Membrane Processes

Molecular Biology and Medical Applications

Membrane Processes

Molecular Biology
and Medical Applications

Edited by
Gheorghe Benga
Harold Baum
Fred A. Kummerow

With 84 Figures

Springer-Verlag
New York Berlin Heidelberg Tokyo

Gheorghe Benga
Department of Cell Biology
Faculty of Medicine
Medical and Pharmaceutical Institute
Cluj-Napoca, Roumania

Harold Baum
Department of Biochemistry
Chelsea College
University of London
London, United Kingdom

Fred A. Kummerow
Department of Food Science
Burnside Research Laboratory
University of Illinois
Urbana, Illinois 61801
U.S.A.

Library of Congress Cataloging in Publication Data
Main entry under title:
Membrane processes.
Includes bibliographies and index.
1. Membranes (Biology)—Addresses, essays, lectures.
I. Benga, Gheorghe. II. Baum, Harold. III. Kummerow, Fred.
QH601.M46855 1984 574.87'5 83-20379

Softcover reprint of the hardcover 1st edition 1984

Typeset by MS Associates, Champaign, Illinois.

9 8 7 6 5 4 3 2 1

ISBN-13: 978-1-4613-8276-8 e-ISBN-13: 978-1-4613-8274-4
DOI: 10.1007/978-1-4613-8274-4

Preface

The burgeoning interest in biomembranes in recent years has been such that "membranology" is now virtually a subject in its own right, cutting vertically, as it were, through the strata of conventional disciplines from mathematics and physics, through chemistry, to biology. The very scope of the topic is thus so daunting that it is tempting to treat it only at one stratum of this hierarchy, be it the biophysics of phospholipid bilayers or the biochemistry of interactions at the cell surface.

Such an approach is entirely valid, particularly among specialists with common interests. However, this approach does present a distorted perspective to the newcomer to the field, and, more significantly, it fails to stimulate cross fertilization of ideas among workers at the various disciplinary levels. For example, as in all areas of molecular biology, the clinicians are frequently unaware of the contributions to their problems that might be made by the application of more basic knowledge and techniques. Conversely, biochemists or biophysicists may be ignorant of the existing practical problems to which they might address their expertise.

The present collection of chapters is intended as a contribution to such cross fertilization. It is by no means a comprehensive treatise on all aspects of membranology, rather it is a sampling of the status of selected topics at different levels, selected to illustrate the interconnections that become apparent between basic and applied biology when a common theme is recognized. The volume provides contributions for reference purposes at the professional level and aims broadly at biologists, biochemists, biophysicists, physicians, and so forth; that is, those who are active investigators working on cell membranes. We hope it will also be of great help to teachers and students at both the undergraduate and graduate levels.

Contents

Contributors

I. Baciu, Department of Physiology, Faculty of Medicine, Bucharest, Roumania

H. Baum, Department of Biochemistry, Chelsea College, University of London, United Kingdom

Gh. Benga, Department of Cell Biology, Faculty of Medicine, Medical and Pharmaceutical Institute, Cluj-Napoca, Roumania

R.F.G. Booth, Department of Biochemistry, Chelsea College, University of London, United Kingdom

D.W. Deamer, Department of Zoology, University of California, Davis, California 95616, U.S.A.

C. Ganea, Department of Biophysics, Faculty of Medicine, Bucharest, Roumania

R.P. Holmes, Department of Food Science and Burnside Research Laboratory, University of Illinois, Urbana, Illinois 61801, U.S.A.

E. Katona, Department of Biophysics, Faculty of Medicine, Bucharest, Roumania

F.A. Kummerow, Department of Food Science and Burnside Research Laboratory, University of Illinois, Urbana, Illinois 61801, U.S.A.

J.A. Lucy, Department of Biochemistry, Royal Free Hospital Medical School, University of London, United Kingdom

V.V. Morariu, Institute of Isotopic and Molecular Technology, Cluj-Napoca, Roumania

P.H. Naccache, Department of Physiology and Pathology, University of Connecticut Health Center, Farmington, Connecticut 06032, U.S.A.

L. Packer, Membrane Bioenergetics Group, Lawrence Berkeley Laboratory, Department of Physiology/Anatomy, University of California, Berkeley, California 94720, U.S.A.

C.A. Pasternak, Department of Biochemistry, St. George's Hospital Medical School, University of London, United Kingdom

A. Popescu, Department of Biophysics, Faculty of Medicine, Bucharest, Roumania

A.T. Quintanilha, Membrane Bioenergetics Group, Lawrence Berkeley Laboratory, Department of Physiology/Anatomy, University of California, Berkeley, California 94720, U.S.A.

R.I. Sha'afi, Department of Physiology and Pathology, University of Connecticut Health Center, Farmington, Connecticut 06032, U.S.A.

S. Smith, Department of Biochemistry, Chelsea College, University of London, United Kingdom

T.L. Smith, Department of Food Science and Burnside Research Laboratory, University of Illinois, Urbana, Illinois 61801, U.S.A.

V. Vasilescu, Department of Biophysics, Faculty of Medicine, Bucharest, Roumania

J.M. Wrigglesworth, Department of Biochemistry, Chelsea College, University of London, United Kingdom

C. Zaciu, Department of Biophysics, Faculty of Medicine, Bucharest, Roumania

Part I

Structural and Functional Relationships in Cell Membranes

1
Reconstituted Membrane Systems

J.M. Wrigglesworth

This chapter is concerned with describing recent advances in the study of membrane systems by reconstitution methods. Particular emphasis is placed on the use of artificial phospholipid structures in the reconstitution of vectorial reactions catalyzed by membrane-associated enzymes.

A widely used approach to the study of enzyme systems is to fractionate and isolate the individual components of the system for further investigation. The resolved components can then be analyzed in isolation from other components to give specific information about individual catalytic reactions. A major disadvantage of this approach is that functional interactions among the different components are lost, especially in membrane-associated systems in which vectorial characteristics are important. In reconstitution studies, an attempt is made to reconstitute the original physiological process from the isolated components. In this way interactions among the different components and the importance of asymmetry and orientation in the membrane can be investigated. This approach is now being used on an increasing number of membrane-associated enzyme systems (Table 1.1), but probably its greatest success has been in investigations into the mechanism of coupling electron transfer reactions in mitochondria to the synthesis of adenosine triphosphate (ATP). In this chapter, we will mainly be concerned with examples from this area.

The success of the reconstitution approach for membrane systems has depended experimentally on the availability of suitable artificial membranes. With this in mind, the development of the liposome concept will first be reviewed together with the development of other, less widely used, artificial membrane systems. Before the full potential of membrane reconstitution could be realized, however, it is necessary to have a theoretical framework covering the concept of vectorial reactions. This was provided by the chemiosmotic theory of Mitchell (1961), and various consequences of this theory relevant to reconstitution studies will also be discussed. Finally, the question of how vectorial reactions in reconstituted systems relate to current ideas on the organization of membrane systems in vivo will be explored.

Table 1.1. Some Membrane Associated Enzyme Systems Investigated by Reconstitution Methods Using Model Membrane Structures

System	Membrane	References
H^+-ATPase	Liposome	Kagawa and Racker (1971)
	Planar bilayer	Drachev et al. (1974)
	Collagen membrane	Blanchy et al. (1979)
NAD-CoQ reductase	Liposome	Ragan and Hinkle (1975)
$CoQH_2$-cytochrome c reductase	Liposome	Leung and Hinkle (1975)
Cytochrome oxidase	Liposome	Hinkle et al. (1972); Jasaitis et al. (1972)
	Planar bilayer	Drachev et al. (1974)
Ca^{2+}-ATPase	Liposome	Racker and Eytan (1973)
Na^+/K^+-ATPase	Liposome	Goldin and Tong (1974)
Rhodopsin	Liposome	Davoust et al. (1979)
Bacteriorhodopsin	Liposome	Racker and Stoeckenius (1974)
	Planar bilayer	Kayushin and Skulachev (1974)
α-glutamyltranspeptidase	Liposome	Sikka and Kalra (1978)
Microsomal electron transfer components	Liposome	Strittmatter et al. (1978)
Chlorpolast pigments	Liposome	Mangel (1976)
	Planar bilayer	Ilani and Berns (1972)
Bacteriochlorophyll	Planar bilayer	Barsky et al. (1976)
Glucose transport	Liposome	Fairclough et al. (1979)
Band III anion transporter	Liposome	Rothstein et al. (1975)
Mitochondrial transhydrogenase	Liposome	Rydström et al. (1975)

Liposomes

Bangham (1963) was among the first to demonstrate that when phospholipids above a certain chain length are exposed to a large excess of aqueous medium they spontaneously arrange themselves into multilamellar concentric bilayer vesicles (Fig. 1.1) termed liposomes. Liposomes have been extensively studied as models for biological membranes (Bangham et al. 1965a, 1965b, 1965c; Bangham and Papahadjopoulos 1966; Bangham et al. 1967; Johnson and Bangham 1969a, 1969b), but more recently other approaches have been developed. Their potential to act as drug carriers for therapeutic use has been explored (Tyrrell et al. 1976; Colley and Ryman 1976; Finkelstein and Weissman 1978; Papahadjopoulos 1978). The possibility of introducing surface receptors into liposomes to target entrapped materials to specific cells in the body has been investigated (Gregoriadis and Neerunjun 1975; Juliano and Stamp 1976), and liposomal membranes have provided a framework for a multitude of reconstituted enzyme systems (Table 1.1).

Not all phospholipids are suitable for liposome formation. The ability of

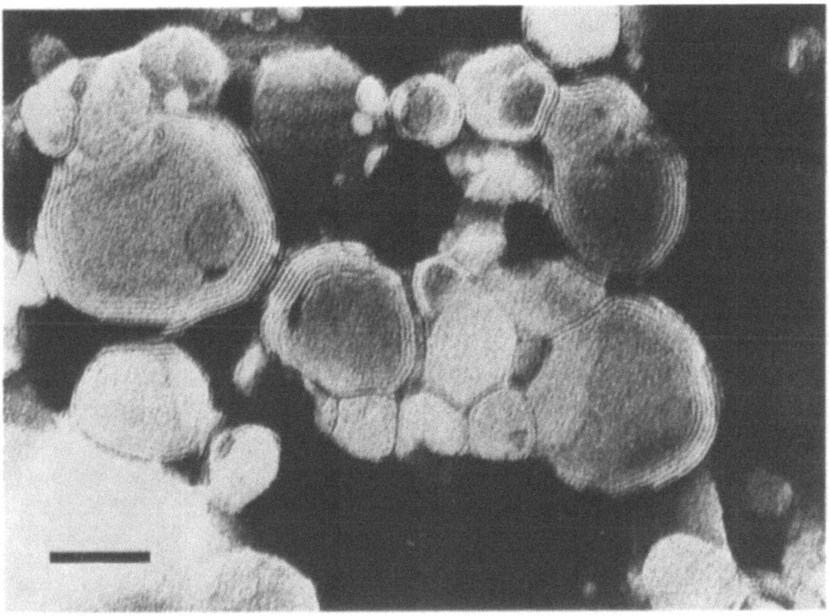

Figure 1.1. Multilamellar bilayer structures formed by mechanical agitation of phospholipids in salt solution. A mixture of phosphatidylcholine and phosphatidic acid (3:1) in a chloroform–ethanol solution was dried under nitrogen and hydrated by the addition of 50 m*M* potassium phosphate with mechanical agitation followed by brief (< 1 min) sonication. Two percent phosphotungstate, pH 7, was used for negative staining. Bar indicates 100 nm.

amphiphiles to form a "mesomorphic phase," or bilayer, is dependent on certain geometrical limitations (Israelachvili and Mitchell 1975; Israelachvili et al. 1976). Single-chained phospholipids generally form micellar structures, whereas double-chain phospholipids above a critical chain length form bilayers. This ability correlates with the very low critical micelle concentration of biological phospholipids compared to most single-chained amphiphiles. In addition, the hydrocarbon chains on the phospholipid have to exist in a liquid state (i.e., above their transition temperature) for successful bilayer formation (Chapman and Fluck 1966). Phospholipids bearing a net charge generally have to be "diluted" with neutral lipid. With these constraints satisfied, the ability to form lamellar structures appears general among the various phospholipids.

Mechanically shaken dispersions of suitable phospholipids form multilamellar liposomes, but prolonged sonication will form an aqueous dispersion of mainly unilamellar vesicles heterogeneous in size ranging in diameter from 250 to 1500 Å (Fig. 1.2). A preparation of uniform size can be isolated by molecular-sieve chromatography (Huang 1969) or by ultracentrifugation procedures (Barenholz et al. 1977). The smaller vesicles approach a lower limit of size where

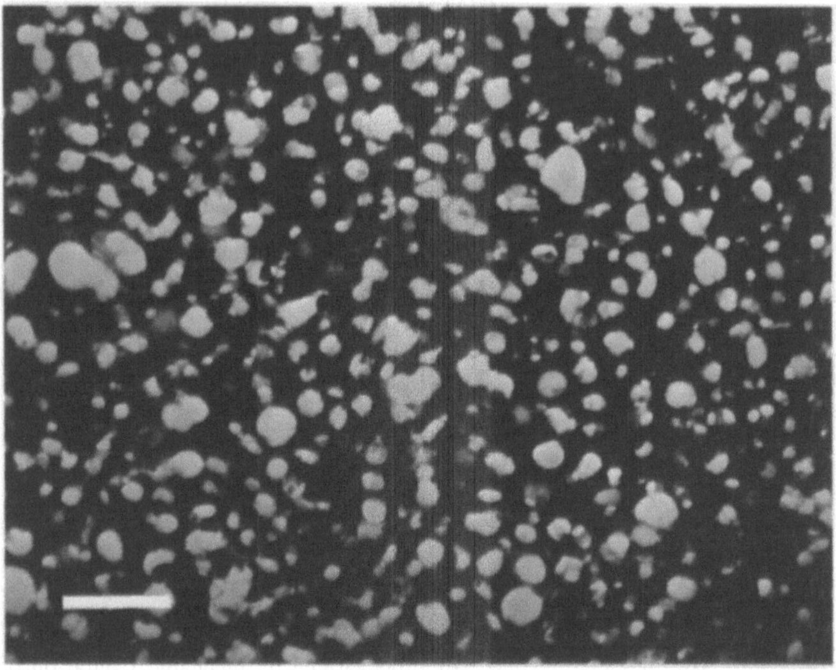

Figure 1.2. Unilamellar vesicles of phospholipid formed by prolonged sonication. Phosphatidylcholine and phosphatidic acid (2:1) were hydrated in a solution of 10 mM sodium chloride pH 6 and subjected to intermittent sonication for a total time period of 20 min in a probe sonicator on ice. Four percent ammonium molybdate, pH 7, was used for negative staining. Bar indicates 200 nm.

packing constraints prevent any further vesicle shrinkage (Cornell et al. 1980). Minimal size phosphatidylcholine vesicles have an average Stokes radius of 11 nm, with approximately 1100 phospholipid molecules in the inner layer of the bilayer and 1900 in the outer layer (Brouillette et al. 1982). The distribution of charged phospholipids between the two membrane surfaces is strongly dependent on the size of the vesicle. In small vesicles, the internal radius of curvature causes strong electrostatic repulsion between charged headgroups, and charged phospholipids appear to partition in the outer layer preferentially (Michaelson et al. 1973). However, this effect strongly depends on the mol% composition of the mixture (Barsukov et al. 1980) and would be expected to be strongly influenced by vesicle size and the presence of protein. It is interesting to note that for these small vesicles, the internal volume of 1.3×10^{-24} m^3 (or 1.3×10^{-21} L/vesicle) would allow for only one molecule of solute/vesicle at 1 mM solute concentration.

Methods other than sonication or mechanical dispersion have been used successfully to form phospholipid vesicles. Many of these rely on exposing an organic solution of the lipid to an excess of aqueous medium. Large (up to 1 μm in diameter) unilamellar vesicles have been produced by this method (Mueller and Rudin 1968; Deamer and Bangham 1976; Nichols et al. 1980). Dispersing the phospholipids in an aqueous medium with detergent followed by dilution (Racker et al. 1975) or removal of the detergent by column chromatography or dialysis (Kagawa et al. 1973) can result in a relatively uniform dispersion of vesicles. A suitable detergent has to be chosen. Cholate or the nonionic detergent octyl glucoside are often used since their relatively high critical micellar concentrations allow for easy removal by dialysis, but other detergents such as Triton X-100 (Wolosin 1980) and lysophosphatidylcholine (Eytan et al. 1975) have been used.

A critical factor in vesicle formation is the detergent-to-lipid concentration. This must be sufficiently high to disperse the phospholipid fully before dialysis. Mimms et al. (1981) found that dialysis of egg lecithin and octyl glucoside dispersions resulted in unilamellar vesicles when the detergent:lipid ratio was kept above 5:1. When lower ratios were used, much of the lipid was found in multilamellar aggregates. This is consistent with the results of Jackson et al. (1982) who followed the solubilization of large unilamellar egg phosphatidylcholine vesicles on the addition of octyl glucoside. Full conversion of bilayers into mixed micelles was only found above a detergent:lipid ratio of 3:1.

In the presence of hydrophobic proteins, many of the physical properties of liposomes are altered. To the author the most perplexing observation, often seen when membrane proteins are incorporated into phospholipid bilayers, is the influence of the protein in "directing" the phospholipid to form uni- rather than multilamellar vesicles. Kagawa et al. (1973) report that multilayered myelin figures are produced by the cholate-dialysis method when only phospholipids are present; when certain hydrophobic proteins are included in the detergent mixture, dialysis often produces smaller unilamellar vesicles. Similar findings have been reported by MacDonald and MacDonald (1975) in which the assembly of phospholipid vesicles incorporating sialoglycoprotein from the erythrocyte membrane was shown to result in unilamellar structures in contrast to multilamellar vesicles obtained with lipid alone. Liposomal systems incorporating different membrane proteins are found to respond differently to the various reconstitution procedures (Racker 1979), and the assembly of vesicular structures in the presence of hydrophobic protein is also reported to be strongly influenced by the composition of the phospholipid (Kagawa et al. 1973). Variations in detergent:lipid ratios may account for some of these effects (Mimms et al. 1981), but the conclusion from these and other studies must be that at present it is prudent to take an empirical approach for each multicomponent system under investigation.

Phospholipid vesicles, however prepared, are freely permeable to water and therefore subject to osmotic forces. Large liposomes will change volume in response to the concentration gradient of impermeable solutes (Bangham et al. 1967), but volume changes in preparations of the smaller size of vesicles are restricted by packing density and surface tension considerations. The mechanical strength of the smallest of the membrane vesicles may enable them to support quite large differences in osmotic pressure (Walter 1975; but see Tanford 1979).

In view of the importance of ion transport concentration gradients in cellular bioenergetics, it is of interest to investigate whether artificial systems can serve as good models for such studies. Biological membranes appear to have high permeability coefficients for cations compared to most model systems (Deamer 1982), probably arising from different membrane mechanisms for ion translocation. Model systems have relatively low sodium and potassium ion permeability coefficients (10^{-13}-10^{-14} cm/s), but the passive permeability of artificial lipid bilayers, including liposomes, to protons is the subject of some disagreement. Values ranging from 10^{-4} cm/s (Nichols et al. 1980; Nichols and Deamer 1980; Clement and Gould 1981; Biegel and Gould 1981) to 10^{-9} cm/s (Nozaki and Tanford 1981; Gutknecht and Walter 1981) have been reported. However, Deamer (1982) points out that despite these conflicting data, the values are all still lower than the measured proton permeability of natural membranes. At a first approximation, it would seem that model systems do provide a sufficient barrier to proton flux for useful reconstitution experiments.

Black-Lipid Membranes and Other Model Membrane Systems

Black-lipid membranes can be formed by stroking a brush loaded with a solution of suitable lipids over a small hole in a hydrophobic barrier between two aqueous compartments (Bangham 1968). The term "black-lipid membrane" arises from the reflectance properties of the thin lipid film. The physical properties of these membranes have been extensively studied (Mueller et al. 1962; Mueller et al. 1964; Finkelstein and Cass 1968). They have a bimolecular leaflet structure similar to the liposome but usually also contain significant amounts of organic solvent, especially in the boundary region in which the film thickens as it attaches to the solid hydrophobic support (Henn and Thompson 1967; Pagano et al. 1972). The organic solvent can affect the electrical properties of the bilayer (Haydon et al. 1977) and also may affect the properties of any incorporated protein system. Black-lipid membranes are not nearly as stable as liposome bilayers; but despite the many difficulties associated with their use, they have been the chosen method for many successful studies on bilayer permeability. With this system it is possible to measure electrical potential differences and transport of material across a phospholipid bilayer separating two bulk aqueous phases.

Other hydrophobic material can easily be introduced into the bilayer. Mueller and colleagues (1962) were the first to study the ion permeability effects of added protein to black-lipid membranes. They showed that the addition of suitable proteins to the bilayer can induce voltage-dependent conductance properties similar to those found in biological membranes. These studies have been extended by Mueller and other workers (Mueller and Rudin 1968; Ting-Beall et al. 1979). Reconstitution studies of membrane-associated enzyme systems using black-lipid membranes have allowed direct measurements to be made of electric current generation by cytochrome oxidase, adenosine triphosphatase (ATPase), and bacteriorhodopsin (Drachev et al. 1974; Dancshazy and Karvaly 1976; Issaurat et al. 1980).

Other phospholipid systems have proved less useful in reconstitution studies. The limitations of using monolayers for investigating vectorial reactions is obvious, but they can be used to provide details of specific protein–lipid and lipid–lipid interactions that cannot be measured in less well-defined structures (Redwood and Patel 1974; Wooster and Wrigglesworth 1976; Quinn and Esfahani 1980). The use of lipid-impregnated filters as an artificial framework for reconstitution studies has been explored with some success by Skulachev and co-workers (Skulachev 1976; Konstantinov et al. 1980). These filters provide the advantages of the black-lipid membrane system without the instability problems inherent in unsupported bilayers.

Electrochemical Gradients and Ionophores

The importance of electrochemical gradients in cell physiology has long been known from studies on the cation balance in cells, especially the nerve cell. However, a shift in paradigm was necessary before the general role of vectorial reactions in metabolism could be realized. This was provided by Mitchell in his chemiosmotic theory (Mitchell 1961). The theory sets out to explain how certain chemical reactions can be linked by a suitable vectorial arrangement of the reactants and products (Fig. 1.3). According to the theory, the coupling between redox reactions and ATP synthesis occurs via the intermediate of an electro-

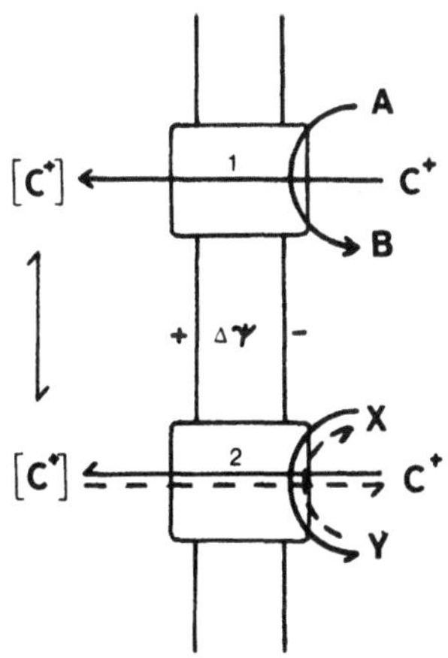

Figure 1.3. Coupling of scalar chemical reactions by a vectorial flow of reactants across a membrane. Both reactions 1 and 2 can translocate compound C across the membrane. If either or both of the reactions are reversible, then the gradient of C formed by one reaction can be used to reverse the other. When C is charged, the electrochemical gradient of C will depend on the concentration difference of C across the membrane and the membrane potential ($\Delta\psi$).

chemical gradient of protons across the coupling membrane (Mitchell 1966; Greville 1969).

The electrochemical potential difference of an ion ($\Delta\tilde{\mu}_i$) between two phases separated by a membrane can be written as

$$\Delta\tilde{\mu}_i = ZF\Delta\psi + RT\ln\frac{c_1}{c_2}$$

where Z is the charge number, F the Faraday constant, $\Delta\psi$ the membrane potential, R the gas constant, and c the concentration of the ion in the two phases assuming activity coefficients of one. For protons we have

$$\Delta\tilde{\mu}_{H^+} = F\Delta\psi - 2.3\,RT\Delta\,\text{pH}$$

From these equations it can be seen that investigations of vectorial reactions often require the measurement of membrane potential and concentration differences. The availability of specific ion transporters, or ionophores, has been a powerful aid in such investigations.

Several antibiotics have been shown to promote the transport of inorganic cations across an aqueous-lipid phase boundary and to induce conductance in lipid bilayers (for reviews, see Gómez-Puyou and Gómez-Lojero 1977; Reed 1979). These ionophores stabilize the ion by replacing its hydration shell. Lipid solubility is promoted because of externally exposed lipophilic groups on the complex (Pressman 1968). There are various types of ionophores (Table 1.2), but two main classes can be defined on the basis of whether the ionophoric action is electrogenic or electroneutral. Electrogenic ionophores form charged complexes with selected ions, and the movement of the complex across the membrane is directed by membrane potentials. Ion transport by these carriers will collapse any preformed potentials ($\Delta\psi = 0$), but bulk movement of ions

Table 1.2. Types of Ionophores Commonly Used in Reconstitution Studies

Classification	Examples	Specificity	References
Electrogenic	Valinomycin	$Rb^+ > K^+ >> Na^+$	Pressman (1968)
Electroneutral	Nigericin	K^+/H^+ exchange	Pressman (1968)
	X-573A	Alkali ion/H^+ exchange Divalent cation/H^+ exchange	Celis et al. (1974)
	A23187	Divalent cation/H^+ exchange	Pfeiffer et al. (1974)
Uncouplers	Lipid-soluble weak acid anions	H^+	Hanstein (1976); Heytler (1979)
	Anionic detergents	H^+	Abramson and Shamoo (1979)
Channel formers	Gramicidin	Alkali ion/H^+ exchange	Bamberg and Lauger (1973)
	Alamethicin	Alkali ion/H^+ exchange	Pressman (1968)

will not happen in an electroneutral system unless an equivalent charge can be moved in the opposite direction. Electrogenic ionophores therefore cannot be used alone to alter chemical gradients across an electrically neutral bilayer. Electroneutral ionophores are lipid insoluble in the absence of any complexing ions but form lipid-soluble neutral complexes with specific ions. The movement of the complex is directed by chemical gradients alone. Nigericin, for example, transports alkali cations as electrically neutral dipoles and protons in their electrically neutral, undissociated form (Pressman 1968). Thus at equilibrium in the presence of nigericin with only K^+ and H^+ cations present, we have

$$-\Delta\tilde{\mu}_{H^+} = \Delta\tilde{\mu}_{K^+}$$

and

$$-\Delta\,pH = \Delta\,pK$$

Montal et al. (1970) have used this property of nigericin to estimate the value

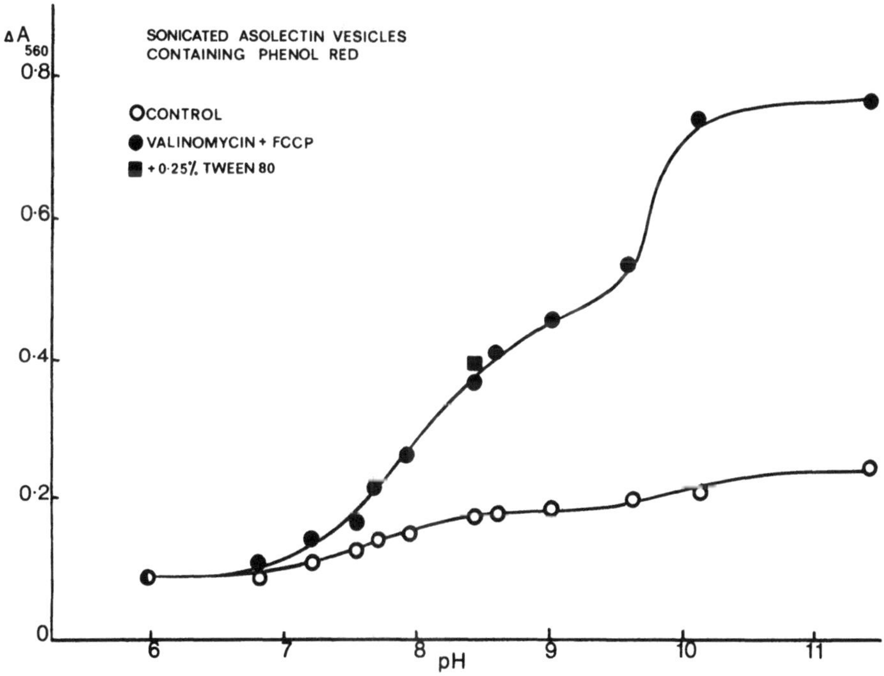

Figure 1.4. Effect of ionophores on the pH equilibrium across a liposomal membrane. Liposomes containing the pH indicator dye phenol red were prepared at pH 6 according to Wrigglesworth (1978). Variation of the pH of the external medium has little effect on internal pH, as assessed by changed in phenol red absorbance, unless valinomycin plus the uncoupler FCCP are present. In the presence of nigericin or the detergent 0.25% Tween-80, the pH internal and external to the bilayer is in equilibrium (Wrigglesworth and Nicholls, unpublished work).

of the pH gradient across the mitochondrial membrane by pH measurements at different potassium concentrations.

Proton carriers fall into a class of their own under the general classification of uncouplers. Most of these are lipid-soluble weak acids that can cross the membrane in a protonated as well as an unprotonated form (Hanstein 1976; McLaughlin and Dilger 1980). Their inability to transport ions other than protons, and thus mediate an exchange of charge, makes their action electrogenic in nature. Other ionophores can facilitate transport by the formation of transmembrane aqueous channels (Table 1.2).

A useful model to illustrate the action of ionophores in relation to their use in reconstitution studies is described by Hinkle (1973, 1979). Liposomes can be prepared in the presence of a pH or redox indicator dye and once formed can be separated from excess dye by dialysis or passage down a gel filtration column. For example, the internal pH of vesicles as indicated by changes in phenol red absorption can be conveniently monitored in a spectrophotometer and can be seen to be insensitive to external pH in the absence of suitable ionophores (Fig. 1.4). Separate addition of electrogenic ionophores such as carbonyl cyanide *m*-fluorophenylhydrazone (FCCP) or valinomycin does not allow immediate pH equilibration to occur, but nigericin, or FCCP and valinomycin in combination, dissipate any pH gradients and the entrapped phenol red then responds directly to changes in external pH (Fig. 1.4).

Lipophilic Ions as Indicators of $\Delta\psi$

Certain lipophylic ions can be used to monitor membrane potentials. A lipid-permeable ion will move across a membrane in response to an applied potential. At equilibrium, the electrochemical potential of the ion will be zero, and from the previous equation we find

$$\Delta\psi = \frac{RT}{ZF} \ln \frac{c_{\text{in}}}{c_{\text{out}}}$$

The membrane potential can thus be calculated from a knowledge of the concentration ratio of the permeant ion at equilibrium. Jasaitis and colleagues (1972) have used the phenyl dicarbaundecaborane anion (PCB^-) as a penetration anion probe to estimate $\Delta\psi$ in liposomes inlayed with cytochrome oxidase or with mitochondrial ATPase. PCB^- is accumulated by vesicles having a positive internal potential, and the drop in external concentration can be measured using a black-lipid membrane system as a selective electrode for the ion. Changes in the external concentration of lipophylic cations such as *N,N*-dibenzyl *N,N*-dimethyl ammonium (DDA^+) have been used in the same way to detect negative internal potentials in reconstituted systems (Grinius et al. 1970; Skulachev 1971, 1979).

Gutweniger and co-workers (1977) describe a spectrophotometric method of monitoring $\Delta\psi$ by using the positive dye safranine. Reconstituted phospholipid vesicles incorporating cytochrome oxidase concentrate the dye when an internal negative potential is generated. Aggregation of safranine molecules inside the vesicles causes a decrease in the absorbance of the monomeric form at 520 nm, whereas another absorption band characteristic of the multimeric form appears at shorter wavelengths. The dye has been used successfully to monitor membrane potentials in intact mitochondria (Akerman and Wikström 1976; Akerman 1979) and *Escherichia coli* vesicles (Schuldiner and Kabak 1975). Changes in the fluorescence of various probe molecules, mainly negatively charged, have also been used to indicate changes in membrane potential (Cohen et al. 1974; Aiuchi et al. 1977; Bashford et al. 1979). However, the response of most of these molecules is critically dependent on the local environment of their binding site, making detailed interpretation difficult (Bashford and Smith 1979).

Reconstitution Procedures

As mentioned previously, each membrane-associated enzyme system appears to differ in its requirements for optimal reconstitution (Racker 1976, 1979). Phospholipid requirements for reconstitution of a number of membrane-associated enzymes are illustrated in Table 1.3. It is necessary therefore to have an assay for the enzyme that is specific for the correct incorporation of the enzyme into

Table 1.3. Phospholipid Requirements for Optimal Reconstitution of Some Membrane-Associated Enzyme Systems

System	Reconstitution Procedure	Optimum Phospholipid Composition	References
Mitochondrial transhydrogenase	Cholate dialysis	PC[a]	Rydström et al. (1975)
$(Mg^{2+} + Ca^{2+})$ATPase	Cholate dialysis	PC	Warren et al. (1974)
	Cholate dialysis sonication	PE[b]:PC:cardiolipin (3:1:0.67)	Knowles et al. (1976)
		PE:PC:cardiolipin (3:1:0.5)	Racker and Eytan (1973)
Cytochrome oxidase	Cholate dilution	PE:PC (4:1)	Carroll and Racker (1977)
	direct incorporation	PE:PC:PS[c] (2.4:0.8:3)	Eytan et al. (1976)
Oligomycin sensitive ATPase	Cholate dilution	PE:PC (1:1)	Racker et al. (1975)
	cholate dialysis	PE:PC (4:1)	Kagawa et al. (1973)

[a]PC = phosphatidylcholine.
[b]PE = phosphatidylethanolamine.
[c]PS = phosphatidylserine.

the membrane so that the degree of reconstitution can be continually assessed. In many cases the catalytic activity of an enzyme in the reconstituted system leads to the translocation of ions or charge across the membrane, and the resulting electrochemical gradient often has an inhibitory effect on turnover. The effect of specific ionophores on the activity of the enzyme can then be used to assess the degree of reconstitution. For example, the incorporation of cytochrome oxidase into phospholipid vesicles can be assessed by the degree of stimulation of electron transfer from cytochrome *c* to oxygen by uncouplers or suitable combinations of ionophores (Fig. 1.5). Electron transfer can be monitored by oxygen uptake measurements using an oxygen electrode or by following the steady-state reduction level of cytochrome *c* at 550 nm (Wrigglesworth 1978; Hansen et al. 1978).

Once reconstitution can be conveniently monitored, then most of the methods described earlier for the preparation of liposomes can be used with suitable modification to allow for the inclusion of membrane proteins in the mixture. The first successful method for incorporating protein into a phospholipid bilayer was that described by Mueller and colleagues (1962), who incorporated an unidentified heat-stable factor from egg white into a bilayer (black) phospholipid membrane by adding the protein to the aqueous phase. Incorporation was monitored by the induction of a gating reaction to direct current electrical stimulation. The formation of planar bilayer membranes from a mixture of phospholipids and isolated enzyme systems in an organic solvent has been used by Skulachev and co-workers to reconstitute the electrogenic function of bacteriorhodopsin, cytochrome oxidase, H^+-ATPase (Drachev et al. 1974, 1976a) and the bacteriochlorophyll reaction center complexes from *Rhodospirillum rumbrum* (Drachev et al. 1976b). The sidedness of the proteins in the bilayer is random, but asymmetry can be obtained by adding nonpermeable substrates to one side of the compartment only.

Simple addition of hydrophobic protein to an aqueous suspension of preformed liposomes can often result in a successful incorporation, provided the liposomes are prepared from an appropriate mixture of phospholipids (Racker 1979). According to Eytan et al. (1976), some acidic phospholipids need to be present for the incorporation of cytochrome oxidase, QH_2 or $CoQH_2$ reduced coenzyme Q-cytochrome *c* reductase, and the oligomycin-sensitive ATPase. The incorporation of these enzymes into preformed liposomes also needs to be carried out at room temperature or above and is accelerated in the presence of Mg^{2+}. One advantage of the direct incorporation procedure is that detergents or sonication treatments that may be inhibitory to the enzyme are avoided. In addition, a sequential addition of different proteins, which can provide information on protein-protein interactions in the bilayer, is possible (Eytan and Racker 1977). A greater control of the orientation of the protein in the membrane has been reported (Eytan et al. 1976). Cytochrome oxidase appears to preferentially incorporate into preformed liposomes with the cytochrome *c* reaction site exposed to the external medium, in contrast to the random sidedness found when sonication or detergent dialysis procedures are used (Wrigglesworth 1978; Wrigglesworth and Nicholls 1980).

The detergent dialysis method is one of the most widely used reconstitution procedures and has proved successful on a variety of membrane-associated systems. Essentially, a sonicated suspension of phospholipids in an appropriate de-

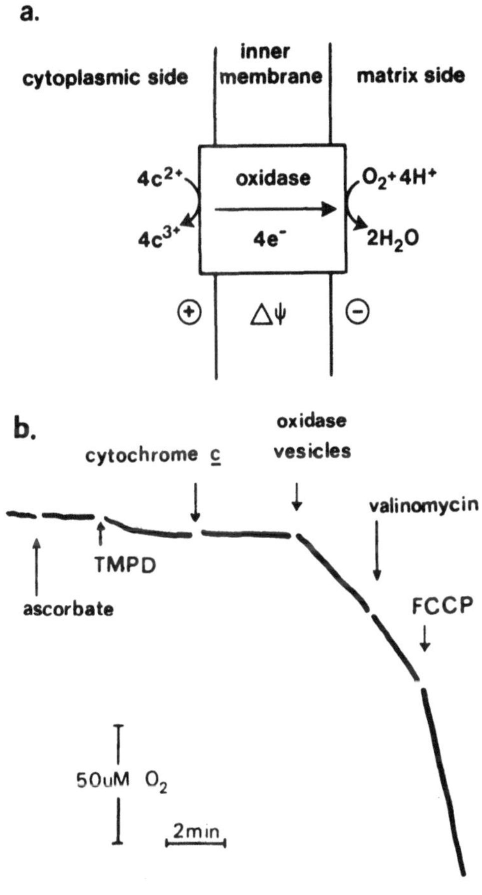

Figure 1.5. Respiratory control in reconstituted systems containing cytochrome oxidase. (*a*) Electron transfer from cytochrome *c* to oxygen induces an electric potential across the coupling membrane. [The arrangement of reactants in this figure is one of the simplest to explain the observed potential generation, but many other arrangements of the components of the oxidase can be envisaged to give the same result (e.g., see King 1978; Nicholls et al. 1980).] (*b*) Oxygen uptake by vesicles inlayed with cytochrome oxidase oxidizing externally added cytochrome *c* is stimulated by valinomycin plus FCCP. The degree of stimulation can be used to monitor the incorporation of the enzyme into the bilayer of closed vesicles. TMPD, N,N,N^1,N^1-tetramethyl-p-phenylenediamine.

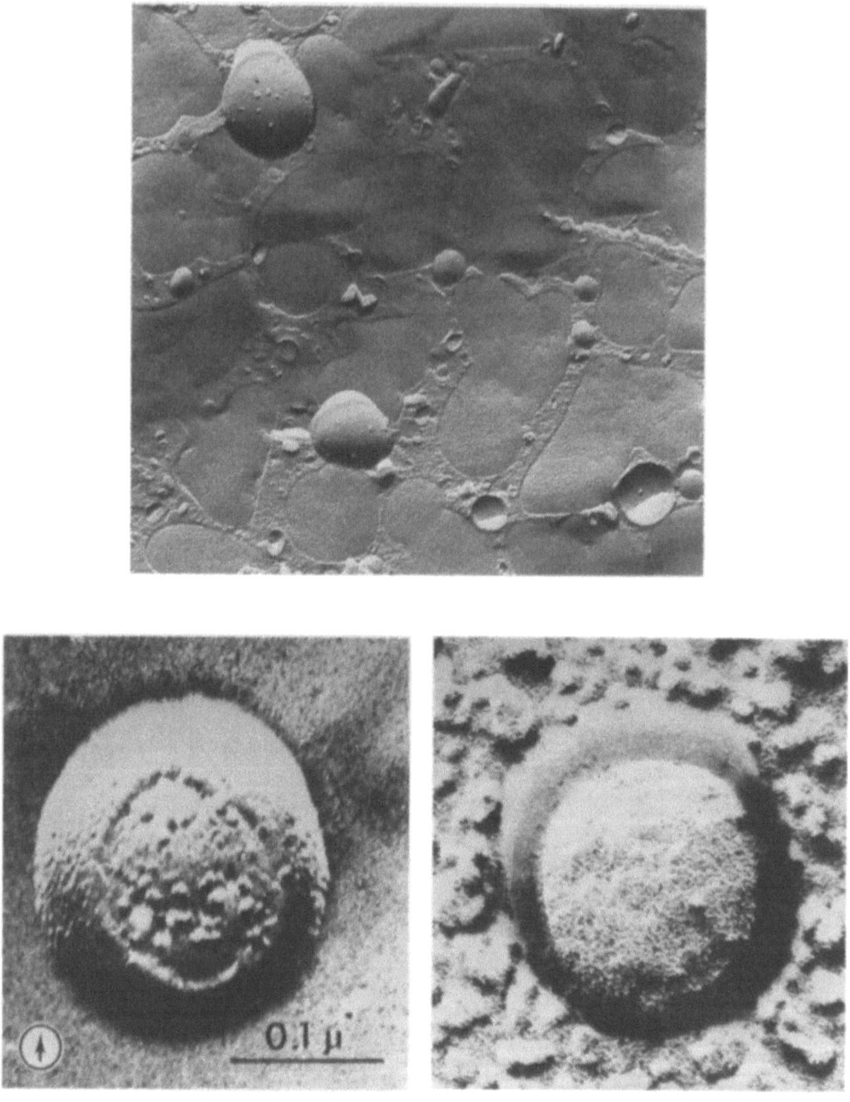

Figure 1.6. Freeze-fracture electron microscopy of (*top*) a population of reconstituted cytochrome oxidase vesicles and (*bottom, left*) a submitochondrial particle and a reconstituted vesicle incorporating cytochrome oxidase. The high lipid:protein ratio used for successful reconstitution studies with cytochrome oxidase reduces the number of protein units in the bilayer. Reconstituted vesicles with more than two to three intramembrane particles in the fracture face are seldom seen compared to the high density of intramembrane proteins in the natural membrane. Conditions of preparation of sample were described in Wrigglesworth et al. (1970).

tergent is mixed with the enzyme system. The detergent is then slowly removed by dialysis over a period of 24 h. Details on the use of cholate (Racker 1979) and octyl glucoside (Mimms et al. 1981) in this method are described. An alternative method of removing the detergent is by absorption or column chromatography. A Bio-Bead column has been used to remove Triton X-100 in a reconstituted Ca^{2+}-ATPase system (Chiesi et al. 1978). Dialysis was not possible because of the low critical micelle concentration of the detergent. Even simple dilution of the mixture often appears to lower the concentration of detergents sufficiently for reconstituted vesicles to form (Racker et al. 1975). The detergent dialysis procedure is reported to form unilamellar vesicles (Kagawa et al. 1973; Mimms et al. 1981), but this cannot be taken to be the general rule for all enzyme systems.

Sonication of a suitable mixture of phospholipids and protein can result in active reconstituted systems (Racker 1973). It is necessary to sonicate the the protein-phospholipid mixture to clarity, and care has to be taken to avoid overheating. Nevertheless, several enzymes remain sensitive to sonic treatment, and turnover in the final system can be markedly lowered. Some protection against damage may be afforded by using high [$\sim$20 (wt/wt)] lipid:protein ratios. This may be true for reconstituted systems of cytochrome oxidase, although high lipid:protein ratios are also found to be necessary for this enzyme in the cholate dialysis procedure. Carroll and Racker (1977) suggest an explanation may be that cytochrome oxidase is preferentially incorporated into a subpopulation of liposomes in reconstitution procedures. When reconstituted vesicles of cytochrome oxidase are fractionated on a Ficoll gradient, most of the cytochrome oxidase activity is found associated with only 10-20% of the total phospholipid (Carroll and Racker, 1977). Even so, examination by freeze-fracture electron microscopy rarely reveals more than two to three oxidase molecules per vesicle (Fig. 1.6).

Reconstitution Studies on Respiratory Chain Components

Membrane potential and proton movements have been monitored in several reconstituted liposome systems incorporating respiratory chain components. The results are summarized in Table 1.4. Asymmetry in most of these systems has been imposed by an asymmetric arrangement of reactants. For example, liposomes incorporating cytochrome oxidase develop a membrane potential, negative inside, when an impermeant mixture of ascorbate and cytochrome *c* are added to the suspending medium in the presence of oxygen. If the cytochrome oxidase liposomes contain entrapped cytochrome *c*, they generate a membrane potential, positive inside, when a permeant reductant such as phenazine methosulphate is added. The movement of the lipophilic anion PCB^- across the liposomal membrane can be used to monitor the membrane potential generation

Table 1.4. Membrane Potential Generation by Reconstituted Vesicles of Phospholipid Incorporating Various Electrogenic Enzyme Systems

System	Experimental Observations: Negative Ion Movement	Experimental Observations: H^+ Production[a]	Internal Potential ψ	References
H^+-ATPase	In	In	Positive	Kagawa and Racker (1971) Jasaitis et al. (1972)
Cytochrome oxidase				
c inside	In		Positive	Jasaitis et al. (1972)
c outside	Out	Out	Negative	Hinkle et al. (1972)
Micochondrial complex I	In	In	Positive	Ragan and Hinkle (1975)
Mitochondrial complex III		Out	Negative	Leung and Hinkle (1975)
Bacteriorhodopsin	In	In	Positive	Racker and Stoeckenius (1974)

[a] In presence of valinomycin.

(Jasaitis et al. 1972). A direct measurement of the movement of protons and potassium ions in suspensions of cytochrome oxidase vesicles has demonstrated that a respiration-dependent release of protons and uptake of potassium ions occurs when reduced cytochrome *c* is in the external medium (Hinkle et al. 1972; Wikström et al. 1981). Valinomycin has to be present to collapse $\Delta\psi$ and maximize ion movements. In a similar way, Kagawa et al. (1973) have demonstrated proton translocation into vesicles inlayed with mitochondrial ATPase when ATP is added to the external medium. PCB^- uptake occurs in this system (Jasaitis et al. 1972). Reconstituted vesicles containing complex I (NADH-CoQ reductase)* and complex III (CoQ-cytochrome *c* reductase) have also been investigated in the same way (Ragan and Hinkle 1975; Leung and Hinkle 1975). The direction of ion movements and the sidedness of membrane potential generation for reconstituted liposomes containing these complexes is indicated in Table 1.4. Bacteriorhodopsin has unusual characteristics not shared by most of the other proteins used in reconstitution experiments. It appears to incorporate into liposomal bilayers in a unidirectional manner (Racker and Stoeckenius 1974). This fortuitous property allows asymmetric ion movements and membrane potential generation to occur on illumination of the vesicles.

The strength of the reconstitution approach is emphasized by the predictions that can be made about the properties of liposomes containing a mixture of the

*NADH: Reduced nicotinamide adenine dinucleotide.

separate systems described in Table 1.4. Thus, a mixed cytochrome oxidase-ATPase liposome (Racker and Kandrach 1971) will catalyze ATP synthesis if the reductant for the oxidase is presented internally to the enzyme ($\Delta\psi$ internal positive) and adenosine diphosphate (ADP) and phosphate are present in the external medium. The electrochemical gradient of protons generated by the oxidase acts to couple the two enzyme systems (see Fig. 1.3) and reverse the ATP hydrolysis reaction. Cytochrome *c* added to the external medium inhibits ATP synthesis. A classic combination of two separate electrogenic systems coupled by an electrochemical gradient of protons is the bacteriorhodopsin and mitochondrial ATPase mixed liposome (Racker and Stoeckenius 1974). As could be predicted from the sidedness of the membrane potential generated by these two systems alone (Fig. 1.4), illumination of a liposome containing a combination of the two results in phosphorylation of ADP in the external medium. Without the theoretical framework provided by Mitchell and the experimental techniques developed by Bangham, Racker, and others, the successful prediction of the properties of reconstituted systems such as these would have been inconceivable.

References

Abramson, J.J. and Shamoo, A.E. (1979). Anionic detergents as divalent cation ionophores across black lipid membranes. *J. Membr. Biol. 50,* 241–255.

Aiuchi, T., Kamo, N., Kurihara, K., and Kobatake, Y. (1977). Significance of surface potential in interaction of 8-anilino-1-naphthalenesulfonate with mitochondria. *Biochemistry 16,* 1626–1630.

Akerman, K.E.O. (1979). Qualitative measurements of the mitochondrial membrane potential in situ in Ehrlich ascites tumour cells using the safranine method. *Biochim. Biophys. Acta 546,* 341–348.

Akerman, K.E.O. and Wikström, M.K.F. (1976). Safranine as a probe of the mitochondrial membrane potential. *FEBS Lett. 68,* 191–197.

Bamberg, E. and Lauger, P. (1973). Channel formation kinetics of gramicidin A in lipid bilayer membranes. *J. Membr. Biol. 11,* 177–194.

Bangham, A.D. (1963). Physical structure and behavior of lipids and lipid enzymes. *Adv. Lipid Res. 1,* 65–104.

Bangham, A.D. (1968). Membrane models with phospholipids. *Prog. Biophys. Mol. Biol. 18,* 31–95.

Bangham, A.D. and Papahadjopoulos, D. (1966). Biophysical properties of phospholipids 1. Interaction of phosphatidylserine monolayers with metal ions. *Biochim. Biophys. Acta 126,* 181–184.

Bangham, A.D., Standish, M.M., and Watkins, J.C. (1965a). Diffusion of univalent ions across the lamellae of swollen phospholipids. *J. Mol. Biol. 13,* 238–252.

Bangham, A.D., Standish, M.M., and Weissmann, G. (1965b). The action of steroids and streptolysin S on the permeability of phospholipid structures to cations. *J. Mol. Biol. 13,* 253–259.

Bangham, A.D., Standish, M.M., and Miller, N. (1965c). Cation permeability of phospholipid model membranes: Effect of narcotics. *Nature 208,* 1295–1297.

Bangham, A.D., DeGier, J., and Grenville, G.D. (1967). Osmotic properties and water permeability of phospholipid liquid crystals. *Chem. Phys. Lipids 1,* 225–246.

Barenholz, Y., Gibbes, D., Litman, B.J., Goll, J., Thompson, T.E., and Carlson, F.D. (1977). A simple method for the preparation of homogeneous phospholipid vesicles. *Biochemistry 16,* 2806–2810.

Barsky, E.L., Dancshazy, Dh., Drachev, L.A., Il'ina, M.D., Jasaitis, A.A., Kondrashin, A.A., Samailov, V.D., and Skuhchev, V.P. (1976). Reconstitution of biological molecular generators of electric current: Bacteriochlorophyll and plant chlorophyll complexes. *J. Biol. Chem. 251,* 7066–7071.

Barsukov, L.I., Victorov, A.V., Vasilenko, I.A., Evstigneeva, R.P., and Bergelson, L.D. (1980). Investigation of the inside–outside distribution intermembrane exchange and transbilayer movement of phospholipids in sonicated vesicles by shift reagent NMR. *Biochim. Biophys. Acta 598,* 153–168.

Bashford, C.L. and Smith, J.C. (1979). The use of aptical probes to monitor membrane potential. In *Methods in Enzymol,* Fleischer, S. and Packer, L., eds. Academic Press, New York, San Francisco, London, vol. LV, pp. 569–586.

Bashford, C.L., Chance, B., and Prince, R.C. (1979). Oxonol dyes as monitors of membrane potential. Their behaviour in photosynthetic bacteria. *Biochim. Biophys. Acta 545,* 46–55.

Biegel, C.M. and Gould, J.M. (1981). Kinetics of hydrogen ion diffusion across phospholipid vesicle membranes. *Biochemistry 20,* 3474–3479.

Blanchy, B., Godinot, C., and Gautheron, D.C. (1979). Reconstitution of ATP synthetase on a Collagen membrane that can synthesize ATP using a pH gradient. In *Methods in Enzymol,* Fleischer, S. and Packer, L., eds. Academic Press, New York, San Francisco, London, vol. LV, pp. 742–748.

Brouillette, C.G., Segrest, J.P., Ng, T.C., and Jones, J.L. (1982). Minimal size phosphatidylcholine vesicles: Effects of radius of curvature on head group packing and conformation. *Biochemistry 21,* 4569–4575.

Carroll, R.C. and Racker, E. (1977). Preparation and characterization of cytochrome *c* oxidase vesicles with high respiratory control. *J. Biol. Chem. 252,* 6981–6990.

Celis, H., Estrada-O, S., and Montal, M. (1974). Model translocation for divalent and monovalent ion transport in phospholipid membranes. *J. Membr. Biol. 18,* 187–199.

Chapman, D. and Fluck, D.J. (1966). Physical studies of phospholipids. III *J. Cell Biol. 30,* 1–11.

Chiesi, M., Peterson, S.Q., and Acuto, O. (1978). Reconstitution of a Ca^{2+}-transporting ATPase system from Triton X-100 soluhilized sarcoplasmic reticulum. *Arch. Biochem. Biophys. 189,* 132–136.

Clement, N.R. and Gould, J.M. (1981). Pyranine as a probe of internal aqueous hydrogen ion concentration in phospholipid vesicles. *Biochemistry 20,* 1534-1538.

Cohen, L.B., Salzberg, B.M., Davila, H.V., Ross, W.N., Landowne, D., Waggoner, A.S., and Wang, C.H. (1974). Changes in azon fluorescence during activity: Molecular probes of membrane potential. *J. Membr. Biol. 19,* 1-36.

Colley, C.M. and Ryman, B.E. (1976). The liposome: From membrane model to therapeutic agent. *Trends Biochem. Sci. 1,* 203-205.

Cornell, B.A., Middlehurst, J., and Separovic, F. (1980). The molecular packing and stability within highly curved phospholipid bilayers. *Biochim. Biophys. Acta 598,* 405-410.

Dancshazy, Z. and Karvaly, B. (1976). Incorporation of bacteriorhodopsin into a bilayer lipid membrane: A photoelectric-spectroscopic study. *FEBS Lett. 72,* 136-138.

Davoust, J., Schoot, B., and Devaux, P.F. (1979). Rhodopsin-phosphatidylcholine vesicles. *Proc. Natl. Acad. Sci. USA 76,* 2755-2759.

Deamer, D.W. (1982). Proton permeability in biological and model membranes. In *Intracellular pH: Its Measurement, Regulation and Utilization in Cellular Functions.* Nuccitelli, E. and Deamer, D.W., eds. Alan R. Liss, New York.

Deamer, D.W. and Bangham, A.D. (1976). Large volume liposomes by an ether vaporization method. *Biochim. Biophys. Acta 443,* 629-634.

Drachev, L.A., Jasaitis, A.A., Kaulen, A.D., Kondrashin, A.A., Liberman, E.A., Nemecek, I.B., Ostroumov, S.A., Semenov, A.Yu., and Skulachev, V.P. (1974). Direct measurement of electric current generation by cytochrome oxidase, H^+-ATPase and bacteriorhodopsin. *Nature 249,* 320-324.

Drachev, L.A., Frolov, V.N., Kaulen, A.D., Liberman, E.A., Obstroumov, S.A., Plakunova, V.G., Semenov, A.Y., and Skulachev, V.P. (1976a). Reconstitution of biological molecular generators of electric current: Bacteriorhodopsin. *J. Biol. Chem. 251,* 7059-7065.

Drachev, L.A., Frolov, V.N., Kaulen, A.D., Kondrashin, A.A., Samuilov, V.D., Semenov, A.Y., and Skulachev, V.P. (1976b). Generation of electric current by chromatophores of rhodospirillum rubrum and reconstitution of electrogenic function. *Biochim. Biophys. Acta 440,* 637-660.

Eytan, G.D. and Racker, E. (1977). Selective incorporation of membrane proteins into proteoliposomes of different compositions. *J. Biol. Chem. 252,* 3208-3213.

Eytan, G., Matheson, M.J., and Racker, E. (1975). Incorporation of biologically active proteins into liposomes. *FEBS Lett. 57,* 121-125.

Eytan, G.D., Matheson, M.J., and Racker, E. (1976). Incorporation of mitochondrial membrane proteins into liposomes containing acidic phospholipids. *J. Biol. Chem. 251,* 6831-6837.

Fairclough, P., Malathi, P., Preiser, H., and Crane, R.K. (1979). Reconstitution into liposomes of glucose active transport from the rabbit renal proximal tubule. *Biochim. Biophys. Acta 553,* 295-304.

Finkelstein, M. and Weissman, G. (1978). The introduction of enzymes into cells by means of liposomes. *J. Lipid Res. 19,* 289-303.

Finkelstein, A. and Cass, A. (1968). Permeability and electrical properties of thin lipid membranes. *J. Gen. Physiol. 52*(2), 145–172.

Goldin, S.M. and Tong, S.W. (1974). Reconstitution of active transport catalyzed by the purified sodium and potassium ion-stimulated ATPase from canine renal medulla. *J. Biol. Chem. 249,* 5907–5915.

Gómez-Puyou, A. and Gómez-Lojero, C. (1977). The use of ionophores and channel formers in the study of the function of biological membranes. In *Current Topics in Bioenergetics,* Sanadi, D.R. ed. Academic Press, New York, San Francisco, London, vol. 6, pp. 221–257.

Gregoriadis, G. and Neerunjun, E.D. (1975). Homing of liposomes to target cells. *Biochem. Biophys. Res. Commun. 65,* 537–544.

Greville, G.D. (1969). A scrutiny of Mitchell's Chemiosmotic Hypothesis of Respiratory Chain and Photosynthetic Phosphorylation. In *Current Topics in Bioenergetics,* vol. 3, pp. 1–78.

Grinius, L.L., Jasaitis, A.A., Kadziauskas, Y.P., Liberman, E.A., Skulachev, V.P., Topali, V.P., Tsofina, L.M., and Vladimirova, M.A. (1970). Conversion of biomembrane-produced energy into electrical form. I. Submitochondrial particles. *Biochim. Biophys. Acta 216,* 1–12.

Gutknecht, J. and Walter, A. (1981). Transport of protons and hydrochloric acid through lipid bilayer membranes. *Biochim. Biophys. Acta 641,* 183–188.

Gutweniger, H., Massari, S., Beltrame, M., Colonna, R., Veronese, P., and Ziche, B. (1977). Cation transport in cytochrome oxidase reconstituted vesicles. *Biochim. Biophys. Acta 459,* 216–224.

Hansen, F.B., Miller, M., and Nicholls, P. (1978). Control of respiration in proteoliposomes containing cytochrome aa. *Biochim. Biophys. Acta 502,* 385–399.

Hanstein, W.G. (1976). Uncoupling of oxidative phosphorylation. *Biochim. Biophys. Acta 456,* 129–148.

Haydon, D.A., Hendry, B.M., Levinson, S.R., and Requena, J. (1977). Anaesthesia by the n-Alkanes. A comparative study of nerve impulse blockage and the properties of black lipid bilayer membranes. *Biochim. Biophys. Acta 470,* 17–34.

Henn, F.A. and Thompson, T.E. (1967). Properties of lipid bilayer membranes separating two aqueous phases: Composition studies. *J. Mol. Biol. 31,* 227–235.

Heytler, P.G. (1979). Uncouplers of oxidative phosphorylation. In *Methods in Enzymology,* Fleischer, S. and Packer, L., eds. Academic Press, New York, San Francisco, London, vol. LV, pp. 462–472.

Hinkle, P. (1973). Electron transfer across membranes and energy coupling. *Fed. Proc. 32,* 1988–1992.

Hinkle, P.C. (1979). A model system for transmembrane redox reactions. In *Methods in Enzymology,* Fleischer, S. and Packer, L., eds. Academic Press, New York, San Francisco, London, vol. IV, pp. 541–544.

Hinkle, P.C., Kim, J.J., and Racker, E. (1972). Ion transport and respiratory control in vesicles formed from cytochrome oxidase and phospholipids. *J. Biol. Chem. 247,* 1338–1339.

Huang, C. (1969). Studies on PC vesicles: Formation and physical characteristics. *Biochemistry 8,* 344–351.

Ilani, A. and Berns, D.S. (1972). Photoresponse of chlorophyll-containing bileaflet membranes and effect of phycocyanin as extrinsic membrane protein. *J. Membrane Biol. 8,* 333–356.

Issaurat, B., Amblard, G., and Garach, C. (1980). Electrical manifestations of ion transport across black lipid membranes generated by a Na^+/K^+-ATPase. *FEBS Lett. 115,* 148–150.

Israelachvili, J.N. and Mitchell, D.J. (1975). A model for the packing of lipids in bilayer membranes. *Biochim. Biophys. Acta 389,* 13–19.

Israelachvili, J.N., Mitchell, D.J., and Ninham, B.W. (1976). Theory of self assembly of hydrocarbon amphiphiles into micelles and bilayers. *J. Chem. Soc. Faraday Trans. II 72,* 1525–1568.

Jackson, M.L., Schmidt, C.F., Lichtenberg, D., Litman, B.J., and Albert, A.D. (1982). Solubilization of phosphatidylcholine bilayers by octyl glucoside. *Biochemistry 21,* 4576–4582.

Jasaitis, A.A., Nemecek, I.B., Severina, I.I., Skulachev, V.P., and Smirnova, S.M. (1972). Membrane potential generation by two reconstituted mitochondrial systems: Liposomes inlayed with cytochrome oxidase or with ATPase. *Biochim. Biophys. Acta 275,* 485–490.

Johnson, S.M. and Bangham, A.D. (1969a). Potassium permeability of single compartment liposomes with and without valinomycin. *Biochim. Biophys. Acta 193,* 81–91.

Johnson, S.M. and Bangham, A.D. (1969b). The action of anaesthetics on phospholipid membranes. *Biochim. Biophys. Acta 193,* 92–104.

Juliano, R.L. and Stamp, D. (1976). Lectin-mediated attachment of glycoprotein-bearing liposomes to cells. *Nature 261,* 235–238.

Kagawa, Y. and Racker, E. (1971). Partial resolution of the enzymes catalyzing oxidative phosphorylation. *J. Biol. Chem. 246,* 5477–5487.

Kagawa, Y., Kandrach, A., and Racker, E. (1973). Partial resolution of the enzymes catalysing oxidative phosphorylation XXVI specificity of phospholipids required for energy transfer reaction. *J. Biol. Chem. 248,* 676–684.

Kayushin, L.P. and Skulachev, V.P. (1974). Bacteriorhodopsin as an electrogenic proton pump: Reconstitution of bacteriorhodopsin proteoliposomes generating $\Delta\psi$ and ΔpH. *FEBS Lett. 39,* 39–42.

King, T.E. (1978). Hydrogen and electron transfer in mitochondria. In *Membrane Proteins,* Nicholls, P., Moller, J.V., Jorgensen, P.L., and Moody, A.J., eds. Proc. 11th FEBS Meeting, Pergamon Press, Oxford, vol. 45, pp. 17--31.

Knowles, A.F., Eytan, E., and Racker, E. (1976). Phospholipid-protein interactions in the Ca^{2+}-ATPase of sarcoplasmic reticulum. *J. Biol. Chem 251,* 5161–5165.

Konstantinov, A., Skulachev, V.P., and Smirnova, I.A. (1980). Membrane potential generation by SMP associated with a lipid impregnated filter. *FEBS Lett. 114,* 302-306.

Leung, K.H. and Hinkle, P.C. (1975). Reconstitution of ion transport and respiratory control in vesicles formed from reduced CoQ-cytochrome c reductase and phospholipids. *J. Biol. Chem. 250,* 8467-8471.

MacDonald, R.J. and MacDonald, R.C. (1975). Assembly of phospholipid vesicle vesicles bearing sialoglycoprotein from Erythrocyte membrane. *J. Biol. Chem. 250,* 9206-9214.

Mangel, M. (1976). Properties of liposomes that contain chloroplast pigments: Photosensitivity and efficiency of energy conversion. *Biochim. Biophys. Acta 430,* 459-466.

McLaughlin, S.G.A. and Dilger, J.P. (1980). Transport of protons across membranes by weak acids. *Physiol. Rev. 60,* 825-863.

Montal, M., Chance, B., and Lee, C-P. (1970). Ion transport and energy conservation in submitochondrial particles. *J. Membr. Biol. 2,* 201-234.

Michaelson, D.M., Horwitz, A.F., and Klein, M.P. (1973). Transbilayer asymmetry and surface homogeneity of mixed phospholipids in cosonicated vesicles. *Biochemistry 12,* 2637-2645.

Mimms, L.T., Zampiggi, G., Nozaki, Y., Tanford, C., and Reynolds, J.A. (1981). Phospholipid vesicle formation and transmembrane protein incorporation using octyl glucoside. *Biochemistry 20,* 833-840.

Mitchell, P. (1961). Coupling of phosphorylation to electron and hydrogen transfer by a chemi-osmotic type of mechanism. *Nature 191,* 144-148.

Mitchell, P. (1966). *Chemiosmotic Coupling in Oxidative and Photosynthetic Phosphorylation.* Glynn Research Ltd., Bodmin, Cornwall, England.

Mueller, P. and Rudin, D.O. (1968). Resting and action potentials in experimental bimolecular lipid membranes. *J. Theor. Biol. 18,* 222-258.

Mueller, P., Rudin, D.O., Tien, H.T., and Wescott, W.C. (1962). Reconstitution of cell membrane structure *in vitro* and its transformation into an excitable system. *Nature 194,* 979-980.

Mueller, P., Rudin, D.O., Tien, H.T., and Wescott, W.C. (1964). Formation and properties of bimolecular lipid membranes. In *Recent Progress in Surface Science,* Danielli, J.F., Pankhurst, K.G.A., and Riddiford, A.C., eds. Pergamon Press, Oxford, vol. I, pp. 379-393.

Nichols, J.W. and Deamer, D.W. (1980). Net proton-hydroxyl permeability of large unilamellar liposomes measured by an acid-base titration technique. *Proc. Natl. Acad. Sci. USA 77,* 2038-2042.

Nichols, J.W., Hill, M.W., Bangham, A.D., and Deamer, D.W. (1980). Measurement of net proton-hydroxyl permeability of large unilamellar liposomes with the fluorescent pH probe α-animoacridine. *Biochim. Biophys. Acta 596,* 393-403.

Nicholls, P., Hildebrandt, V., and Wrigglesworth, J.M. (1980). Orientation and reactivity of cytochrome aa_3 in proteoliposomes. *Arch. Biochem. Biophys. 204,* 533-543.

Nozaki, Y. and Tanford, C. (1981). Proton and hydroxide ion permeability of phospholipid vesicles. *Proc. Natl. Acad. Sci. USA 78,* 4324-4328.

Pagano, R.E., Ruysschaert, J.M., and Miller, I.R. (1972). The molecular composition of some lipid bilayer membranes in aqueous solution. *J. Membr. Biol. 10,* 11-30.

Papahadjopoulos, D. (ed) (1978). Liposomes and their uses in biology and medicine. *Ann. N.Y. Acad. Sci. 308.*

Pfeiffer, D.R., Reed, P.W., and Lardy, H.A. (1974). U.V. and fluorescent spectral properties of the divalent cation ionophore A 23187 and its metal ion complexes. *Biochemistry 13,* 4007–4014.

Pressman, B.C. (1968). Ionophorous antibiotics as models for biological transport. *Fed. Proc. 27,* 1283–1288.

Quinn, P.J. and Esfahani, M.A. (1980). Ubiquinones having surface properties suited to transport electrons and protons across membranes. *Biochem. J. 185,* 715-722.

Racker, E. (1973). A new procedure for the reconstitution of biologically active phospholipid vesicles. *Biochem. Biophys. Res. Commun. 55,* 224–230.

Racker, E. (1976). *A New Look at Mechanisms in Bioenergetics.* Academic Press, New York, San Francisco, London.

Racker, E. (1979). Reconstitution of membrane processes. In *Methods in Enzymology,* Fleischer, S. and Packer, L., eds. Academic Press, New York, San Francisco, London, vol. LV, pp. 699–711.

Racker, E. and Kandrach, A. (1971). Reconstitution of the third site of oxidative phosphorylation. *J. Biol. Chem. 246,* 7069–7071.

Racker. E. and Eytan, E. (1973). Reconstitution of an efficient calcium pump without detergents. *Biochem. Biophys. Res. Commun. 55,* 174–178.

Racker, E. and Stoeckenius, W. (1974). Reconstitution of purple membrane vesicles catalyzing light-driven proton uptake and adenosine triphosphate formation. *J. Biol. Chem. 249,* 662–663.

Racker, E., Chien, T-F., and Kandrach, A. (1975). A cholate-dilution procedure for the reconstitution of the Ca^{2+} pump ^{32}P-ATP exchange and oxidative phosphorylation. *FEBS Lett. 57,* 14–18.

Ragan, C.I. and Hinkle, P.C. (1975). Ion transport and respiratory control in vesicles formed from NADH-CoQ reductase and phospholipids. *J. Biol. Chem. 250,* 8472–8476.

Redwood, N.R. and Patel, B.C. (1974). Binding of a solubilized membrane ATPase to phospholipid bilayers. *Biochim. Biophys. Acta 363,* 70–85.

Reed, P.W. (1979). Ionophores. In *Methods in Enzymology,* Fleischer, S. and Packer, L., eds. Academic Press, New York, San Francisco, London, vol. LV, pp. 435–454.

Rothstein, A., Cabantchik, Z.I., Balshin, M., and Juliano, R. (1975). Enhancement of anion permeability in lecithin vesicles by hydrophobic proteins extracted from red blood cell membranes. *Biochem. Biophys. Res. Commun. 64,* 144–150.

Rydström, J., Kanner, N., and Racker, E. (1975). Resolution and reconstitution of mitochondrial nicotinamide nucleotide transhydrogenase. *Biochem. Biophys. Res. Commun. 67,* 831-839.

Schuldiner, S. and Kaback, H.R. (1975). Membrane potential and active transport in membrane vesicles from *Escherichia coli. Biochemistry 14,* 5451–5461.

Sikka, S.C. and Kalra, V.K. (1978). Glutamyl transpeptidase-mediated transport of amino acid in lecithin vesicles. *J. Biol. Chem. 255,* 4399–4402.

Skulachev, V.P. (1971). Energy transformations in the respiratory chain. In *Current Topics in Bioenergetics,* Sanadi, D.R., ed. Academic Press, New York, London, vol. 4, pp. 127–190.

Skulachev, V.P. (1979). Membrane potential and reconstitition. In *Methods in Enzymology,* Fleischer, S. and Packer, L., eds. Academic Press, New York, San Francisco, London, vol. LV, pp. 586–603.

Strittmatter, P., Enoch, H.G., and Fleming, P. (1978). Incorporation of microsomal electron-transfer components into liposomes. In *Methods in Enzymology,* Fleischer, S. and Packer, L., eds. Academic Press, New York, San Francisco, London, vol. LII, pp. 206–211.

Tanford, C. (1979). Hydrostatic pressure in small phospholipid vesicles. *Proc. Natl. Acad. Sci. USA 76,* 3318–3319.

Ting-Beall, H.P., Lees, M.B., and Robertson, J.D. (1979). Interaction of Folch-Lees proteolipid apoprotein with planar lipid bilayers. *J. Membr. Biol. 51,* 33–46.

Tyrrell, D.A., Heath, T.D., Colley, C.M., and Ryman, B.E. (1976). New aspects of liposomes. *Biochim. Biophys. Acta 547,* 259–302.

Walker, N.A. (1975). Uncoupling in particles and intact chloroplasts by amines and nigericin–A discussion of the role of swelling. *FEBS Lett. 50,* 98–101.

Warren, G.B., Toon, P.A., Birdsall, N.J.M., Lee, A.G., and Metcalfe, J.C. (1974). Reconstitution of a calcium pump using defined membrane components. *Proc. Natl. Acad. Sci. USA 71,* 622–626.

Wikström, M., Krab, K., and Saraste, M. (1981). *Cytochrome Oxidase–A Synthesis.* Academic Press, London.

Wolosin, J.M. (1980). A procedure for membrane-protein reconstitution and functional reconstitution of the anion transport system of the human-erythrocyte membrane. *Biochem. J. 189,* 35–44.

Wooster, M.S. and Wrigglesworth, J.M. (1976). Adsorption of glyceraldehyde 3-phosphate dehydrogenase on condensed monolayers of phospholipid. *Biochem. J. 153,* 93–100.

Wrigglesworth, J.M. (1978). Control of cytochrome *c* oxidase activity in reconstituted systems. In *Membrane Proteins,* Nicholls, P., ed. 11th FEBS Meeting, Pergamon Press, Oxford, New York, vol. 45, pp. 95–103.

Wrigglesworth, J.M. and Nicholls, P. (1979). Turnover and vectorial properties of cytochrome c oxidase in reconstituted vesicles. *Biochim. Biophys. Acta 547,* 36–46.

Wrigglesworth, J.M., Packer, L., and Branton, D. (1970). Organisation of mitochondrial structure as revealed by freeze-etching. *Biochim. Biophys. Acta 205,* 125–135.

2
Molecular Aspects of Membrane Fusion

J.A. Lucy

The importance of the stability of the phospholipid bilayer is often emphasized, particularly in relation to theories of the origin of life and to the formation of the first membrane-bound, living structures. Indeed, one of the arguments put forward by Danielli early in research on biological membranes, when the basic organization of the phospholipid molecules was still a matter of some debate, was that the lipid bilayer is a thermodynamically stable structure. He proposed that the lipid bilayer is a natural unit of structure in the same sense that the α helix and the double-stranded helix of deoxyribonucleic acid (DNA) are natural units (Danielli 1967).

Although life for higher organisms begins with cell fusion between sperm and egg, it is less commonly acknowledged that an equally important property of biological membranes is their incipient instability. A full understanding of the behavior of biological membranes, in fact, requires not only a study of their integrity but also of their controlled dispersion. Thomas (1971) has made the following reflective comments on the instability of biological membranes:

> The ultimate challenge to the notion of separateness is cell fusion itself. . . . This tendency to join together, to incorporate, to flow into a single system, seems to me a new and dominant fact of life. Nature seems to like itself, even if we sometimes have our minds set against it. If there were no mechanisms around to counter the drift towards unity, we might end up in fact as a single organism, with all kinds of unworkable problems.

Membrane fusion is thus a macromolecular event of fundamental importance in biology. Cell fusion, which represents one class of membrane fusion phenomena, occurs not only in fertilization but also between myoblasts in the development of skeletal muscle and in macrophages, myxomycetes, and fungi. More numerous and of no less importance are the widespread membrane fusion events at the subcellular level. Again, fertilization provides examples such as in the acrosome reaction occurring in sperm cells and in the cortical granule reaction of egg cells. Other instances of membrane fusion at the subcellular level of organization are endocytosis (both pinocytosis and phagocytosis), fusion between lysosomes and other organelles [compare the recent interesting papers by Amano and Mizuno (1980) and Kielian and Cohn (1980)], and secretion by exocytosis.

These and many other examples of membrane fusion have been reviewed in the multiauthor text edited by Poste and Nicholson (1978).

The fusion of somatic cells in the laboratory has been utilized extensively in recent years as a valuable tool in the study of animal cells and in work on higher plants involving the fusion of protoplasts. The uses of cell fusion and hybrid cells obtained by somatic cell fusion are legion and include somatic cell genetics, analysis of the recessive-dominant character of malignancy, studies on protein synthesis, and studies on the structures of membranes and membrane-bound hormone receptors. One of the most important practical developments in this field is the production of monoclonal antibodies by hybrid cells, termed *hybridoma cells* (Melchers et al. 1978), which can be used to yield, for example, a monoclonal antibody to human acute lymphoblastic leukemia antigen (Ritz. et al. 1980). An additional practical use of membrane fusion is the introduction of enzymes, metabolites, hormones, drugs, etc. into cells in vitro or in vivo by the fusion of liposomes with cells (Gregoriadis and Allison 1980; Ryman and Tyrrell 1980). Fused cells can also be used as recipients to test the biological activity of nucleic acids; polykaryons have been reported, for example, to support gene expression of injected Epstein-Barr virus DNA (Graessmann et al. 1980).

Model Systems for the Study of Membrane Fusion

The complexities of membrane fusion reactions in natural systems are such that many investigators have concentrated their studies on membrane fusion occurring in model systems in the hope that information obtained from relatively simple models may facilitate an understanding of fusion reactions in vivo.

Lipid Vesicles and Bilayer Membranes

Simple lipid systems have received considerable attention since, as reviewed by Papahadjopoulos et al. (1979). lipid vesicles and lipid bilayers provide a means of investigating mechanisms of membrane fusion at the molecular level. Papahadjopoulos et al. have summarized their advantages as follows. Phospholipid vesicles and black-lipid films are stable structures with a limited permeability to small ions and large molecules. Their chemical composition can vary considerably from pure one-component systems to mixtures of various phospholipids differing in headgroup or hydrocarbon chain configuration. Questions relating to headgroup specificity, bilayer fluidity, phase transitions, and separations can therefore be studied in detail. In addition, other membrane components such as cholesterol, glycolipids, and various proteins or glycoproteins can be incorporated into the phospholipid bilayer, and their effects on fusion can be studied under controlled conditions. Finally, molecules that are membrane active and drugs that are known to alter the fusion of natural membranes can be added and their influence on the mechanism of fusion assessed. Papahadjopoulos and colleagues have also drawn attention to the fact that a wide range of methods has been used to detect fusion in model systems. Evidence for fusion has been based on the transfer of labeled lipid molecules between vesicles, changes in nuclear magnetic resonance (NMR) spectra of sonicated vesicles, increase in vesicle size, increase in optical density of vesicle suspensions, mixing of lipid molecules from two vesicle populations, transfer of fluorescent probes or ionophores between vesicles and black-lipid films, mixing and reactivity of two membrane proteins incorporated initially into different lipid vesicles, optical microscopic observations on spherical bilayer black-lipid films, and electrical conductance measurements between apposed hemispherical bilayer black-lipid films. From experiments on the fusion of phospholipid membranes in model systems, Papahadjopoulos and colleagues (Papahadjopoulos 1978; Papahadjopoulos et al. 1979) have constructed a detailed hypothesis for the mechanism of membrane fusion. They propose that the crucial event responsible for triggering most and probably all membrane fusion phenomena is a Ca^{2+}-induced separation of acidic phospholipids (such as phosphatidylserine) into rigid crystalline domains with fusion occurring between domain boundaries on closely apposed membranes.

Despite their numerous advantages, phospholipid vesicles and phospholipid bilayers suffer from a major deficiency as models for biological membranes. They contain no protein. As a consequence, interpretations of phenomena observed with these simple phospholipid membranes may not be adequate to account for the behavior of biological membranes. Hence, although many instances of membrane fusion are "triggered" in vivo by a rise in the intracellular concentration of Ca^{2+}, it does not necessarily follow, as is argued by Papahadjopoulos et al. (1979), that the only role of the entering Ca^{2+} is to induce the phase separation of acidic phospholipids in biological membranes.

In relation to exocytosis, Papahadjopoulos et al. (1979) have proposed that when the intracellular concentration of Ca^{2+} in the vicinity of organelle and plasma membranes is raised sufficiently to trigger a phase separation, the membranes will immediately become fusion susceptible. To Lucy at least, this interpretation appears to be the precise antithesis of a hypothesis, which was previously developed by Poste in a series of papers (Poste and Allison 1971; Poste 1972; Poste and Allison 1973) in which it was proposed that fusion is inhibited when membranes are in a Ca^{2+}-associated state and that membranes become "fusion susceptible" when Ca^{2+} is displaced from the membrane.

Erythrocytes

As indicated previously (Lucy 1977), the erythrocyte membrane, especially the hen erythrocyte membrane, has much to recommend it as a model system for studying membrane fusion and cell fusion in particular. First, the cell does not divide and hence there is no possibility of mistaking a cell that is dividing for two cells that are undergoing fusion as may occur, for example, in experiments with fibroblasts. Second, under normal circumstances, erythrocytes do not fuse. As a result, there is no background of spontaneous cell fusion that may complicate the interpretation of experimentally induced fusion occurring under the influence of viral or chemical agents. Also, about 54% of the total membranous material of the hen erythrocyte is plasma membrane, with nearly 46% being nuclear membrane (Zentgraf et al. 1971). Mitochondrial, lysosomal, and endoplasmic reticulum membranes represent less than 1% of the total membrane. It can therefore be concluded with reasonable confidence that any fusion of erythrocyte membranes observed results from a direct modification of the plasma membrane, not from the release of lysosomal enzymes.

In addition to these advantages, it is relatively simple to determine whether or not fusion has occurred in erythrocytes by comparison with model lipid systems for which sophisticated measurements are needed that are not always straightforward in their interpretation. Phase-contrast light microscopy allows the investigator to see two or more nuclei moving in a common cytoplasm when hen erythrocytes are fused. Furthermore, agents that are damaging to cell membranes as well as being fusogenic can readily be identified and compared by measuring the leakage of hemoglobin. Hemolytic studies have shown that the perturbation of erythrocyte membranes that occurs when the cells are fused by lipid-soluble agents is usually accompanied by membrane damage. Thus in the vitamin A series the fusogens, retinol, α-retinoic acid, and iso-13-retinol caused lysis that was similar to that produced by oleic acid and by oleoylglycerol. Interestingly, however, retinaldehyde was markedly less hemolytic although equally fusogenic (Goodall et al. 1980). Water-soluble fusogens by comparison, particularly polyethylene glycol (PEG), are generally considerably less damaging to membranes.

Fusogenic Agents

Viruses

Several reviews on the fusion of cells induced by viruses are available elsewhere (Hosaka and Shimizu 1977; Poste and Pasternak 1978; see also Chapter 8), and it will not be discussed in detail here.

The study of virus-induced cell fusion as a facet of the cytopathic effects of virus infection is of much importance, but use of the Sendai virus as an experimental tool for the production of cell hybrids has disadvantages that have been summarized by Pontecorvo (1975) as follows:

> Virus treatment seems to be ineffective with some kinds of cells, production of the virus is laborious, the activity of various batches is variable and liable to decay, and the possibility of introducing fragments of viral information into the treated cells is not to be ignored. For all these reasons the search for chemical "fusogens" effective on mammalian cells has gone on for some years.

Poly(ethylene glycol), which is discussed below, is currently the agent of choice for the production of cell hybrids. Its use has been paramount in the production of hybridomas since Galfre et al. (1977) fused antibody-producing B lymphocytes with myeloma cells using poly(ethylene glycol) to yield hybrid cells, which individually continued to secrete the specific antibody of the parent B cells but also grew like the myeloma cells as permanent cell lines.

For the above reasons and because it is less controllable, virus-induced cell fusion is arguably less suitable than chemically induced cell fusion as an experimental model for membrane fusion occurring in vivo. It is also important to note that various pieces of evidence indicate that the molecular mechanisms of cell fusion induced by viruses and by chemicals are not necessarily the same. For example, the two systems respond very differently to thiol reagents. We have observed that the fusion of hen erythrocytes by Sendai virus is more than doubled by *N*-ethylmaleimide. This enhancement was maximal after preincubation of erythrocytes with the agent for 30 min at 37°C (Hart et al. 1975), and it therefore appeared that *N*-ethylmaleimide was acting primarily on erythrocyte proteins rather than on the virus. A similar enhancement of cell fusion was obtained with *N*-phenylmaleimide and with iodoacetamide. Lalazar et al. (1977) also found *N*-ethylmaleimide promotes fusion in virus-agglutinated human erythrocyte ghosts. By contrast, *N*-ethylmaleimide inhibited the fusion of hen erythrocytes induced by oleoylglycerol, glycerol, poly(ethylene glycol), and the ionophore A23187 (Hart et al. 1975). *N*-ethylmaleimide at a higher concentration (11 mM rather than 1 mM) also inhibited the fusion of human erythrocytes by glycerol monooleate but less effectively than with hen erythrocytes. Recently, as discussed in the final section of this chapter, work on thiol reagents has been extended to the fusion of rat erythrocytes by benzyl alcohol, which is inhibited by *N*-ethylmaleimide, tetrathionate, iodoacetamide, and cystamine (Ahkong et al. 1980).

On the basis of experiments undertaken with a "1-day" Sendai virus, Knutton and Pasternak (1979) have proposed that cell fusion induced by viruses and chemicals can be separated into the following consecutive stages. (1) cell agglutination, (2) membrane fusion between adjacent cells, and (3) osmotic swelling. They consider that cell swelling occurs after cell fusion because cells that were fused via small cytoplasmic bridges on treatment with 1-day Sendai virus were seen to form spherical polykaryons on subsequent exposure to hypotonic media. Knutton and Pasternak have commented that although cell swelling has been observed before in virus-induced cell fusion (Volsky and Loyter 1978) and in chemically induced fusion (Lucy 1978), the swelling had been assumed to precede formation of cytoplasmic bridges between cells. In fact, however, we have observed cell swelling to occur both before and after cell fusion, depending on the conditions used (Ahkong et al. 1972; Ahkong et al. 1973a, 1973b). Ahkong et al. (1972) found that the cytoplasms of hen erythrocytes were initially joined by narrow bridges when the cells were incubated with lysophosphatidyl choline dispersed in a lipid emulsion. Homokaryons obtained in this way subsequently rounded up into spherical cells. By contrast, hen erythrocytes treated with oleoylglycerol were first swollen and then fused (Ahkong et al. 1973a, 1973b). With oleoylglycerol, fusion occurred between adjacent spherical cells and oval-shaped unswollen cells were never observed to fuse. Other workers have made comparable observations. Kosower et al. (1978) have reported that the membrane mobility agent, A_2C, promotes the fusion of erythrocytes from several strains of leghorn hens with a high efficiency, and they have described the following "typical" sequence of events: Oval cells change to round cells, round cells approach and stick to other round cells, the cells then fuse into binucleated cells that often participate in additional fusion steps to give multinucleated cells (see also Kosower et al. 1977). Under the same conditions, erythrocytes from white rock hens neither changed shape nor fused on treatment with A_2C. Maggio et al. (1978) observed that hen erythrocytes fused by polysialogangliosides became round after 1.5 h, and that bi- and multinucleated cells appeared after 2-3.5 h of incubation. When fibroblasts were incubated with oleoylamine, Bruckdorfer et al. (1974) similarly found that fusion was preceded by rounding and swelling of the cells.

Hence, rather than the mechanism of cell-to-cell fusion proposed by Knutton and Pasternak being universally applicable to virally and chemically induced cell fusion, it seems much more likely that it applies only to those instances in which fusion is induced by a particulate agent that is not in itself especially hemolytic, for example, 1-day Sendai virus particles or small lipid droplets containing a chemical fusogen. On the other hand, when cell surfaces are treated relatively uniformly with a chemical fusogen that increases membrane permeability, the available evidence indicates that cell swelling precedes cell fusion. Indeed, as discussed below, it appears that many instances of chemically induced cell fusion are initiated by an increase in the permeability of their plasma membranes to Ca^{2+}.

Water-Soluble Chemicals

Lysophosphatidyl Choline

The first simple chemical that we investigated as a potential chemical agent for the induction of cell fusion in erythrocytes was lysophosphatidyl choline (Lucy 1978). Lucy had previously suggested that under appropriate circumstances some of the lipids of biological membranes may be organized in globular micelles that are in dynamic equilibrium with the bimolecular leaflet structure (Lucy 1964a) and that a primary requirement for the fusion of two membranes might be that the membranes involved need to have a relatively high proportion of their phospholipid molecules in the micellar configuration (Lucy 1970). As a consequence of its wedge shape, lysophosphatidyl choline causes bilayers of phosphatidyl choline to break down into globular micelles. The presence of lysophosphatidyl choline in membranes was therefore anticipated to facilitate membrane fusion. Under suitable experimental conditions, it will in fact induce the fusion of a variety of cell types (reviewed in Lucy 1978). Cell fusion induced by lysophosphatidyl choline is, however, accompanied by drastic membrane damage, and it is therefore not particularly useful as an experimental tool for producing hybrid cells. Furthermore, no substantial experimental evidence has been forthcoming to support the idea that transiently produced lysophosphatidyl choline may be involved in membrane fusion occurring in vivo (compare Lucy 1978).

Although it appears probable that globular micelles may be involved in fusion caused by lysophosphatidyl choline, it now seems that membrane fusion induced by other fusogenic chemicals proceeds via different mechanisms. However, Breisblatt and Ohki (1975) have suggested that a semimicelle membrane configuration of the hydrocarbon chains of phospholipids may participate in the fusion of spherical bilayers occurring with and without lysophosphatidyl choline. For the alamethicin-mediated fusion of phosphatidyl choline vesicles, Lau and Chan (1975) have also suggested that transient rearrangements in the local structure of the lipid may occur, with the formation of an inverted micelle of alamethicin.

Cryoprotectant Molecules

High concentrations of dimethyl sulfoxide and glycerol can be used to induce the fusion of cells. For example, incubation of hen erythrocytes with 5 M dimethyl sulfoxide for 5 min at 37°C caused the cells to become spherical, and binucleate cells were then observed after 25-30 min. The number of fused cells and the number of nuclei per cell increased on further incubation (Ahkong et al. 1975a). Dimethyl sulfoxide has been used to enhance the fusion of human diploid cells in vitro induced by poly(ethylene glycol), and the extent of fusion is then directly proportional to the concentration of both chemicals (Norwood et al. 1976).

Multinucleated cells were also produced by the treatment of hen erythrocytes with sorbitol; numerous multinucleated cells were present after 2 h of incubation in 2.5 or 3 M sorbitol. Higher and lower concentrations of sorbitol were less effective. Multinucleated cells were also observed with mannitol (1.25 M, 6 h incubation), ethylene glycol (20%, 5 h incubation), and sucrose (1 M, 5 h incubation) (Ahkong et al. 1975a).

Poly(ethylene Glycol)

Of particular importance among the water-soluble chemicals that induce cell fusion is the polymer poly(ethylene glycol), which was initially used to fuse plant protoplasts (Power et al. 1978). Since poly(ethylene glycol) is relatively nontoxic and yields an exceptionally high incidence of fused cells, the scope for using it as a chemical fusogen is very great. The wide range of cells that are susceptible to fusion by the polymer is another of its remarkable properties. It has been used, for example, to fuse bacterial protoplasts (Fodor and Alfölde 1976; Shaeffer et al. 1976) and plant protoplasts with animal cells (Ahkong et al. 1975b; Dudits et al. 1976; Jones et al. 1976), as well as animal cells in tissue culture; its use has been responsible for the development of hybridomas and monoclonal antibodies (Melchers et al. 1978; Edwards 1981; McMichael and Fabre 1982).

Why is poly(ethylene glycol) so effective in fusing cells? A clue to its mode of action is to be found in the fact that high concentrations of the polymer are necessary for the effective induction of cell fusion. Fifty percent poly(ethylene glycol) is commonly used in the production of hybrid cells. The cells are normally treated with the concentrated solution of the polymer for a short time, usually 1 min, because a longer exposure leads to cell damage. Cell fusion is then observed minutes or hours after removal of most of the poly(ethylene glycol). With preparations of poly(ethylene glycol) having molecular weights ranging from 200 to 20,000, we found that the optimum concentration for cell fusion varied with the molecular weight of the polymer when Lesch-Nyhan (LN) fibroblasts were treated for 1 min (Blow et al. 1978). As the molecular weight was increased, the concentration at which the polymer was most effective decreased, from 60% for poly(ethylene glycol)-200 to 45% for poly(ethylene glycol)-20,000.

To obtain information on the molecular mechansisms by which poly(ethylene glycol) induces the fusion of cells, the state of the water in solutions of the polymer was investigated by means of differential scanning calorimetry. It appeared that the ability of the fusogen to bind water and to decrease the free water content of its aqueous solutions is important in causing cells to fuse. Table 2.1 relates the maximum percentage of cell fusion observed with LN fibroblasts to the concentration of free water in the polymer solutions used and shows that the free water (%/wt/wt) in the solutions that induced maximum cell fusion was zero or, with poly(ethylene glycol)-20,000, extremely low. More dilute solutions, which induce fibroblasts to fuse only when the cells are treated for longer time periods, contained increasing quantities of free water. For example, a 35% wt/wt solution of poly(ethylene glycol)-1500 (giving 10% fusion in 30 min) contained approximately 20% wt/wt free water (Blow et al. 1978). Subsequent work, which is discussed below, has shown that exposure of erythrocytes to fusogenic concentrations of poly(ethylene glycol) increases their permeability to Ca^{2+}, and this is thought to be an important feature of the fusion process (Blow et al. 1979). In addition, poly(ethylene glycol) increases the permeability of phospholiposomes in a comparable manner, and it has been proposed that cell fusion induced by the polymer may occur at sites of dehydration-induced discontinuities in the phospholipid bilayer of two closely adjacent cells (Aldwinckle et al. 1982).

Recently, Honda et al. (1981) have reported that the fusion of human erythrocytes induced by poly(ethylene glycol)-6000 is decreased or even completely prevented when the polymer is recrystallized from chloroform and diethyl ether and/or dialyzed against distilled water. These workers have suggested that

Table 2.1. Fusion of LN Fibroblasts and the Binding of Water by Polyethylene Glycol

PEG (mol wt)	Maximum Cell Fusion[a] Observed (%)	Concentration of PEG[b] (% wt/wt) for Maximum Cell Fusion	Concentration of Free Water[c] (% wt/wt) for Maximum Cell Fusion
200	17.8 (2)	60	0.0
400	73.0 ± 4.6 (4)	55	0.0
600	37.2 ± 4.2 (4)	55	0.0
1,500	27.1 (2)	50	0.0
4,000	25.2 (2)	50	0.0
6,000	13.1 ± 2.4 (4)	50	0.0
20,000	5.2 ± 1.2 (4)	45	0.6

SOURCE: Blow et al. (1978). *FEBS Lett. 94,* 305–310. Reproduced with permission.

[a]Cell fusion was measured after a 30–60-s treatment of the cells with PEG. The cells were seeded at 2 × 10^5 cells onto 50 mm diam Petri dishes, each containing one 2 × 2 mm glass cover slip, and the values given are means of the fusion observed in replicate dishes, ± standard error of the mean (SEM) where appropriate, with the number of dishes shown in parentheses.

[b]Cells were treated with PEG solutions of 0, 30, 40, 45, 50, 55, and 60% wt/wt.

[c]The free water content (percent sample weight) was obtained from the differential scanning calorimetry trace by measuring the area of the peak due to the melting of ice and relating this to a standard curve of peak area against weight of water. The area of each melting peak was measured in triplicate with a planimeter. Samples were cooled and heated three times, and the mean values were obtained.

diethyl ether-soluble and water-soluble contaminants are present in commercial poly(ethylene glycol), and they have further proposed that poly(ethylene glycol) itself merely induces cell aggregation, and, hence, when its contaminants are removed, the polymer does not induce cell fusion. The possibility that poly(ethylene glycol) may not be intrinsically fusogenic is of considerably practical importance and interest since, as indicated above, the polymer has been used for a number of years as a laboratory tool to induce cell fusion and it is currently widely used to fuse lymphocytes with myeloma cells in the preparation of monoclonal antibodies. We have therefore investigated poly(ethylene glycol)-6000 from several different commercial sources and, when appropriate conditions for the fusion of erythrocytes were used, have observed that purification did not decrease the ability of four of the five preparations of poly(ethylene glycol) studied to fuse the cells (Smith et al. 1982). Thus it appears that polymeric poly(ethylene glycol) is itself able to fuse cells, but that some commercial preparations, for example, poly(ethylene glycol) (Wako), have enhanced fusogenic properties resulting from the presence of contaminating substances. The extent of cell fusion obtained in some applications of the polymer may therefore depend to a greater or lesser degree on the enhancement of fusion by contaminant molecules. Effects of this kind may be partially responsible for the variability found in the behavior of poly(ethylene glycol) in different laboratories in the production of hybridomas.

High pH and Calcium Phosphate

Toister and Loyter (1971) reported that avian erythrocytes could be fused by Ca^{2+} at pH 10.5. Agglutination, lysis, and cell fusion were induced by preincubation of the cells at pH 10.5 and 37°C, a subsequent addition of Ca^{2+} in the cold to cause agglutination, and a final incubation of the agglutinated cells at 37°C to induce fusion. It was proposed that cell fusion occurring in this way might arise from a production of lysophosphatidyl choline.

Higher plant protoplasts are also fused by incubation in buffer at high pH, raised temperature, and a high concentration of Ca^{2+} (Keller and Melchers 1973). Ward et al. (1979) subsequently coupled this procedure with a finding by Hartmann et al. (1976), who observed that the fusion of erythrocytes by poly(ethylene glycol) was enhanced on pretreating the cells with proteolytic enzymes in a new method for the induction of interkingdom fusion. Thus preincubation, in a protease solution, of carrot protoplasts with a kidney cell line derived from *Xenopus* tissue enhanced subsequent interkingdom fusion using the high pH-high Ca^{2+} procedure. Potentially viable heterokaryons were repeatedly obtained in this way at high frequencies (consistently greater than 10%), which far exceeded those previously achieved using poly(ethylene glycol). Interestingly, although the method gave a high incidence of interkingdom fusion, intraspecific fusion was enhanced to a comparatively small degree. More recent experiments using these techniques, together with immunofluorescence, have demonstrated the presence and distribution of animal-specific membrane components within the plasma membranes of *Xenopus*-carrot heterokaryons. An inhibitor of cellulose synthesis, 2,6-dichlorobenzonitrile, was used to impair cell wall regeneration so that the plasma membranes of cultured heterokaryons would remain accessible to antibodies. In this way, *Xenopus*-specific surface components were demonstrated in the heterokaryons after 14 days of culture (Ward et al. 1980).

Several reports have been published on chemically induced fusion that involves the precipitation of a calcium salt. It was early noted that hen erythrocytes are fused on the addition of fluoride ions to cells in the presence of Ca^{2+} (Lucy et al. 1971). Zakai et al. (1976) have described in detail the fusion of human erythrocyte ghosts that is promoted by the combined action of calcium and phosphate ions: All of the bivalent metal ions that formed precipitates with Ca^{2+} were found to cause the agglutination of erythrocyte ghosts. However, fusion was distinct from agglutination since Zn^{2+} and Mn^{2+} in phosphate buffer did not promote fusion, although they were active in agglutination. Recently it has been reported that intact human erythrocytes (Baker and Kalra 1979) and hen erythrocytes (Majumdar and Baker 1980) are fused by calcium and phosphate ions. Also, incubation of human erythrocytes with either uranyl ions or the ions of rare earth metals at 37°C for 30-45 min results in cell fusion (Majumdar et al. 1980).

Lipid-Soluble Fusogens

In early work, a variety of lipid-soluble substances were tested for their ability to induce the fusion of erythrocytes (Ahkong et al. 1973b). The authors found that about 30 different fat-soluble substances (100 μg/ml) were able to cause the formation of multinucleated cells, using a suspension of 3×10^8 hen erythrocytes/ml. The most effective fusogenic lipids, e.g., oleoylglycerol, induced fusion within 5-10 min at 37°C, but some lipids required about 1 h to cause a significant incidence of cell fusion. Fusion induced by saturated fatty acids usually occurred only with acids containing 10-14 carbon atoms.

Since low-melting lipids, e.g., oleoylglycerol and oleic acid, induced the fusion of hen erythrocytes, whereas the saturated stearic acid was inactive, Ahkong et al. (1973a) suggested that unsaturated fatty acids and their derivatives may induce erythrocytes to fuse by increasing the proportion of hydrocarbon chains in the membrane that are in a relatively liquid state (Ahkong et al. 1973a). Kennedy and Rice-Evans (1976) have undertaken a spectrofluorometric study of the interactions of oleoylglycerol with human erythrocyte ghosts, the results of which support the idea that the role of oleoylglycerol in inducing cell fusion is at least partially a fluidizing one. They observed a red shift in the wavelength of maximum emission of the fluorescent probe 1-anilino-8-naphthalene sulfonate (ANS) when the probe interacted with membrane ghosts that had been perturbed by glycerol monooleate; a decrease in quantum yield was also found. These workers suggested that their findings, and related observations with *N*-phenyl-1-naphthylamine, indicated that oleoylglycerol produces a less-ordered structure in the membrane, decreasing the constraint around the environment of the fluorescent probes. Kosower et al. (1975), who have observed that the membrane mobility agent, A_2C, promotes the fusion of hen erythrocytes, have concluded that it seems reasonable to suppose that some increase in local fluidity in membranes favors cell fusion. The fusion of phospholipid vesicles with one another similarly requires the lipids of the interacting membranes to be in a fluid state, that is, at a temperature above the transition temperature of the phospholipid present (Papahadjopoulos et al. 1973).

Retinol, for which the term ***membrane active*** was originally proposed (Lucy 1964b), rapidly induces the fusion of hen erythrocytes (Ahkong et al. 1973b), and it has recently been found that retinol needs to be present in the erythrocyte membrane in an approximate equimolar ratio to the membrane phospholipids to induce cell fusion (Goodall et al. 1979). It appears that such molecules are potent fusogens because they partition into the cell membrane very efficiently, producing the high intramembranous concentration that is apparently required for fusion even though their aqueous concentration is relatively low. Specific interactions with retinol, oleolyglycerol, and oleic acid have been observed (at similar molar ratios) with membrane phospholipids in lipid monolayers (Maggio and Lucy 1975), in phospholipid vesicles (Howell et al. 1973), and between oleic acid and erythrocyte ghosts (Cullis and Hope 1978). These observations are all consistent with our suggestion that membrane fusion is facilitated by the insertion into membranes of molecules such as oleic acid and oleolyglycerol that increase the proportion of hydrocarbon chains in the membrane in a relatively liquid state (Ahkong et al. 1973a). Unfortunately, the concept that an increase in the overall fluidity of the lipid region of membranes occurs when they contain a high proportion of exogenous lipids with fluid hydrocarbon chains has been misinterpreted by some workers as implying that exogenous lipophilic substances can modify the intrinsic fluidity of the endogenous phospholipid molecules of membranes. Model experiments with dimethyl sulfoxide and poly(ethylene glycol) in phospholipid vesicles (Lyman et al. 1976; Tilcock and Fisher 1979) and in phospholipid monolayers (Maggio and Lucy 1978) have, in fact, indicated that whether or not the transition temperature of a phospholipid is increased or decreased by the presence of (water-soluble) fusogenic molecules depends on the parameters of the system in question.

When the interactions of fusogenic lipids with phospholipids were investigated using the negative-staining technique of electron microscopy, mixtures of phosphatidyl choline with fusogenic lipids–including oleoylglycerol and oleic acid–occasionally exhibited a hexagonal phase with a repeat distance of 8 nm (Howell et al. 1973). It was therefore suggested that a primary effect of low-melting lipids such as oleoylglycerol and oleic acid in causing cell fusion is on the organization of the phospholipids of the plasma membrane. More recently, Cullis and Hope (1978) have interpreted changes in the ^{31}P NMR spectra of erythrocyte ghosts treated with oleic acid in terms of a progressive conversion of the membrane phospholipids from the bilayer phase to the hexagonal (H_{II}) phase. They have proposed a model for cell fusion induced by oleic acid in which the two outer monolayers of the fusing cells combine to make an intermediate hexagonal (H_{II}) phase that is subsequently restabilized to a bilayer structure to complete the fusion process.

It has been briefly reported that the hydrocarbons tetradecane, hexadecane, and octadecane (0.1 mg/ml), which have melting points at 5.5, 20, and 28°C, respectively, induce hen erythrocytes to fuse (Ahkong et al. 1974). Recently, a new class of fusogenic agents, the *n*-alkyl bromides, has been described and has been reported to fuse both photoreceptor membrane vesicles (Mason et al. 1979) and phospholipid vesicles (Mason et al. 1980). Although fusion induced by alkyl bromides appears to differ in some respects from that caused by other fusogenic agents, it nevertheless appears that the lipids involved need to be above their transition temperature for fusion to occur.

Proteinases

Hartmann et al. (1976) showed that the pretreatment of hen erythrocytes with a proteinase preparation greatly increased the incidence of cell fusion on the subsequent addition of the water-soluble fusogen, poly(ethylene glycol). More recently Laster et al. (1979) have reported that although the lipid-soluble fusogen, oleoylglycerol, fails to fuse human erythrocyte ghosts, they can be fused by this agent if they are pretreated with either neuraminidase or Pronase. Pronase-treated human erythrocyte ghosts fused within 1-2 min of the addition of oleoylglycerol at 37°C, whereas about 60-90 min at 37°C were needed for oleoylglycerol to fuse Pronase-treated intact human erythrocytes. Ghosts treated with Pronase under isotonic conditions were fused by Sendai virus, but those treated with the enzyme under hypotonic conditions were agglutinated but not fused by the virus.

Experiments on the fusion of human erythrocytes fused by oleoylglycerol, which are discussed below, have indicated that membrane-bound proteinases might be involved in cell fusion induced by this agent. Ahkong et al. (1978) therefore argued that it should be possible, under appropriate circumstances, to induce human erythrocytes to fuse by treating them with an exogenous proteinase. They incubated these cells with Pronase and chymotrypsin, but only a limited degree of cell fusion was seen. Subtilisin (Carlsburg) and nagarse were considerably more effective, although fused cells were found only after several hours of treatment at 37°C. At 47°C, however, extensive cell fusion was noted after 60 min, particularly with erythrocytes from human blood that had been stored in the cold for 4 wk. Both aggregation and cell fusion were inhibited in these experiments by the proteinase inhibitor phenylmethanesulfonyl fluoride (PMSF); ethyleneglycoltetraacetate (EGTA) inhibited fusion but not aggregation. The intramembranous particles of the *P*-fracture face of the enzyme-treated cells were clustered on cooling to 0°C before freeze fracturing, and this behavior closely resembled that of hen erythrocytes when cell fusion was induced by treatment with the ionophore A23187 and Ca^{2+} (Vos et al. 1976; Volsky and Loyter 1977). PMSF and EGTA inhibited the cold-induced clustering of intramembranous particles, as well as cell fusion, indicating that freedom of the intramembranous particles to move in the plane of the membrane is associated with the cell fusion process. Similar observations have been made in related experiments with hen erythrocytes (Lucy 1980).

Electrically Induced Cell Fusion

A new technique–electrically induced cell fusion–has recently been developed in a number of centers (Zimmerman et al. 1981; Zimmerman 1982; Neumann et al. 1980; Teissie et al. 1982). Electrically induced cell fusion consists experimentally of two consecutive stages. In the first, the membranes of the cells to be fused are brought into contact by dielectrophoresis; in the second, cell fusion is induced by a reversible electrical breakdown of the plasma membranes of cells in contact. *Dielectrophoresis* refers to the movement of cells in an inhomogenous electric field that results from the field exerting a net force on the field-induced dipoles in the cells. The direction of movement is not reversed when the field is reversed. Hence, cells move to one electrode, attach themselves to one another, and can even line up in a chain, approximately normal to the surface of the electrode.

After the cells have been brought in contact by dielectrophoresis in an alternating field, they are fused by the application of an electric pulse, the intensity of which exceeds that for the (highly localized) electrical breakdown of the plasma membrane. However, a prerequisite for reversible electrical breakdown, and hence for cell fusion, is that the critical membrane voltage is reached fairly rapidly. For mammalian cells, the optimum time for the pulse is between 1 and 4 μs. The electrical technique has been used to fuse a variety of plant and animal cells, as well as liposomes. In an interesting development, Veinken and Zimmerman (1982) have reported that it is possible to fuse electrically a single murine myeloma cell with a single lymphocyte and then to recover the fused cell. In this way they obtained a heterokaryon that was uncontaminated by parent cells, thus in principle avoiding the necessity to select hybridoma cells by culturing them in hypoxanthine, aminopterin, thymidine (HAT) medium. These workers observed that the hybridoma cells were viable and that they will grow in a nutrient medium, but there has so far not been any report of monoclonal antibodies being obtained from hybridomas produced electrically.

Behavior of Membrane Lipids and Proteins in Chemically Induced Cell Fusion

The possibility that chemical fusogens allow exogenous Ca^{2+} to enter erythrocytes has been investigated by Blow et al. (1979) in the light of the fusogenic behavior of the ionophore A23187 (Vos et al. 1976). It was found that both water-soluble and lipid-soluble fusogens [e.g., poly(ethylene glycol) and oleoylglycerol] allowed Ca^{2+} to enter hen erythrocytes within 1 min at 37°C, the entry of Ca^{2+} reaching a maximum in 5-10 min. Closely related but nonfusogenic chemicals, e.g., stearoyl glycerol, did not significantly alter the permeability of the cells. It has been suggested by Blow et al. (1979) that fusogenic compounds increase the permeability of erythrocyte membranes to ions by interacting with membrane phospholipids (compare Maggio and Lucy 1975; Maggio and Lucy 1976) and that the consequent increase in intracellular Ca^{2+} initiates or facilitates events that lead to cell fusion. It is relevant that although excess ethylenediaminetetraacetate (EDTA) inhibits the fusion of hen erythrocytes by poly(ethylene glycol) when it is present throughout the experimental procedure, the chelator is unable to inhibit fusion when it is added after the cells have been incubated with poly(ethylene glycol) and Ca^{2+} ions (Maggio et al. 1976). This finding is consistent with an initial increase in membrane permeability to Ca^{2+} occurring on the addition of chemical fusogens to erythrocytes, since it appears that exogenous Ca^{2+} is not required for fusion to be completed once events leading to cell fusion have been initiated by treatment of cells with poly(ethylene glycol).

Possible involvements of membrane proteins in the fusion of human erythrocyte ghosts have been indicated by related findings recently reported in two papers. Sekiguchi and Asano (1978) found that washed human erythrocyte ghosts can be induced to fuse by the addition of the Sendai virus after the ghosts have been loaded with bovine serum albumin and resealed, and that an antispectrin antibody inhibited both the fusion reaction and virus-induced aggregation of intramembranous particles. Lalazar and Loyter (1979) have observed that membrane fusion can be induced in Sendai virus-agglutinated human erythrocyte ghosts by the addition of trypsin, papain, and Pronase. When erythrocyte ghosts were reacted with antispectrin antiserum, however, both the induction of fusion and proteolysis of membrane spectrin were inhibited.

The behavior of membrane proteins in human erythrocytes during cell fusion induced by oleoylglycerol has been investigated by Quirk et al. (1978). Gel electrophoresis of membrane proteins obtained from ghosts of fused human erythrocytes exhibited a production of material of very high molecular weight, development of a new component in the band-3 region, increased staining of bands 4.3 and 4.5, and a new component moving slightly faster than band 6. Bands 2.1-2.3 were also altered, band 3 was decreased, and band 4.1 was lost. Most, but not all, of these changes in membrane proteins appeared to result from the entry of Ca^{2+} into the treated cells. As the proteinase inhibitor TPCK* partially inhibited both cell fusion and the associated decrease in band-3 protein, it was suggested that perturbation of the structure of the plasma membrane by oleoylglycerol may conceivably expose a normally latent Ca^{2+}-dependent proteinase that is located in the interior of the membrane or on its cytoplasmic surface. It was further proposed that the activation of proteolytic activity in the membrane may result in an increased freedom of movement of integral membrane proteins and that this could be an important feature of membrane fusion.

This approach has been pursued further by Ahkong et al. (1980) in work on rat erythrocytes fused by incubation with benzyl alcohol. The membranes of rat erythrocytes treated with this lipid-soluble fusogen in the presence of Ca^{2+} exhibited particularly extensive changes in the polypeptides that correspond on gel electrophoresis to bands 2 and 3 in human cells. Cell fusion and changes in these proteins were markedly inhibited by *N*-ethylmaleimide and by other thiol reagents, as well as by EGTA (Fig. 2.1). The intramembranous particles of the *P*-fracture faces of cells treated with benzyl alcohol to induce fusion were susceptible to cold-induced aggregation, and this phenomenon also was markedly

---

*1-chloro-4-phenyl-3-L-toluene-*p*-sulfonamidobutan-2-one.

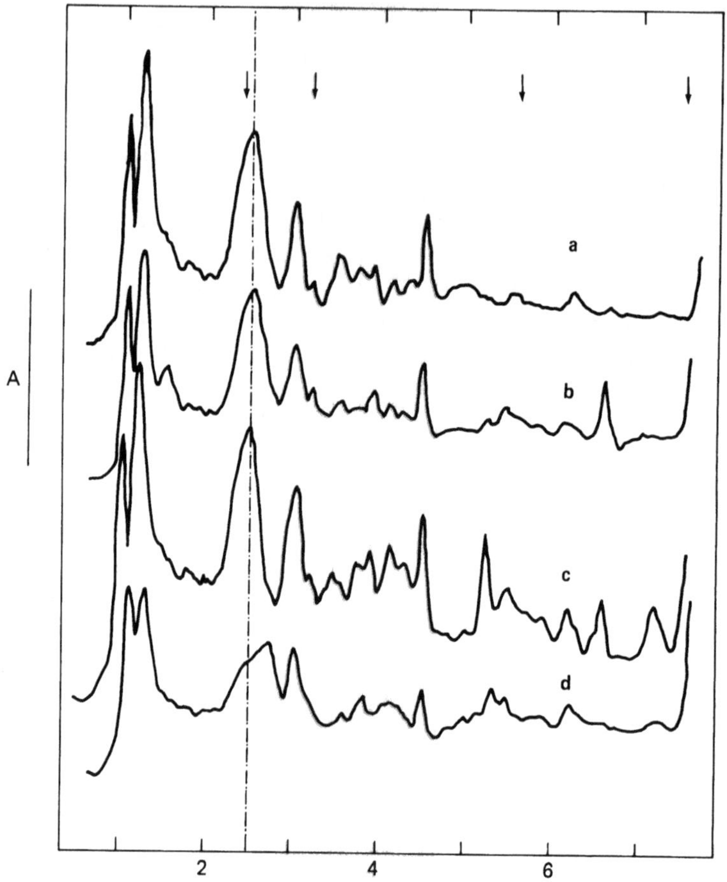

Figure 2.1. Densitometric scans of sodium dodecyl sulfate/polyacrylamide-gel electrophoresis of ghosts obtained from rat erythrocytes treated with benzyl alcohol. The bar (A) in the absorbance scale represents an absorbance increment of 1.0 for each scan. (*a*) The ghosts subjected to electrophoresis were from cells incubated with benzyl alcohol (100 mM) and EGTA (4 mM). (*b*) Cells were preincubated with Ca^{2+} (4 mM) in the absence of benzyl alcohol. (*c*) Cells were preincubated for 10 min at 37°C with *N*-ethylmaleimide (1 mM) and were then incubated with benzyl alcohol (100 mM) and Ca^{2+} (4 mM). (*d*) Cells were incubated with benzyl alcohol (100 mM) and Ca^{2+} (4 mM). All incubations were 15 min at 37°C. Arrows at the top of the figure indicate the positions of molecular weight markers from left to right as follows: phosphorylase *a* (mol. wt. 94,000), bovine serum albumin (68,000), carbonic anhydrase (29,000), and cytochrome *c* (12,384). (Ahkong et al. 1980. *Biochem. J. 192,* 829–836. Data reproduced with permission.)

inhibited by EGTA and partially inhibited by TLCK* and *N*-ethylmaleimide. The ability of the intramembranous particles to move and the altered behavior of membrane proteins on gel electrophoresis support the hypothesis that degradation of membrane proteins may be of major importance in membrane fusion reactions. In human erythrocytes, band 2.1 connects molecules of spectrin, at the cytoplasmic surface, to the major transmembrane protein, band 3 (Bennett and Stenbuck 1979). Losses of band 2.1 in human erythrocytes (Quirk et al. 1978) and of material in the band-2 region in rat erythrocytes together with the production of a polypeptide at or near the front of band 3 (Ahkong et al. 1980) are associated with cell fusion induced by lipid fusogens. It therefore seems likely that the degradation of spectrin-binding proteins in the band-2 region may be responsible for the increased freedom of intramembranous particles in erythrocytes treated with lipid-soluble fusogens and that the subsequent development and interaction of protein-free areas of lipid bilayer in closely adjacent cells may permit cell fusion to occur as suggested earlier (Ahkong et al. 1975a).

References

Ahkong, Q.F., Cramp, F.C., Fisher, D., Howell, J.I., and Lucy, J.A. (1972). Studies on chemically induced cell fusion. *J. Cell Sci. 10,* 769–787.

Ahkong, Q.F., Cramp, F.C., Fisher, D., Howell, J.I., Tampion, W., Verrinder, W., and Lucy, J.A. (1973a). Chemically-induced and thermally-induced cell fusion: Lipid–lipid interactions. *Nature 242,* 215–217.

Ahkong, Q.F., Fisher, D., Tampion, W., and Lucy, J.A. (1973b). The fusion of erythrocytes by fatty acids, esters, retinol and α-tocopherol. *Biochem. J. 136,* 147–155.

Akhong, Q.F., Howell, J.I., Tampion, W., and Lucy, J.A. (1974). The fusion of hen erythrocytes by lipid-soluble and surface-active immunological adjuvants. *FEBS Lett. 41,* 206–210.

Ahkong, Q.F., Fisher, D., Tampion, W., and Lucy, J.A. (1975a). Mechanisms of cell fusion. *Nature 253,* 194–195.

Ahkong, Q.F., Howell, J.I., Lucy, J.A., Safwat, F., Davey, M.R., and Cocking, E.C. (1975b). Fusion of hen erythrocytes with yeast protoplasts induced by polyethylene glycol. *Nature 255,* 66–67.

Ahkong, Q.F., Blow, A.M.J., Botham, G.M., Launder, J.M., Quirk, S.J., and Lucy, J.A. (1978). Proteinases and cell fusion. *FEBS Lett. 95,* 147–152.

Ahkong, Q.F., Botham, G.M., Woodward, A.W., and Lucy, J.A. (1980). Calcium-activated, thiol proteinase activity in the fusion of rat erythrocytes induced by benzyl alcohol. *Biochem. J. 192,* 829–836.

Aldwinckle, T.J., Ahkong, Q.F., Bangham, A.D., Fisher, D., and Lucy, J.A. (1982). Effects of poly(ethylene glycol) on liposomes and erythrocytes

*Tos-Lys-CH_2Cl, 7-amino-1-chloro-3-L-tosylamidoheptan-2-one.

permeability changes and membrane fusion. *Biochim. Biophys. Acta 689,* 548–560.

Amano, F. and Mizuno, D. (1980). F_c-receptor modulates the fusion between phagocytic vesicles and lysosomes in guinea pig polymorphonuclear leukocytes. *FEBS Lett. 115,* 193–196.

Baker, R.F. and Kalra, V.K. (1979). Chemically induced fusion of fresh human erythrocytes. *Biochem. Biophys. Res. Commun. 86,* 920–928.

Bennett, V. and Stenbuck, P.J. (1979). The membrane attachment protein for spectrin is associated with band 3 in human erythrocyte membranes. *Nature 280,* 468–473.

Blow, A.M.J., Botham, G.M., Fisher, D., Goodall, A.H., Tilcock, C.P.S., and Lucy, J.A. (1978). Water and calcium ions in cell fusion induced by poly(ethylene glycol). *FEBS Lett. 94,* 305–310.

Blow, A.M.J., Botham, G.M., and Lucy, J.A. (1979). Calcium ions and cell fusion: Effects of chemical fusogens on the permeability of erythrocytes to calcium and other ions. *Biochem. J. 182,* 555–563.

Briesblatt, W. and Ohki, S. (1975). Fusion in phospholipid spherical membranes I. Effect of temperature and lysolecithin. *J. Membr. Biol. 23,* 385–401.

Bruckdorfer, K.R., Cramp, F.C., Goodall, A.H., Verrinder, M., and Lucy, J.A. (1974). Fusion of mouse fibroblasts with oleoylamine. *J. Cell Sci. 15,* 185–199.

Cullis, P.R. and Hope, M.J. (1978). Effects of fusogenic agent on membrane structure of erythrocyte ghosts and the mechanism of membrane fusion. *Nature 271,* 672–674.

Danielli, J.F. (1967). The formation, physical stability and physiological control of paucimolecular membranes. In *Formation and Fate of Cell Organelles,* K.B. Warren, ed. Academic Press, New York, pp. 239–253.

Dudits, D., Raskó, I., Hadlaczky, G.Y., and Lima-de-Faria, A. (1976). Fusion of human cells with carrot protoplasts induced by polyethylene glycol. *Hereditas 82,* 121–124.

Edwards, P.A.W. (1981). Some properties and applications of monoclonal antibodies. *Biochem. J. 200,* 1–10.

Fodor, K. and Alfölde, L. (1976). Fusion of protoplasts of *Bacilles megaterium. Proc. Natl. Acad. Sci. USA 73,* 2147–2150.

Galfre, G., Howe, S.C., Milstein, C., Butcher, G.W., and Howard, J.C. (1977). Antibodies to major histocompatibility antigens produced by hybrid cell lines. *Nature 266,* 550–552.

Goodall, A.H., Galloway, M.J., Fisher, D., and Lucy, J.A. (1979). The interactions of dispersions of lipid-soluble fusogens with hen erythrocytes. *Biochem. Soc. Trans. 7,* 937–939.

Goodall, A.H., Fisher, D., and Lucy, J.A. (1980). Cell fusion, haemolysis and mitochondrial swelling induced by retinol and derivatives. *Biochim. Biophys. Acta 595,* 9–14.

Graessmann, A., Wolf, M., and Bornkamm, G.W. (1980). Expression of Epstein-Barr virus genes in different cell types after micro-injection of viral DNA. *Proc. Natl. Acad. Sci. USA 77,* 433–436.

Gregoriadis, G. and Allison, A.C. (1980). *Liposomes in Biological Systems.* John Wiley and Sons, Chichester.

Hart, C.A., Ahkong, Q.F., Fisher, D., Hallinan, T., Quirk, S.J., and Lucy, J.A. (1975). Effects of thiol reagents on virus- and chemically induced fusion of erythrocytes. *Biochem. Soc. Trans. 3,* 734-736.

Hartman, J.X., Galla, J.D., Emma, D.A., Kao, K.N., and Gamborg, O.L. (1976). The fusion of erythrocytes by treatment with proteolytic enzymes and polyethylene glycol. *Can. J. Genet. Cytol. 18,* 503-512.

Honda, K., Maeda, Y., Sasakawa, S., Ohno, H., and Tsuchida, E. (1981). The components contained in polyethylene glycol of commercial grade (PEG-6,000) as cell fusogen. *Biochim. Biophys. Acta 101,* 165-171.

Hosaka, Y. and Shimizu, K. (1977). Cell fusion by Sendai virus. In *Virus infection and the cell surface. Cell Surface Reviews,* Poste, G. and Nicholson, G.L., eds. North-Holland Publishing Co., Amsterdam, vol. 2, pp. 129-155.

Howell, J.I., Fisher, D., Goodall, A.H., Verrinder, M., and Lucy, J.A. (1973). Interactions of membrane phospholipids and fusogenic lipids. *Biochim. Biophys. Acta 332,* 1-10.

Jones, C.W., Mastrangelo, I.A., Smith, H.A., Liu, H.Z., and Meck, R.A. (1976). Interkingdom fusion between human (HeLa) cells and tobacco hybrid (GGLL) protoplasts. *Science 193,* 401-403.

Keller, W.A. and Melchers, G. (1973). The effect of high pH and calcium on tobacco leaf protoplast fusion. *Z. Naturforsch. 28c,* 737-741.

Kennedy, A. and Rice-Evans, C. (1976). A spectrofluorimetric study of cell interaction of glycerol mono-oleate with human erythrocyte ghosts. *FEBS Lett. 69,* 45-50.

Kielian, M.C. and Cohn, Z.A. (1980). Phagosome-lysosome fusion. Characterization of intracellular membrane fusion in mouse macrophages. *J. Cell Biol. 85,* 754-765.

Knutton, S. and Pasternak, C.A. (1979). The mechanism of cell-cell fusion. *Trends Biochem. Sci. 4,* 220-223.

Kosower, N.S., Kosower, E.M., and Wegman, P. (1975). Membrane mobility agents. II. Active promoters of cell fusion. *Biochim. Biophys. Acta 401,* 530-534.

Kosower, E.M., Kosower, N.S., and Wegman, P. (1977). Membrane mobility agents. IV. The mechanism of particle-cell and cell-cell fusion. *Biochim. Biophys. Acta 471,* 311-329.

Kosower, N.S., Wegman, P.O., Neiman, T., and Kosower, E.M. (1978). Membrane mobility agents. V. Genetic variability in the fusibility of hen red cells. *Exp. Cell Res. 116,* 454-456.

Lalazar, A. and Loyter, A. (1979). Involvement of spectrin in membrane fusion: Induction of fusion in human erythrocyte ghosts by proteolytic enzymes and its inhibition by antispectrin antibody. *Proc. Natl. Acad. Sci. USA 75,* 1740-1744.

Lalazar, A., Michaeli, D., and Loyter, A. (1977). Restoration of the fusion capacity of human erythrocyte ghosts by SH-blocking reagents. *Exp. Cell Res. 107,* 77-98.

Laster, Y., Lalazar, A., and Loyter, A. (1979). Viral and non-viral induced fusion of pronase-digested human erythrocyte ghosts. *Biochim. Biophys. Acta 551,* 282–294.

Lau, A.L.Y. and Chan, S.I. (1975). Alamethicin-mediated fusion of lecithin vesicles. *Proc. Natl. Acad. Sci. USA 72,* 2170–2174.

Lucy, J.A. (1964a). Globular lipid micelles and cell membranes. *J. Theor. Biol. 7,* 360–373.

Lucy, J.A. (1964b). Membrane permeability and the control of cellular function. Nat. Cancer Inst. Monograph No. 13. 93–107.

Lucy, J.A. (1970). The fusion of biological membranes. *Nature 227,* 814–817.

Lucy, J.A. (1977). The membrane of the hen erythrocyte as a model for studies on membrane fusion. In *Structure of Biological Membranes,* Abrahamsson, S. and Pascher, I., eds. Plenum Press, New York, pp. 275–291.

Lucy, J.A. (1978). Mechanisms of chemically induced cell fusion. In *Membrane Fusion. Cell Surface Review,* Poste, G. and Nicholson, G.L., eds. North-Holland Publishing Co., Amsterdam, vol. 5, pp. 267–304.

Lucy, J.A. (1980). Biochemical and ultrastructural studies on chemically induced cell fusion. In *Membrane-Membrane Interactions,* Gilula, N.B., ed. Raven Press, New York, pp. 81–98.

Lucy, J.A., Ahkong, Q.F., Cramp, F.C., Fisher, D., and Howell, J.I. (1971). Cell fusion without viruses. *Biochem. J. 124,* 46P–47P.

Lyman, G.K., Preisler, H.D., and Papahadjopoulos, D. (1976). Membrane action of DMSO and other chemical inducers of Friend leukemic cell differentiation. *Nature 262,* 360–363.

Maggio, B. and Lucy, J.A. (1975). Studies on mixed monolayers of phospholipids and fusogenic lipids. *Biochem. J. 149,* 597–608.

Maggio, B. and Lucy, J.A. (1976). Polar-group behaviour in mixed monolayers of phospholipids and fusogenic lipids. *Biochem. J. 155,* 353–364.

Maggio, B. and Lucy, J.A. (1978). Interactions of water-soluble fusogens with phospholipids and monolayers. *FEBS Lett. 94,* 301–304.

Maggio, B., Ahkong, Q.F., and Lucy, J.A. (1976). Poly(ethylene glycol) surface potential and cell fusion. *Biochem. J. 158,* 647–650.

Maggio, B., Cumar, F.A., and Caputto, R. (1978). Induction of membrane fusion by polysialogangliosides. *FEBS Lett. 90,* 149–152.

Majumdar, S. and Baker, R.F. (1980). Phosphate-calcium induced fusion of chicken erythrocytes. *Exp. Cell Res. 126,* 175–182.

Majumdar, S., Baker, R.F., and Kalra, V.K. (1980). Fusion of human erythrocytes induced by uranyl acetate and rare earth metals. *Biochim. Biophys. Acta 598,* 411–416.

Mason, W.T., Hladky, S.B., and Haydon, D.A. (1979). Fusion of photoreceptor membrane vesicles. *J. Membr. Biol. 46,* 171–181.

Mason, W.T., Lane, N.J., Miller, N.G.A., and Bangham, A.D. (1980). Fusion of liposome membranes by the n-alkyl bromides. *J. Membr. Biol. 55,* 69–79.

McMichael, A.J. and Fabre, J.W. (1982). *Monoclonal Antibodies in Clinical Medicine.* Academic Press, London.

Melchers, F., Potter, M., and Warner, N.L., eds. (1978). *Lymphocyte hybridomas. Current Topics in Microbiology and Immunology.* Springer-Verlag, New York, vol. 81.

Neumann, E., Gerisch, G., and Opatz, K. (1980). Cell fusion induced by high electric impulses applied to *Dictyostelium. Naturwissenschaften 67,* 414-415.

Norwood, T.H., Zeigler, C.J., and Martin, G.M. (1976). Dimethyl sulphoxide enhances polyethylene glycol-mediated somatic cell fusion. *Somatic Cell Genet. 2,* 263-270.

Papahadjopoulos, D. (1978). Calcium-induced phase-phase changes and fusion in natural and model membranes. In *Membrane Fusion. Cell Surface Reviews,* Poste, G. and Nicholson, G.L., eds. North-Holland Publishing Co., Amsterdam, vol. 5, pp. 765-790.

Papahadjopoulos, D., Poste, G., and Schaeffer, B.E. (1973). Fusion of mammalian cells by unilamellar lipid vesicles: Influence of lipid surface charge, fluidity and cholesterol. *Biochim. Biophys. Acta 323,* 23-42.

Papahadjopoulos, D., Poste, G., and Vail, W.J. (1979). Studies on membrane fusion with natural and model membranes. In *Methods in Membrane Biology,* Korn, E.D., ed. Plenum Press, New York, vol. 10, pp. 1-121.

Pontecorvo, G. (1975). Production of mammalian somatic cell hybrids by means of polyethylene glycol treatment. *Somatic Cell Genet. 1,* 397-400.

Poste, G. (1972). Mechanisms of virus-induced cell fusion. *Int. Rev. Cytol. 33,* 157-252.

Poste, G. and Allison, A.C. (1971). Membrane fusion reaction: A theory. *J. Theor. Biol. 32,* 165-184.

Poste, G. and Allison, A.C. (1973). Membrane fusion. *Biochim. Biophys. Acta 300,* 421-465.

Poste, G. and Nicholson, G.L. (eds.) (1978). *Membrane Fusion. Cell Surface Reviews.* North-Holland Publishing Co., Amsterdam, vol. 5.

Poste, G. and Pasternak, C.A. (1978). *Virus Induced Cell Fusion. Membrane Fusion. Cell Surface Reviews,* Poste, G. and Nicholson, G.L., eds. North-Holland Publishing Co., Amsterdam, vol. 5, pp. 305-367.

Power, J.B., Evans, P.K., and Cocking, E.C. (1978). Fusion of plant protoplasts. In *Cell Surface Reviews,* Poste, G. and Nicholson, G.L., eds. North-Holland Publishing Co., Amsterdam, vol. 5, pp. 369-385.

Quirk, S.J., Ahkong, Q.F., Botham, G.M., Vos, J., and Lucy, J.A. (1978). Membrane proteins in human erythrocytes during cell fusion induced by oleoylglycerol. *Biochem. J. 176,* 159-167.

Ritz, J., Pesando, J.M., Notis-McConarty, J., Lazarus, H., and Scholssman, S.F. (1980). A monoclonal antibody to human acute lymphoblastic leukaemia antigen. *Nature 283,* 583-585.

Ryman, B.E. and Tyrrell, D.A. (1980). Liposomes–Bags of potential. *Essays Biochem. 16,* 49-98.

Schaeffer, P., Cami, B., and Hotchkiss, R.D. (1976). Fusion of bacterial protoplasts. *Proc. Natl. Acad. Sci. USA 73,* 2151-2155.

Sekiguchi, K. and Asano, A. (1978). Participation of spectrin in Sendai virus-induced fusion of human erythrocyte ghosts. *Proc. Natl. Acad. Sci. USA 75,* 1740–1744.

Smith, C.L., Ahkong, Q.F., Fisher, D., and Lucy, J.A. (1982). Is purified poly-(ethylene glycol) able to induce cell fusion? *Biochim. Biophys. Acta 692,* 109–114.

Teissie, J., Knutson, V.P., Tsong, T.Y., and Lane, M.D. (1982). Electric pulse-induced fusion of 3T3 cells in monolayer culture. *Science 216,* 537–538.

Thomas, L. (1971). Adaptive aspects of inflammation. In *Immunopathology of Inflammation,* Forscher, B.K. and Houck, J.C., eds. Excerpta Medica, Amsterdam, pp. 1–10.

Tilcock, C.P.S. and Fisher, D. (1979). Interaction of phospholipid membranes with poly(ethylene glycol). *Biochim. Biophys. Acta 577,* 53–61.

Toister, Z. and Loyter, A. (1971). Ca^{2+}-induced fusion of avian erythrocytes. *Biochim. Biophys. Acta 241,* 719–724.

Vienken, J. and Zimmermann, U. (1982). Electric field-induced fusion: Electrohydraulic procedure for production of heterokaryon cells in high yield. *FEBS Lett. 137,* 11–13.

Volsky, D. and Loyter, A. (1977). Rearrangement of intramembranous particles and fusion promoted in chicken erythrocytes by intracellular Ca^{2+}. *Biochim. Biophys. Acta 471,* 243–259.

Volsky, D.J. and Loyter, A. (1978). Role of Ca^{++} in virus-induced membrane fusion. Ca^{++} accumulation and ultrastructural changes induced by Sendai virus in chicken erythrocytes. *J. Cell Biol. 78,* 465–479.

Vos, J., Ahkong, Q.F., Botham, G.M., Quirk, S.J., and Lucy, J.A. (1976). Changes in the distribution of intramembranous particles in hen erythrocytes during cell fusion induced by the bivalent-cation ionophore A23187. *Biochem. J. 158,* 651–653.

Ward, M., Davey, M.R., Mathias, R.J., Cocking, E.C., Clothier, R.H., Balls, M., and Lucy, J.A. (1979). Effects of pH, Ca^{2+}, temperature and protease pre-treatment on interkingdom fusion. *Somatic Cell Genet. 5,* 529–536.

Ward, M., Davey, M.R., Cocking, E.C., Balls, M. Clothier, R.H., and Lucy, J.A. (1980). Animal-specific membrane components visualised on the surface of animal/plant heterokaryons. *Protoplasma 104,* 75–80.

Zakai, N., Kulka, R.G., and Loyter, A. (1976). Fusion of human erythrocyte ghost promoted by the combined action of calcium and phosphate ions. *Nature 263,* 696–699.

Zentgraf, H., Deumling, B., Jarasch, E-D., and Franke, W.W. (1971). Nuclear membranes and plasma membranes. I. Isolation, characterisation and comparison. *J. Biol. Chem. 246,* 2986–2995.

Zimmermann, U., Scheurich, P., Pilwat, G., and Benz, R. (1981). Cells with manipulated functions: New perspectives for cell biology, medicine and technology. *Angew. Chem. Int. Ed. Eng. 20,* 325–344.

Zimmermann, U. (1982). Electric field-mediated fusion and related electrical phenomena. *Biochim. Biophys. Acta 694,* 227–277.

3

Microheterogeneity in Biological Membranes

R.P. Holmes, T.L. Smith, and F.A. Kummerow

Cellular membranes are often conveniently fractionated into different classes depending on their origin in the cell: plasma membranes, microsomal membranes, mitochondrial membranes, and so on. In some instances more refined fractionations can be made, producing, for example, inner and outer mitochondrial membranes, smooth and rough endoplasmic reticulum membranes, and basolateral and brush-border plasma membranes. Such preparations have routinely been used to characterize the composition and properties of these membranes. These analyses, however, still do not yield a complete picture, as they fail to recognize two apparently characteristic features of biological membranes: first, lipids, proteins, carbohydrates, and other membrane components are asymmetrically distributed between the hemileaflets of the lipid bilayer, the main structural core of membranes; second, evidence is accumulating that within each hemileaflet membrane components segregate into discrete domains that may differ in their size, their composition, and hence their properties. Failure to consider these factors in experiments produces results that are difficult, if not impossible, to interpret fully. The evidence that suggests microheterogeneity is a fundamental property of all biological membranes includes microscopic analyses of cell surfaces, subfractionation of membrane preparations, and biophysical analyses of membranes.

Microscopic Distribution of Cell Surface Components

A morphological examination of the surfaces of most cell types has revealed that they are not smooth and may contain projections, villi, or other protuberances. However, the possibility exists that such morphological variation is imposed on the membrane by intracellular components such as the cytoskeleton. The capping of B lymphocytes in response to antibodies or lectins provides a dramatic demonstration of the interaction between the cytoskeleton and cell surface receptors. The use of fluorescently labeled compounds has made this phenomenon readily visible in the light microscope (Taylor et al. 1971). This movement has been shown to coincide with that of cytoskeletal components, and receptor-cytoskeleton complexes have even been isolated (Flannagan and Koch 1978;

Sheterline and Hopkins 1981) . In another approach, fluorescence recovery after photobleaching experiments have identified a population of immobile components, and their immobility may be caused by cytoskeletal attachments (Edidin 1982). One example could be the β receptors of Chang liver cells that have been shown to be immobile and to exist in patches (Henis et al. 1982). There is evidence from the partitioning of free fatty acids in membranes, which we will discuss in a later section, that receptor-cytoskeletal complexes are apparently localized in specific lipid domains.

There are several good examples of morphological variations within the one cell type. In brush-border-containing epithelial cells, there are gross morphological differences between ventral and dorsal surfaces. The isolated brush-border and basolateral membranes obtained from these surfaces differ both in composition and functional properties (e.g., Del Castillo and Robinson 1982). The sperm cell is another example, as regions of its cell surface are quite clearly structurally and functionally diversified. Cytochemical and morphological analyses have shown that the different surface regions have different compositions and properties (Friend 1982). The hepatocyte is also a morphologically diversified cell. It has three segments in its plasma membrane: sinusoidal, lateral, and bile canicular. These can be readily discerned in isolated plasma membrane fractions (Hubbard et al. 1983). However, the heterogeneity apparent in these cell types is a type of macroheterogeneity, and the question we are really pursuing is whether there is a further heterogeneous distribution of components within these macrodomains. Brush borders, for example, consist of a mixture of highly curved and planar segments, raising the possibility that compositional differences are associated with these ultrastructural variations.

To be observed visually, segregated domains in a cellular membrane, by necessity, have to be examined at the electron microscope level. With this technique, the distribution of surface markers on capillary endothelium has been examined using lectins and cationic ferritin (Simionescu et al. 1981, 1982). These experiments showed a heterogeneous distribution of surface receptors for these compounds apparently related to structural differentiations on the luminal surface. The labeling of the capillary endothelium surface with *Ricinus communis* agglutinin, a galactosyl-specific lectin, is shown in Fig. 3.1. In a similar type of cytochemical approach, using ferritin-labeled antibodies, a nonrandom distribution of surface immunoglobulin molecules was detected on lymphocyte surfaces during microvillus formation (De Petris 1978). These antibodies are preferentially localized in microvilli. A morphological analysis of receptor-mediated endocytosis has also revealed an example of microheterogeneity (Goldstein et al. 1979). Coated pits form on the surface before internalization of the receptor-ligand complex. Many surface proteins are excluded from these pits. The internalized receptors are recycled by reinsertion in the plasma membrane at the cell's leading edge, creating further heterogeneity (Bretscher 1983). Filipin binding to cholesterol that produces membrane deformations when freeze fractures are examined has proven to be a useful cytochemical tech-

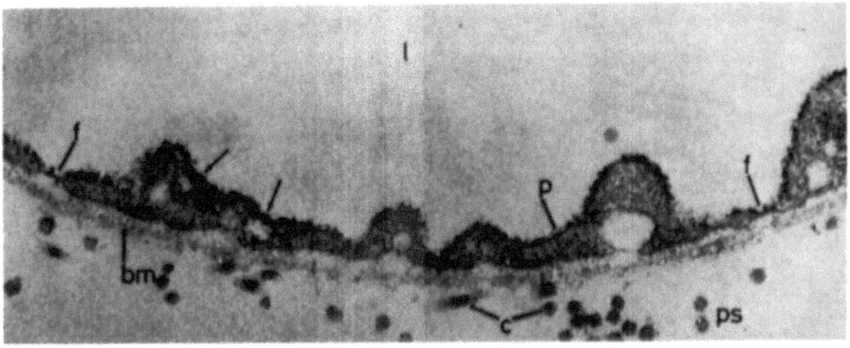

Figure 3.1. Mouse capillary endothelium decorated with the lectin, *R. communis* agglutinin, conjugated to horse radish peroxidase. The labeling on the surface exposed to the lumen is heterogeneous. It is heaviest on stomatal diaphragms (arrows); it occurs in a particulate rather than continuous form on the plasmalemma proper (p); and it is light or absent on fenestral diaphragms (f). bm, basement membrane; c, collagen fibrils; and ps, pericapillary space. Simionescu et al. (1982). *J. Cell Biol. 94,* 406–413. Used with permission of the Rockefeller University Press.

nique for examining the distribution of cholesterol in membranes. Severs (1981), for example, observed a patchy distribution of such deformations in cardiac muscle plasma membranes. An example of the pattern observed is shown in Fig. 3.2. Freeze fracture can also illustrate the distribution of proteins in membranes. An analysis of sperm cells has revealed that within the diverse structural macrodomains of the surface membrane there exists a heterogeneous distribution of particles presumably protein in nature (Friend 1982). Thus, microdomains exist within these macrodomains. A heterogeneous distribution of particles was also observed in nuclear and mitochondrial membranes.

The heterogeneous distribution of membrane components is also evident in the formation of specialized structures in plasma membranes, including neural junctions, junctions between cells, and budding animal viruses. Experiments, for example, with viruses that bud from the plasma membrane have shown that the virus envelope forms from localized lipid regions that do not reflect the average lipid composition of the plasma membrane (Pessin and Glaser 1980). The relationship to the cytoskeleton of the movement of these types of proteins into defined segments of the plasma membrane is uncertain. In some instances, proteins may aggregate because of physical linkages created by transglutamination or some other process.

In model membrane systems there is also microscopic evidence that under certain conditions lipids can segregate into patches differing in lipid composition. A clear demonstration of this is shown in Fig. 3.3, with vesicles consisting of a mixture of phosphatidylserine and dipalmitoylphosphatidylcholine. In these micrographs the corrugated structures are believed to represent dipalmitoyl-

Figure 3.2. Plasma membrane (p face) of a filipin-treated cardiac muscle cell. Filipin-cholesterol deformations occur in clusters, and extensive regions of the membrane remain unaffected by the drug. The bar represents 1 μm. Severs (1981). *Eur. J. Cell Biol.* *25*, 289–299. Used with permission.

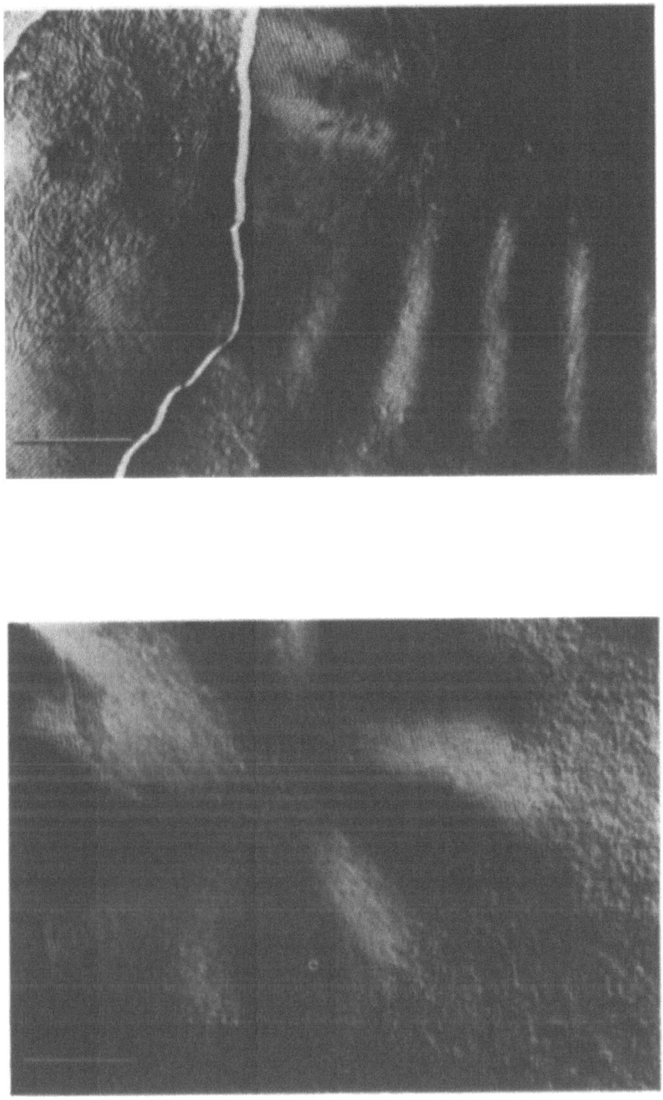

Figure 3.3. Freeze-fracture electron micrographs of lipid dispersions of 33 mol% bovine phosphatidylserine in dipalmitoylphosphatidylcholine quenched from 34°C. Jain (1983) drew attention to the ridges that apparently peeled off from the banded regions and thought they may reflect lipids undergoing cooperative phase changes. The bar represents 0.5 μm. Used with permission.

phosphatidylcholine because the vesicles were quenched at 34°C, which is below the 42°C phase-transition temperature of pure dipalmitoylphosphatidylcholine. The smoother, although still textured, surface is believed to be predominantly phosphatidylserine. Biophysical analyses have been extensively used to study such phase separations in model membrane systems. This evidence will be discussed briefly below. There is no microscopic evidence to suggest that under physiological conditions such lipid segregations occur in biological membranes. There is, however, biophysical evidence that we will outline below.

Isolation of Membrane Subfractions

Several techniques have been successfully used to subfractionate membranes, relying on properties resulting from the heterogeneous distribution of membrane components. Such techniques may rely on charge differences, density differences, or the heterogeneous distribution of specific proteins.

Free-flow electrophoresis has been the main technique used to subfractionate membranes based on a heterogeneous distribution of charged components. Fractions will migrate in an electric current based on the number and type of charged residues. This migration can be manipulated by varying the pH of the electrophoresis buffers. This technique has been successfully applied to the subfractionation of thymocyte plasma membranes (Brunner et al. 1977b). These membranes were separated into three fractions that differed in their cholesterol:phospholipid ratios, enzyme distribution, and the pattern of proteins revealed on sodium dodecyl sulfate (SDS) polyacrylamide gel electrophoresis.

Separation of membrane fractions into subfractions has been observed in density gradients. An example is the separation of "light" and "heavy" lymphocyte plasma membranes on sucrose gradients (Hoessli and Rungger-Brandle 1983). An analysis of surface glycoproteins showed the segregation of surface proteins in these fractions that differed in density because of their different protein contents. Another indication of heterogeneous membrane densities is the spread of a particular membrane marker over many more fractions than could be expected if the membrane fragments being separated all had identical densities. Similarly, the activities of two different marker enzymes for a particular membrane may segregate, as has been observed with the plasma membrane markers, 5′-nucleotidase and Na^+,K^+-adenosine triphosphatase (ATPase), in cultured muscle cells (Schimmel et al. 1973).

Affinity chromatography is proving to be a valuable tool in the separation of heterogeneous membrane fragments. Insulin (Soderman et al. 1973), H2 antibody (Ozer and Wallach 1967), concanavalin A (Brunner et al. 1977a; Schroeder et al. 1982), and a nicotinic receptor (Johansson et al. 1981) are among the affinity ligands that have been successfully used. In several of these studies it was shown that the origin of the different chromatographic behavior was not because the original membrane fractions were a mixture of inside-out and right-side-out vesicles (e.g., Schroeder et al. 1982).

Biophysical Evidence for Microheterogeneity

Model Membrane Systems

Three questions can be asked in model systems concerning membrane microheterogeneity. The first is whether mixtures of different lipids form a homogeneous mixture in lipid bilayers and, if not, what structural forms will this immiscibility create; the second is whether proteins preferentially associate with certain lipids creating heterogeneity; and the third is whether in mixed lipid systems two different proteins will collide with a frequency compatible with completely random movement.

Experiments related to the first question indicate that the coexistence of fluid and gel phases in membranes and the immiscibility of lipids in either the gel or fluid phase can occur (Jain 1983). The degree of phase separation will depend on acyl chain length, the saturation of the acyl chains, the headgroup, the lipid class, and environmental conditions such as temperature, pH, and the concentration of divalent cations. A number of biophysical techniques has been applied to study lipid interactions in mixed bilayers, including differential scanning calorimetry, fluorescence spectroscopy, and electron spin resonance (ESR).

The partitioning of lipophilic probes in liposomes made from lipid mixtures has been particularly useful in studying phospholipid associations in membranes. McConnell and co-workers (Shimshick and McConnell 1973; Wu and McConnell 1975) followed the partitioning of the spin label probe Tempo (2,2,6,6-tetramethylpiperidinyl-1-oxy) in disaturated phospholipid mixtures. Some of their observations have interesting implications for cellular membranes. The first concerns phase transitions. In single-lipid liposomes, at the phase transition temperature, different structural forms of the lipid coexist. At this temperature, fluid and gel domains of the lipid coexist. Properties of membranes, such as the permeability to small molecules, change at the phase transition temperature, possibly because of packing defects at the boundaries between the two phases (Jain 1983; Melchior and Steim 1979). This suggests that if gel and fluid lipid domains coexist in biological membranes, boundary regions may be a vital contributing factor to membrane properties. In binary lipid mixtures, the phase transition temperature was intermediate between those of the single lipids. This indicates that lipid mixing of phases occurs such that at the phase transition temperature the gel and fluid domains each consist of differing proportions of the two lipids.

McConnell and co-workers have been able to construct phase diagrams relating the composition of gel and fluid phases to a particular temperature as shown in Fig. 3.4. Phosphatidylcholines that differed in acyl chain length by two carbons could form a complete range of mixtures. When the headgroup differed in mixtures of phosphatidylcholine and phosphatidylethanolamine, some immiscibility was apparent in the gel phase. Phase diagrams describing the lipid composition of gel and fluid phases for a variety of lipid mixtures have since been studied and have been reviewed by Lee (1977) and Melchior and Steim (1979).

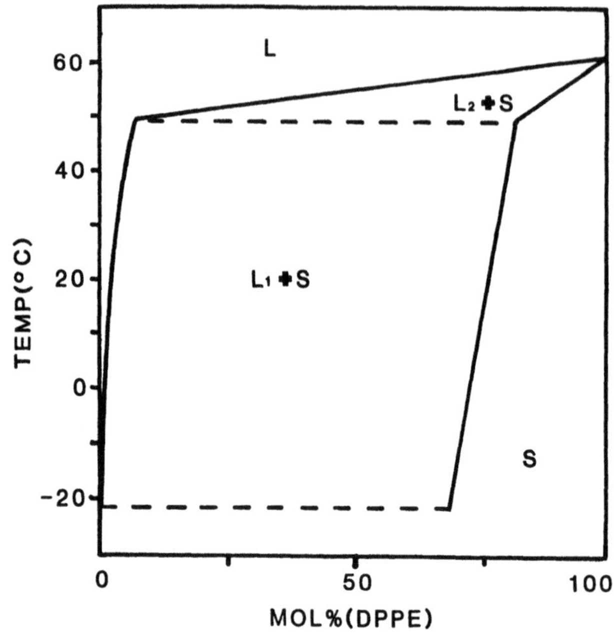

Figure 3.4. Phase diagram for an aqueous dispersion of dioleoylphosphatidylcholine and dipalmitoylphosphatidylethanolamine. S, solid phase, L, L_1, and L_2, liquid phases. Wu and McConnell (1975). *Biochemistry 14,* 847–854. Used with permission of the American Chemical Society.

Wu and McConnell (1975) showed that immiscibility of fluid phase lipids could occur in mixtures of dipalmitoylphosphatidylethanolamine and dielaidoylphosphatidylcholine. Similarly, Galla and Sackmann (1975) showed that dipalmitoylphosphatidylcholine and dipalmitoylphosphatidic acid segregated in the fluid state. In mixtures of dipalmitoylphosphatidylcholine and cholesterol, at temperatures at which dipalmitoylphosphatidylcholine is in the fluid state, two phases form (Recktenwald and McConnell 1981). One has been postulated to be pure dipalmitoylphosphatidylcholine and the other a mixture of cholesterol and dipalmitoylphosphatidylcholine.

Further evidence for the segregation of fluid phase phospholipids has been obtained from an examination of the effects of divalent cations on model membranes. Ca^{2+}, for example, binds tightly to the acidic phospholipids, phosphatidylserine, phosphatidylinositol, cardiolipin, and phosphatidic acid. The addition of Ca^{2+} to membranes containing such acidic phospholipids results in the formation of lipid patches rich in acidic phospholipids (Galla and Sackmann 1975; Ohnishi and Ito 1974; Ito and Ohnishi 1974). In the case of cardiolipin, nonbilayer phases apparently form based on ^{31}P-nuclear magnetic resonance (NMR) spectral analyses and freeze-fracture electron microscopy (Cullis et al. 1978).

The association of lipids with proteins has been a controversial issue since Jost et al. (1973) observed that in protein-lipid mixtures, two populations of lipids were apparent: one associated with proteins and the other with the remaining bulk phase lipids. The lipids associated with proteins were thought to be immobilized, but as this population of lipids is only visible on an ESR time scale (10^{-8} s) and not on an NMR time scale (10^{-5} s), the association is now recognized as being only transitory (Devaux and Davoust 1982). Further confusion exists because the majority of lipid observed to be immobilized on an ESR time scale may represent probe molecules trapped in protein aggregates (Devaux and Davoust 1982). Whether or not this transitory association involves only selected components in a lipid mixture is not known with certainty. It has been suggested, for example, that cholesterol is excluded from those lipids associating with the sarcoplasmic reticulum Ca^{2+}-adenosine triphosphatase (ATPase) (Schneider and Hackenbrock 1982). Such experiments are of profound significance in the consideration of membrane microheterogeneity, as the possibility exists that each membrane protein is transitorily associated with its own unique group of lipid molecules.

If it is accepted that proteins are the main functional units of membranes and that their activity can be influenced by their lipid environment, it becomes reasonable to ask whether different proteins preferentially segregate into specific lipid domains. A consequence of this occurring would be that the frequency of two proteins in different types of lipid domains colliding would be less than two proteins in the same domain. This can be tested by examining a protein activity dependent on protein-protein interactions. Experiments with components of the electron transport chain in the inner mitochondrial membrane indicate that functional associations of segments of the chain occur with a frequency consistent with random movement of the proteins in a single lipid environment (Schneider and Hackenbrock 1982). However, these experiments were carried out with "swollen" membranes, and the relationship of these results to the behavior of proteins in native curved cristal membranes is uncertain.

Thus, studies with model membranes indicate that the types of lipid present in membranes, the presence of integral membrane proteins, the curvature of the membrane, and the interaction of ions with charged headgroups could influence microheterogeneity in biological membranes.

Biological Membranes

As we have already mentioned, morphological analyses and fractionation procedures have provided experimental evidence that lipids in biological membranes segregate into discrete domains. Biophysical analyses with probe molecules have in general supported these observations. One approach has been to measure the lifetime of a fluorescent probe incorporated into membranes (Klausner et al. 1980). More than one lifetime is indicative that the probe has partitioned into more than one lipid domain with differing physical characteristics. In lymphocyte plasma membranes, a heterogeneity analysis of the fluorescence decay of the probe, diphenylhexatriene, demonstrated that 10-20% of the decay was associated with a rapidly decaying lifetime of 2-3 ns, whereas the bulk decayed with a lifetime of 9-10 ns.

The partitioning of other lipophilic probes has yielded some evidence for a heterogeneous distribution of membrane lipids, although advances have not been rapid due to the complex structure and composition of biological membranes. Karnovsky et al. (1982) have reviewed the evidence, primarily from their laboratory, that saturated and unsaturated fatty acids partition into different lipid domains in the membrane: the saturated acids into gellike domains and the unsaturated acids into fluidlike domains. This partitioning is not absolute, and the fatty acids will partition into each type of domain, although preferring one of them. This explanation has been invoked to explain the effect of unsaturated fatty acids on the fluorescence polarization of diphenylhexatriene in biological membranes. These acids appear to function primarily by disordering gel domains in the membrane although they partition mainly in the fluid regions. The addition of unsaturated fatty acids affects membrane phenomena such as capping (Klausner et al. 1980), the movement of membrane-associated cytoskeletal elements (Hoover et al. 1981), and platelet aggregation (MacIntyre et al. 1980). In surveying a range of membrane functional activities affected by the partitioning of saturated and unsaturated fatty acids, it was apparent that many functions were affected in opposing manners and to different extents (Table 3.1). This suggests that the membrane proteins responsible for the expression of the membrane functions were in a range of different lipid domains. Reservations must be held about some of these experiments, as Corps et al. (1980) have found that unsaturated fatty acids affect cellular adenosine triphosphate (ATP) production, apparently by uncoupling mitochondria, whereas saturated fatty acids do not. In studying the lateral diffusion of a variety of fluorescent lipid probes (3,3′dialkylindocarbocyanines) differing in their acyl chain length, Wolf et al. (1981) concluded that the probes they used partitioned into different microenvironments in sea urchin egg plasma membranes and that multiple domains existed in these membranes.

Laggner (1981) examined the interaction of spin-labeled fatty acids in sarcoplasmic reticulum membranes at physiological temperatures. Broadening of the spectra was less in the native membranes than in liposomes prepared from phospholipids extracted from them. This was consistent with the fact that there was a diffusion barrier to lipid movement in native membranes not permitting probe molecules to collide. Laggner estimated that clusters of lipids containing 50 molecules existed in these membranes. It is also possible that in the native membranes fluid and gel domains coexist and that the probe preferentially partitions in the gel regions. There is evidence that the probe, 5-nitroxystearate, partitions in this way (Butler et al. 1974).

Table 3.1. Functional Differentiation of Group A and Group B Free Fatty Acids[a]

Function	Group A[a]	Group B[b]
Temperature agglutination in 3T3 cells		
Wheat germ agglutinin	No effect	Inhibit below 32°C
Concanavalin A	Inhibit below 5°C	Inhibit below 25°C
Adhesion of BHK cells	Inhibit	No effect
Adhesion of mouse lymphocytes	Inhibit	Stimulate
Chloride transport in isolated frog cornea	Stimulate	No effect
D-Glucose transport in isolated plasma membranes from rat adipocytes	Stimulate (—insulin) No effect (+insulin)	No effect (—insulin) Inhibit (+insulin)
Glucose-6-phosphate transport in rat liver membranes	Inhibit	No effect
Brain Na, K-ATPase	Inhibit more	Inhibit less
Fusion of chromaffin granules	Stimulate	No effect
Superoxide production by leukocytes	Stimulate	No efffect
1,25-dihydroxyvitamin D[c] stimulation of Ca^{2+} transport in brush border vesicles isolated from chick intestinal mucosa	Stimulate	No effect

SOURCE: Karnovsky et al. (1982). *J. Cell Biol. 94,* 1-6. The original references may be found in that article. Used with permission.

[a]Group A acids include oleic, *cis*-vaccenic, linoleic, and arachidonic acids.

[b]Group B acids include elaidic, *trans*-vaccenic, stearic, and nonadecanoic.

[c]Fontaine et al. (1981). *Proc. Natl. Acad. Sci. USA 78,* 1751-1754.

Conclusions

An increasing body of evidence suggests that lipids and proteins are not homogeneously mixed in each leaflet of the bilayer. There is some evidence that suggests that proteins may preferentially associated with a particular type of lipid domain in the membrane. For example, β-glucoside and β-galactoside transporters in *Escherichia coli* appear to preferentially partition into fluid domains in the membrane (Thilo et al. 1977). The possible segregation of lipids into discrete domains and the association of proteins with these domains is illustrated in Fig. 3.5. Functionally related proteins may associate in the same domain, facilitating their interaction. The type of lipid in the domain may be intimately associated with the functional activity of the proteins within it. Advances in this area most likely will come from studies on purified membrane proteins reconstituted in liposomes or planar lipid bilayers of known lipid composition and from examination of the relationship between activity and the partitioning behavior of the protein. Other approaches may be developed to elucidate these questions.

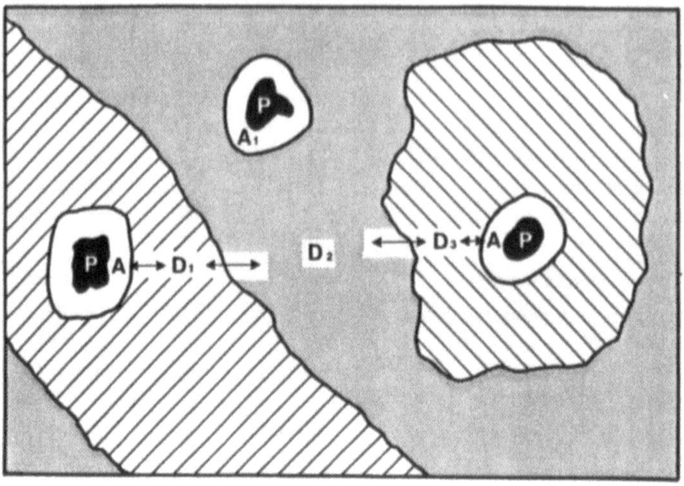

Figure 3.5. The possible arrangement of lipids and proteins in biological membranes. Patches or "islands" of gel-state lipid (D_1 and D_3) are depicted in a "sea" of fluid-state lipid (D_2). When proteins (P) are present they interact with the lipid by transiently immobilizing adjacent lipid molecules and by disturbing their packing (nanosecond time scale). This would create an annulus of lipid (A) affected by the protein. Although the annulus lipid exchanges rapidly with neighboring lipids, a much slower exchange is presumed to occur between the lipids in the various domains. More than one protein may exist in each domain, and the protein density may be greater than that depicted.

An example could be the examination of compositional, functional, and biophysical changes in the membranes of cultured fibroblasts from a poikilothermic organism (flathead minnow) grown at different temperatures (Edidin and Sessions 1983).

Many experiments have attempted to relate a physiological or pathological change in a membrane function to changes in membrane lipid composition and the biophysical properties of the membrane. Such an approach is overly simplistic and ignores bilayer asymmetry in the distribution of lipids, the heterogeneous distribution of lipid in the plane of the membrane, and the actual distribution of molecular species of phospholipids. The reason for not considering these factors is mainly technical. Methods to answer some of these questions have not been completely developed or rigorously tested, and those that have may be difficult, involve expensive instrumentation, and be time consuming. Also, the precise role lipids play in influencing the functional expression of membrane protein activity is uncertain. Studies in which purified membrane proteins are reconstituted in membranes and the lipid composition is systematically varied are still in their infancy (e.g., Churchill et al. 1983).

Concerning the biophysical properties of biological membranes that stem from the physical characteristics of membrane lipids, oversimplification again abounds. Experimental approaches usually have involved examining the spectral properties of fluorescent, spin-label, or isotopic (e.g., deuterated) probes. In the interpretation of most of these experiments, the asymmetric distribution of membrane lipids and the heterogeneous distribution of lipids in each hemileaflet have been ignored. The final spectra obtained then may represent the average of many different lipid domains or may report on one or two lipid environments into which the probe preferentially partitions. Studies analyzing the spectra of probes inserted in liposomes made with phospholipids extracted from biological membranes actually defy interpretation as to their relationship to the state of lipids in the intact membrane. Asymmetry and heterogeneity are most likely abolished. It is our belief that until these considerations are taken into account, the role membrane lipids play in membrane functions will not be clarified.

References

Bretscher, M.S. (1983). Distribution of receptors for transferrin and low density lipoprotein on the surface of giant HeLa cells. *Proc. Natl. Acad. Sci. USA 80,* 454–458.

Brunner, G., Ferber, E., and Resch, K. (1977a). Fractionation of membrane vesicles. I. A separation method for different populations of membrane vesicles of thymocytes by affinity chromatography on Con A-sepharose. *Anal. Biochem. 80,* 420–429.

Brunner, G., Heidrich, H.-G., Golecki, J.R., Bauer, H.C., Suter, D., Pluckhahn, P., and Ferber, E. (1977b). Fractionation of membrane vesicles. II. A method for separation of membrane vesicles bearing different enzymes by free flow electrophoresis. *Biochim. Biophys. Acta 471,* 195–212.

Butler, K.W. Tattrie, N.H., and Smith, I.C.P. (1974). The location of spin probes in two phase mixed lipid systems. *Biochim. Biophys. Acta 363,* 351–360.

Churchill, P., McIntyre, J.O., Eibl, H., and Fleischer, S. (1983). Activation of D-β-hydroxybutyrate apodehydrogenase using molecular species of mixed fatty acyl phospholipids. *J. Biol. Chem. 258,* 208–214.

Corps, A.N., Pozzan, T., Hesketh, T.R., and Metcalfe, J.C. (1980). cis-Unsaturated fatty acids inhibit cap formation on lymphocytes by depleting cellular ATP. *J. Biol. Chem. 255,* 10566–10568.

Cullis, P.R., Verkleij, A.J., and Ververgaert, P.H.J.TH. (1978). Polymorphic phase behavior of cardiolipin as detected by ^{31}P NMR and freeze-fracture techniques. Effects of calcium, dibucaine and chlorpromazine. *Biochim. Biophys. Acta 513,* 11–20.

Del Castillo, J.R., and Robinson, J.W.L. (1982). Preparation of basolateral and brush-border membrane vesicles from guinea-pig intestinal epithelium, and the determination of the orientation of the basolateral vesicles. *Biochim. Biophys. Acta 688,* 45–56.

De Petris, S. (1978). Preferential distribution of surface immunoglobulins on microvilli. *Nature 272,* 66–68.

Devaux, P.F., and Davoust, J. (1982). Current views on boundary lipids deduced from electron spin resonance studies. In *Membranes and Transport,* Martonosi, A.N., ed. Plenum Press, New York, vol. 1, pp. 125–133.

Edidin, M. (1982). Lateral diffusion of membrane proteins. In *Membranes and Transport,* Martonosi, A.N., ed. Plenum Press, New York, vol. 1, pp. 141–144.

Edidin, M. and Sessions, A.V. (1983). Heterogeneity in the plasma membrane lipids of eukaryotic cells. *Ann. N.Y. Acad. Sci. 414,* 8–18.

Flannagan, J. and Koch, G.L.E. (1978). Cross-linked surface Ig attaches to actin. *Nature 273,* 278–281.

Fontaine, O., Matsumoto, T., Goodman, D., and Rasmussen, H. (1981). Liponomic control of Ca^{2+} transport: Relationship to mechanism of action of 1,25-dihydroxyvitamin D_3. *Proc. Natl. Acad. Sci. USA 78,* 1751–1754.

Friend, D.S. (1982). Plasma-membrane diversity in a highly polarized cell. *J. Cell Biol. 93,* 243–249.

Galla, H.-J. and Sackmann, E. (1975). Chemically induced lipid phase separation in model membranes containing charged lipids: A spin label study. *Biochim. Biophys. Acta 401,* 509–529.

Goldstein, J.L., Anderson, R.G.W., and Brown, M.S. (1979). Coated pits, coated vesicles, and receptor-mediated endocytosis. *Nature 279,* 679–685.

Henis, Y.I., Hekman, M., Elson, E.L., and Helmreich, E.J.M. (1982). Lateral motion of β receptors in membranes of cultured liver cells. *Proc. Natl. Acad. Sci. USA 79,* 2907–2911.

Hoessli, D.C. and Rungger-Brandle, E. (1983). Isolation of plasma membrane domains from murine T lymphocytes. *Proc. Natl. Acad. Sci. USA 80,* 439–443.

Hoover, R.L., Fujiwara, K., Klausner, R.D., Bhalla, D.K., Tucker, R., and Karnovsky, M.J. (1981). Effects of free fatty acids on the organization of cytoskeletal elements in lymphocytes. *Mol. Cell Biol. 1,* 939–948.

Hubbard, A.L., Wall, D.A., and Ma, A. (1983). Isolation of rat hepatocyte plasma membranes. I. Presence of the three major domains. *J. Cell Biol. 96,* 217–229.

Ito, T. and Ohnishi, S. (1974). Ca^{2+}-induced lateral phase separations in phosphatidic acid-phosphatidylcholine membranes. *Biochim. Biophys. Acta 352,* 29–37.

Jain, M.K. (1983). Non-random lateral organization in bilayers and biomembranes. In *Membrane Fluidity in Biology,* Aloia, R.C., ed. Academic Press, New York, vol. 1, pp. 1–37.

Johansson, G., Gysin, R., and Flanagan, S.D. (1981). Affinity partitioning of membranes. Evidence for discrete membrane domains containing cholinergic receptor. *J. Biol. Chem. 256,* 9126–9135.

Jost, P.C., Griffith, O.H., Capaldi, R.A., and Vanderkooi, G. (1973). Evidence for boundary lipid in membranes. *Proc. Natl. Acad. Sci. USA 70,* 480-484.

Karnovsky, M.J., Kleinfeld, A.M., Hoover, R.L., and Klausner, R.D. (1982). The concept of lipid domains in membranes. *J. Cell Biol. 94,* 1-6.

Klausner, R.D., Kleinfeld, A.M., Hoover, R.L., and Karnovsky, M.J. (1980). Lipid domains in membranes. Evidence derived from structural perturbations induced by free fatty acids and lifetime heterogeneity analysis. *J. Biol. Chem. 255,* 1286-1295.

Klausner, R.D., Bhalla, D.K., Dragsten, P., Hoover, R.L., and Karnovsky, M.J. (1980). Model for capping derived from inhibition of surface receptor capping by free fatty acids. *Proc. Natl. Acad. Sci. USA 77,* 437-441.

Laggner, P. (1981). Lateral diffusion of lipids in sarcoplasmic reticulum membranes is area limited. *Nature 294,* 373-374.

Lee, A.G. (1977). Lipid phase transitions and phase diagrams. II. Mixtures involving lipids. *Biochim. Biophys. Acta 472,* 285-344.

MacIntyre, D.E., Hoover, R.L., Karnovsky, M.J., and Salzman, E.W. (1980). Inhibition of platelet aggregation by unsaturated fatty acids. *Circulation 62,* 191a.

Melchior, D.L. and Steim, J.M. (1979). Lipid-associated thermal events in biomembranes. *Prog. Surf. Memb. Sci. 13,* 211-296.

Ohnishi, S. and Ito, T. (1974). Calcium-induced phase separations in phosphatidylserine-phosphatidylcholine membranes. *Biochemistry 13,* 881-887.

Ozer, J.H. and Wallach, D.F.H. (1967). H-2 components and cellular membranes: Distinctions between plasma membrane and endoplasmic reticulum governed by the H-2 region in the mouse. *Transplantation 5,* 652-667.

Pessin, J.E. and Glaser, M. (1980). Budding of Rous sarcoma virus and vesicular stomatitus virus from localized lipid regions in the plasma membrane of chicken embryo fibroblasts. *J. Biol. Chem. 255,* 9044-9050.

Recktenwald, D.J. and McConnell, H.M. (1981). Phase equilibria in binary mixtures of phosphatidylcholine and cholesterol. *Biochemistry 20,* 4505-4510.

Schimmel, S.D., Kent, C., Bischoff, R., and Vagelos, P.R. (1973). Plasma membranes from cultured muscle cells: Isolation procedure and separation of putative plasma-membrane marker enzymes. *Proc. Natl. Acad. Sci. USA 70,* 3195-3199.

Schneider, H. and Hackenbrock, C.R. (1982). The significance of protein and lipid mobility for catalytic activity in the mitochondrial membrane. In *Membranes and Transport,* Martonosi, A.N., ed. Plenum Press, New York, vol. 1, pp. 431-435.

Schroeder, F., Fontaine, R.N., and Kinden, D.A. (1982). LM fibroblast plasma membrane subfractionation by affinity chromatography on Con A-sepharose. *Biochim. Biophys. Acta 690,* 231-242.

Severs, N.J. (1981). Plasma membrane cholesterol in myocardial muscle and capillary endothelial cells. Distribution of filipin-induced deformations in freeze-fracture. *Eur. J. Cell Biol. 25,* 289-299.

Sheterline, P. and Hopkins, C.R. (1981). Transmembrane linkage between surface glycoproteins and components of the cytoplasm in neutrophil leukocytes. *J. Cell Biol. 90,* 743–754.

Shimshick, E.J. and McConnell, H.M. (1973). Lateral phase separation in phospholipid membranes. *Biochemistry 12,* 2351–2360.

Simionescu, M., Simionescu, N., and Palade, G.E. (1982). Differentiated microdomains on the luminal surface of capillary endothelium: Distribution of lectin receptors. *J. Cell Biol. 94,* 406–413.

Simionescu, N., Simionescu, M., and Palade, G. (1981). Differentiated microdomains on the luminal surface of the capillary endothelium. I. Preferential distribution of anionic sites. *J. Cell Biol. 90,* 605–613.

Soderman, D.D., Germershausen, J., and Katzen, H.M. (1973). Affinity binding of intact fat cells and their ghosts to immobilized insulin. *Proc. Natl. Acad. Sci. USA 70,* 792–796.

Taylor, R.B., Duffus, P.H., Raff, M.C., and de Petris, S. (1971). Redistribution and pinocytosis of lymphocyte surface immunoglobulin molecules induced by anti-immunoglobulin antibody. *Nature New Biol. 233,* 225–229.

Thilo, L., Trauble, H., and Overath, P. (1977). Mechanistic interpretation of the influence of lipid phase transitions on transport functions. *Biochemistry 16,* 1283–1290.

Wolf, D.E., Kinsey, W., Lennarz, W., and Edidin, M. (1981). Changes in the organization of the sea urchin egg plasma membrane upon fertilization: Indications from the lateral diffusion rates of lipid-soluble fluorescent dyes. *Dev. Biol. 81,* 133–138.

Wu, S.H.-W. and McConnell, H.M. (1975). Phase separations in phospholipid membranes. *Biochemistry 14,* 847–854.

4

Molecular Composition and Functional Properties of Human Liver Mitochondria

Gh. Benga

The liver plays a vital role in human physiology as the major metabolic center for manufacture and interconversion of various metabolites. In the past 3 decades considerable effort has been devoted to the isolation of hepatocyte organelles in order to characterize their composition and function. Such studies, together with ultrastructural approaches, form the basis of our knowledge of how cells function (Fleischer and Fleischer 1977). Liver is a particularly useful tissue for characterization of subcellular organelles since the purified fractions prepared from liver are derived mainly from a single cell type, the hepatocyte. This is not only because hepatocytes constitute 78% of the volume of liver (Blouin et al. 1977) but also because they are the major cell type broken by homogenization of the tissue (Dallner et al. 1966).

A great deal is known about the composition and functional properties of subcellular fractions of rat liver (Rouser et al. 1968). In particular, rat liver mitochondria have been studied extensively (Whittaker and Danks 1978) and the essential role of phospholipids in mitochondrial function has been recognized (Green and Tzagoloff 1966). However, the structure of membranes, in particular the association between lipids and proteins, is a matter of considerable discussion (Benga 1979). The significance of fatty acid composition of lipids is poorly understood. The analysis of the composition of mitochondria from different sources, combined with the knowledge of their properties, should lead to a better understanding of interaction between lipids and proteins in membranes, particularly in regard to their functional significance.

On the other hand, although it was considered that in humans liver is the best enzymologically studied organ, investigations have been performed mainly on homogenates (Schmidt and Schmidt 1970) and have given conflicting results. Studies on subcellular fractions of human liver therefore represent an advancement in this field. The study of human liver mitochondria, for example, is important both for a better knowledge of the metabolic properties of human liver and to illustrate differences in structure and function between liver mitochondria in normal and pathological conditions (cholecystitis, cirrhosis, jaundice, or diseases in which the general metabolism is deranged such as diabetes mellitus and hyperlipidemia).

Isolation of Human Liver Mitochondria

To begin the study of human liver mitochondria we first had to find the optimal conditions for isolation and study of enzymic properties (Benga et al. 1971, 1972). The isolation of human liver mitochondria raises some special problems. Human tissue is obtained by intraoperatory biopsy, whereas the isolation and study of mitochondria must be performed in the laboratory.

Surgical biopsy specimens weighing 0.5-1.0 g were used for isolation of mitochondria as previously described (Benga et al. 1972). The liver biopsy specimen was placed immediately in ice cold 250 mM sucrose, 10mM tris-hydrochloric acid[1] (pH 7.4), and 2 mM sodiumethylenediaminetetraacetate (EDTA) (STE medium) and brought to the laboratory within 15-30 min. All procedures were carried out at 2-4°C. The piece of liver was blotted, quickly weighed, and minced with scissors. After being rinsed in more STE medium to diminish contamination with blood, the material was further homogenized in STE supplemented with 1% defatted serum albumin (bovine or human) in a glass homogenizer with a Teflon pestle. The homogenate (20%) was centrifuged for 10 min at 600 g and the resulting supernate for 10 min at 5500 g. By sedimenting the mitochondria at a lower centrifugal force (5500 g instead of the conventional force of 12,000 g) and by washing the mitochondrial pellet twice, the contamination with lysosomes and microsomes was reduced.

In order to evaluate the optimum conditions for transportation of the tissue from surgery to the laboratory, half the sample was homogenized immediately after its removal, whereas the other half was homogenized only after reaching the laboratory. The transport in bulk is better than that as a homogenate, as revealed by the higher value of respiratory activity obtained in the former case (Benga et al. 1972).

Oxygen uptake was measured spectrophotometrically by the method of Bârzu et al. (1972). The spectrophotometric method is based on utilization of oxyhemoglobin to serve both as a source of oxygen and as an indicator of respiration, the deoxygenation of oxyhemoglobin being accompanied by changes in optical density at suitable wavelengths. The spectrophotometric method for assay of mitochondrial oxygen uptake has been greatly improved in recent years (Bârzu et al. 1978; Mureşan et al. 1980). A typical measurement of oxygen uptake and acceptor control index (ACI) with human liver mitochondria is shown in Fig. 4.1.

The use of defatted serum albumin in the homogenization and washing medium proved to be essential in order to obtain maximum respiratory rates and high values of acceptor control index (ACI) (Fig. 4.2). Table 4.1 shows the oxygen uptake of mitochondria with different substrates; control experiments made by the conventional Warburg technique gave similar results. From citric acid cycle intermediates, isocitrate was oxidized at the fastest rate and pyruvate

[1] Tris = 2-amino-2(hydroxymethyl) propane-1:3 diol

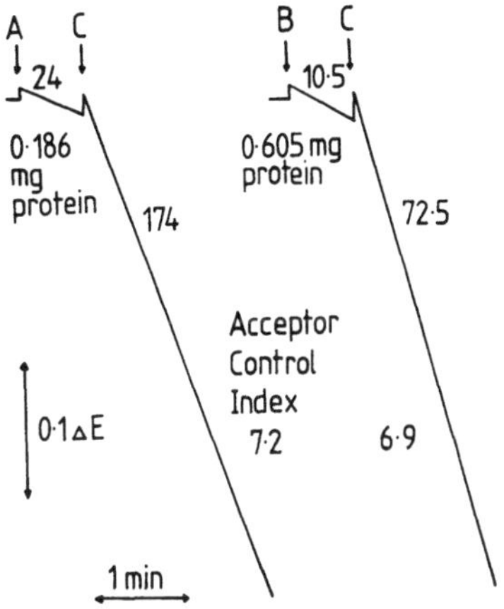

Figure 4.1. Spectrophotometric measurement of oxygen uptake and ACI. Human liver mitochondria were incubated at 37°C in the respiratory medium containing in 1 ml final volume: 180 mM sucrose, 50 mM KCl, 16 mM tris-HCl (pH 7.4), 2.5 mM EDTA, 2.5 mM $MgCl_2$, 5 mM phosphate buffer (pH 7.4), 0.5 mg hexokinase (1 mg = about 3.7 EU), 56 mM glucose, 10 mg defatted bovine serum albumin (BSA), 0.04 meq/l HbO_2. The additions, at the points indicated, were 10 mM succinate (A), 6 mM isocitrate plus 4 mM malate (B) and 0.3 mM adenosine-5′-diphosphate (ADP) (C). The numbers beside the traces indicate the oxygen uptake expressed as nAtoms/min/mg of protein. Data from Benga et al. 1980). *Rev. Roum. Biol. 25,* 147–154.

plus malate at the slowest. The rate of respiration with pyruvate plus malate was the first to be depressed and showed the greatest decrease in the case of liver disease.

In studies on isolated mitochondria there are many differences in the media and conditions used for isolation of organelles and incubation during measurement of enzymic activities. In spite of the greater difficulties entailed, a standardization comparable with that used in the case of determination of serum enzymes is necessary in studies of liver mitochondria isolated from control

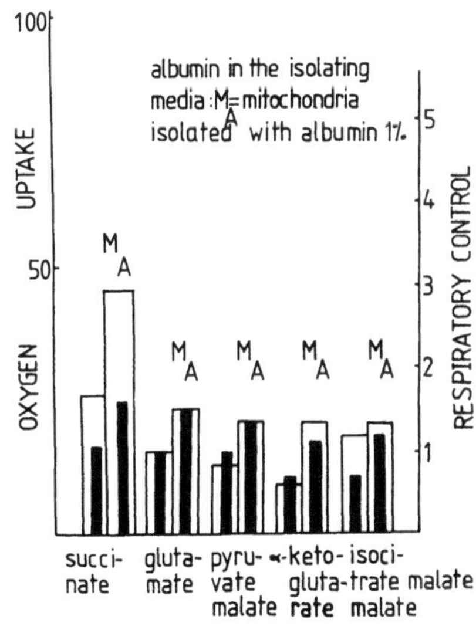

Figure 4.2. Protective role of albumin on the respiratory activity of mitochondria. Effect of defatted BSA on oxygen uptake and respiratory control. The incubation mixture contained the reagents mentioned in the legend to Figure 4.1. The oxygen uptake of mitochondria in state 3 was expressed as nanogram atoms per minute per milligram protein, whereas the respiratory control represents the ratio of the respiratory rate in the presence of ADP (0.2 mM) to the rate before ADP has been added. M_A represent mitochondria isolated with 1% albumin. Data from Benga (1974). *St. Cerc. Biochim. 17,* 123–135.

Table 4.1. Respiratory Activities of Human Liver Mitochondria

Substrate	Additions: Hexokinase Plus Glucose	Additions: NaF	Oxygen Uptake (No. Atoms/min/mg protein)	Acceptor Control Index
Succinate, 10 mM	+	—	134.1 ± 9.9[a]	6.2 ± 0.6[a]
	—	—	102.6 ± 6.2	6.1 ± 0.4
	+	+	123.0 ± 9.3	5.9 ± 0.3
Pyruvate, 6 mM,	+	—	53.1 ± 5.7	5.0 ± 0.6
plus malate,	—	—	30.2 ± 4.6	4.5 ± 0.5
4 mM	+	+	32.7 ± 6.3	4.7 ± 0.4
Isocitrate, 6 mM,	+	—	60.2 ± 4.7	5.6 ± 0.4
plus malate,	—	—	46.6 ± 5.0	5.0 ± 0.4
4 mM	+	+	50.4 ± 3.5	5.1 ± 0.3

Source: Data from Benga et al. (1978). *Eur. J. Biochem. 84,* 625–633.

Note: The oxygen uptake was measured spectrophotometrically (Bârzu et al. 1972). Mitochondria were incubated at 37°C in the respiratory medium containing in 1 ml final volume to 10 mM sucrose, 50 mM KCl, 16 mM tris-HCl (pH 7.4), 2.5 mM EDTA, 2.5 mM $MgCl_2$, 5 mM phosphate buffer (pH 7.4), 10 mg defatted BSA, 0.04 meq oxygenated hemoglobin (HbO_2). Additions were 0.5 mg hexokinase (1 mg = about 3.7 EU) plus 56 mM^2 glucose and 10 mM NaF. Acceptor control index was calculated as the ratio of the respiratory rate in the presence of ADP to the rate before ADP was added.

[a]Mean and standard error of the mean (SEM) for eight determinations.

subjects and those with different diseases. Therefore, we studied different conditions of incubation (Toader et al. 1976). At 37°C the respiratory activity was greater than at 24°C, but it was much greater than expected for enzyme-catalysed, temperature-dependent reactions. This might be explained by the more rapid alterations of mitochondria at 37°C compared with that at 24°C. The effect of the hexokinase-glucose system is dependent on the source of mitochondria. With human liver mitochondria, the addition of hexokinase-glucose results in a stimulation of the respiratory activity. The stability of isolated mitochondria is important for obtaining a confidence time interval for the investigation of different parameters. The respiratory activities of human liver mitochondria were found to be rather stable. However, 3-4 h after isolation, a small decrease of the oxygen uptake and respiratory control index was noted, which was more marked when isocitrate plus malate was the substrate (Toader et al. 1976).

In conclusion, the reproducible procedure for isolation of human liver mitochondria outlined here resulted in a relatively pure fraction exhibiting high values of respiratory activities and respiratory control index. The integrity and purity of the mitochondria was also checked by electron microscopy. As shown in Fig. 4.10(*a*) the isolated human liver mitochondria are in a condensed form, with well-preserved structure.

Lipid Composition of Human Liver Mitochondria

The main characteristic of human liver mitochondria proved to be their lipid composition (Benga et al. 1978). This is illustrated in Table 4.2, which shows the total lipid content as well as the proportion of different lipid classes present in human and rat liver mitochondria. It can be seen that human liver mitochondria contain twice as much lipid as those from the rat; (the difference was statistically significant; $p < 0.001$ by the unpaired t test). This increased lipid content of human liver mitochondria is not associated with an increase in content of a particular lipid class. On the contrary, the proportions between different lipid classes are similar in both human and rat liver mitochondria. Likewise, there were no differences in the proportions of the phospholipid classes between mitochondria from human and rat liver (Table 4.2). However, the fatty acid composition of the total lipids as well as that of the various lipid classes appear to be different in human compared to rat liver mitochondria (Table 4.3). The differences were related to the essential fatty acids; it was thus interesting to notice that the total lipids of human liver mitochondria contain more linoleic acid and less arachidonic acid than those of the rat. These differences appeared

Table 4.2. Lipid Composition of Mitochondria

Lipid	Human Liver μg/mg protein	Human Liver % of total lipid	Rat Liver μg/mg protein	Rat Liver % of total lipid
Total lipids	401.0 ± 29.1 (*N* = 30)	100	205.3 ± 17.2 (*N* = 25)	100
Phospholipids	344.0 ± 16.9 (*N* = 12)	86.0 ± 4.3	180.0 ± 12.5 (*N* = 10)	88.0 ± 3.6[a]
Phosphatidylcholine		40.5 ± 4.0[a]		41.5 ± 5.6[a]
Phosphotidylethanolamine		31.8 ± 8.6[a]		32.7 ± 4.5[a]
Cardiolipin		14.0 ± 1.2[a]		14.2 ± 1.7[a]
Phosphatidylinositol		6.2 ± 1.2[a]		5.6 ± 1.1[a]
Phosphotidylserine		4.0 ± 0.9[a]		3.9 ± 0.7[a]
Sphingomyelin		2.4 ± 0.7[a]		2.1 ± 0.5[a]
Phosphatidic acid		1.0 ± 0.1[a]		0.9 ± 0.1[a]
Triglycerides	20.7 ± 3.3 (N = 9)	5.2 ± 0.5 (n = 9)	12.4 ± 3.0 (N = 5)	6.0 ± 1.2 (N = 5)
Free fatty acids	24.0 ± 5.4 (N = 6)	6.0 ± 1.7 (n = 6)	16.0 ± 7.0 (N = 5)	7.9 ± 2.4 (N = 5)
Cholesterol	11.2 ± 1.9	2.9 ± 0.3	5.4 ± 2.1	2.6 ± 0.3

Note: The measurements were performed as described previously (Benga et al. 1978). The results are expressed as the mean ± SEM; N = the number of experiments.

[a]Percent of total phospholipids (mean and SEM for six experiments).

Table 4.3. Fatty Acid Composition of Rat and Human Liver Mitochondria

Lipid	Source of Mitochondria	Fatty Acids (% of total)[a]						
		14:0	16:0	16:1	18:0	18:1	18:2	20:4
Total lipids	Human (N = 9)	2.2 ± 0.3	19.2 ± 1.3	3.8 ± 0.2	15.0 ± 1.2	17.2 ± 2.1	35.6 ± 4.6	3.6 ± 0.4
	Rat (N = 12)	1.8 ± 0.4	18.0 ± 1.1	3.0 ± 0.5	17.0 ± 2.8	18.1 ± 1.4	26.2 ± 3.1 ($p < 0.025$)	12.0 ± 1.4 ($p < 0.001$)
Phospholipids	Human (N = 7)	1.4 ± 0.7	21.2 ± 3.2	2.9 ± 0.6	14.2 ± 3.9	16.0 ± 1.2	38.1 ± 2.8	4.0 ± 1.4
	Rat (N = 6)	1.1 ± 0.4	17.7 ± 5.0	1.9 ± 0.3	14.0 ± 3.4	16.6 ± 1.6	25.7 ± 6.6 ($p < 0.025$)	14.5 ± 1.7 ($p < 0.01$)
Triglycerides	Human (N = 4)	5.5 ± 1.2	23.0 ± 2.4	11.9 ± 1.8	9.1 ± 1.2	23.5 ± 5.0	23.0 ± 4.5	4.0 ± 1.2
	Rat (N = 4)	6.2 ± 1.2	26.3 ± 2.5	7.0 ± 0.3	10.0 ± 0.7	23.0 ± 1.2	27.0 ± 2.8	4.3 ± 1.1
Free fatty acids	Human (N = 5)	5.4 ± 1.2	29.5 ± 1.4	3.2 ± 1.4	21.2 ± 6.3	16.5 ± 5.7	18.8 ± 3.9	2.0 ± 0.9
	Rat (N = 4)	3.0 ± 0.6	31.1 ± 1.7	4.7 ± 0.9	15.7 ± 3.4	20.5 ± 4.3	22.5 ± 2.8	2.2 ± 0.2

Source: Data from Benga et al. (1978). *Eur. J. Biochem, 84,* 625–633.

[a]Fatty acids are designated as follows: 14:0 myristate, 16:0 palmitate, 16:1 palmitoleate, 18:0 stearate, 18:1 oleate, 18:2 linoleate, 20:4 arachidonate, 22:6 4,7,10,13,16,19-docosahexaenoate. Means ± SEM for the number of determinations (N) indicated; values are expressed in relative percentages of the sum of fatty acids.

Table 4.4. Fatty Acid Composition of Phospholipid Classes in Rat and Human Liver Mitochondria

Phospholipid	Source of Mitochondria	Fatty Acids (% of total)[a]							
		14:0	16:0	16:1	18:0	18:1	18:2	20:4	22:6
Phosphatidyl-choline	Human	1.1	16.5	3.7	19.3	20.4	34.0	5.5	–
	Rat	0.3	18.4	3.0	16.1	22.7	19.8	17.5	2.9
Phosphatidyl-ethanolamine	Human	1.1	21.7	2.6	16.2	18.5	35.0	3.6	–
	Rat	1.0	19.5	2.5	22.6	15.5	16.9	17.0	5.2
Cardiolipin	Human	1.2	10.0	3.7	7.7	7.4	70.2	0.0	–
	Rat	1.0	5.0	1.9	5.0	6.6	77.3	3.2	1.1

Source: Data from Benga et al. (1978). *Eur. J. Biochem. 84*, 625–633.

[a]Mean of three to four determinations.

in the phospholipid fraction of mitochondrial lipids and not in triglyceride and free fatty acid fractions (Table 4.3). As the phospholipids are found predominately in cell membranes, it is clear that the lipid composition of mitochondrial membranes in human and rat livers are different.

The same pattern of fatty acid composition was observed in the major phospholipid classes, that is, more linoleic acid and less arachidonic acid in human compared to rat liver mitochondria (Table 4.4). When the extent of unsaturation of the lipids was determined it was clear that human liver mitochondria contain more saturated lipids than those of the rat (Table 4.5). The degree of saturation of the lipids was quantitated by calculating the unsaturation index and the

Table 4.5. Unsaturation of Fatty Acids in Liver Mitochondria

Lipid	Source of Mitochondria	Unsaturation Index[a]	Σ Unsaturated Fatty Acid / Σ Saturated Fatty Acid
Total lipids	Human	106.6	1.6
	Rat	121.5	1.6
Phospholipids	Human	111.1	1.6
	Rat	152.7	1.8
Phosphatidylcholine	Human	114.1	1.7
	Rat	125.3	1.8
Phosphatidylethanolamine	Human	105.3	1.4
	Rat	151.0	1.3
Cardiolipin	Human	154.3	4.3
	Rat	181.9	8.1

Source: Data from Benga et al. (1978). *Eur. J. Biochem. 84*, 625–633.

[a]Unsaturation index = $\sum_{a=1}^{k}$ [(number of double bonds in a) × (wt% occurrence of a)].

ratio of unsaturated fatty acids to saturated fatty acids in the particular lipid fraction. The unsaturation index of total lipids and of various phospholipid classes was greater in the rat than in the human liver mitochondria. The ratio of unsaturated fatty acids to saturated fatty acids in cardiolipin is also greater in rat liver mitochondria.

The fatty acid composition of mitochondrial phospholipids is a challenging problem. Marked variations in fatty acid composition of mitochondria from different species have been reported (Rouser et al. 1968). A comparison of the fatty acid composition of liver mitochondria obtained from different species gives an interesting picture (Table 4.6). Chicken liver mitochondria contain more oleic acid and less stearic and arachidonic acids than rat liver mitochondria. Fish liver mitochondria contain very little linoleic, linolenic, and arachidonic acids, however, they have a high content of unsaturation because they contain substantial amounts of very long-chain polyenoic acyl residues (20:5, 22:5, 22:6). Although dietary differences may be responsible for some of the species variation in fatty acid composition, it is clear that mitochondria may have widely different fatty acid compositions (Rouser et al. 1968). Richardson et al. (1961, 1962) showed that fish-eating birds retain linoleic and arachidonic acids in mitochondria, although acids of the linolenate family were also found, demonstrating that the diet only influences to some extent the fatty acid composition of mitochondria (Table 4.6).

Human liver mitochondria, with a high linoleate and a low arachidonate con-

Table 4.6. Fatty Acid Composition of Liver Mitochondria of Different Species[a]

Fatty	Human	Rat	Chicken	Salmon	Catfish	Cod	Pelican	Cormorant
14:0	2.2	1.8	0.3	1.2	0.9	1.6	–	–
16:0	19.2	18.0	26.0	19.1	19.1	21.6	17.0	20.2
16:1	3.8	3.0	1.3	2.5	2.5	8.6	4.9	4.9
18:0	15.0	17.0	7.5	11.2	6.1	7.0	19.6	19.5
18:1	17.2	18.1	38.2	19.9	41.2	32.5	19.6	16.7
18:2	35.6	26.2	16.1	0.9	–	0.4	2.2	2.2
18:3	–	–	–	–	–	–	–	0.3
20:3	–	–	–	1.0	–	–	–	–
20:4	3.6	12.0	6.2	4.0	4.5	1.5	8.5	12.2
20:5	–	–	–	16.5	8.4	14.0	14.2	6.8
22:5	–	–	–	6.5	1.8	1.1	–	0.5
22:6	–	–	2.6	15.6	11.8	7.4	11.9	12.7

Sources: Human and rat data from Benga et al. (1978). *Eur. J. Biochem. 84*, 625–633; chicken, salmon, catfish, and cod data from Richardson et al. (1961). *Arch. Biochem. Biophys. 94*, 1–6; pelican and cormorant data from Richardson et al. (1962). *J. Lipid Res. 3*, 344–350.

[a]Percent of total fatty acids.

tent, appear to be another interesting case of species variation in fatty acid composition. It is noteworthy that such a pattern of mitochondrial fatty acids has not been reported for any other species. Moreover, when dietary alterations of the mitochondrial fatty acid composition are induced in the rat, the linoleate and arachidonate increase or decrease in parallel (Hayashida and Portman 1960; Stancliff et al. 1969). The pattern of fatty acid composition of human liver mitochondria reported here does not seem to be related to the diet of the patients investigated; the determinations were performed over a period of several years on subjects with no special diets or alimentary habits. The variations among different subjects were rather small, this is, of the same order of magnitude as the variations for the determinations on rats fed a standard laboratory diet. The higher linoleate and lower arachidonate in human liver mitochondria suggest peculiarities in the phospholipid metabolism of human liver.

Spin-Label Studies

A comparison of human and rat liver mitochondria using spin-labeled fatty acid probe methods was undertaken to determine the effect of the higher lipid content of human liver mitochondria and the differences in fatty acid composition between the two species.

For the spectra that indicated strong impedence of spin-label motion (5 NS[2] in mitochondrial suspensions), the splitting between the low- and high-field extrema ($2\ T_{\parallel}$) was used as an index of probe mobility (see notation in Fig. 4.3). A greater value of the hyperfine splitting $2\ T_{\parallel}$ reflects more restricted rotational motion, therefore indicating a more viscous environment around the spin probe (McConnell and McFarland 1970). As shown in Fig. 4.4 the mobility of 5 NS is higher in liver mitochondria from the human than in those from the rat, at all temperatures studied.

The rotational correlation time (τ_c) is a defined motion parameter that is reasonably accurate for isotropic motion in the fast tumbling region (Kivelson 1960) and can be expressed (Griffith et al. 1965; Mehlhorn and Keith 1972) as follows:

$$\tau_c = 6.5 \times 10^{-10}\, W_0 (\sqrt{h_0/h_1} - 1)$$

where W_0 (in gauss) is the width of the midfield line and h_0 and h_1 are the heights of the mid- and high-field lines on the first derivative spectrum (see notation in Fig. 4.5). A large value of τ_c indicates a rigid microenvironment around the spin label. The rotational correlation time has been used as a motion parameter for the spectra of 16 NS[3] in mitochondrial suspensions. The temperature

[2] 5 NS = *N*-oxyl-4,4′-dimethyloxazolidine derivative of 5 ketostearic acid.
[3] 16 NS = *N*-oxyl-4,4′-dimethyloxazolidine derivative of 16 ketostearic acid.

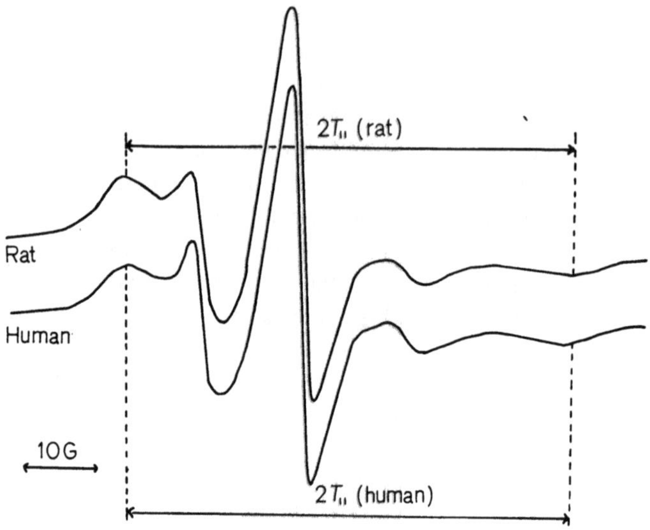

Figure 4.3. The electron spin resonance (ESR) spectra of rat and human liver mitochondria labeled with 5 NS. A modulation amplitude of 1.6 G, a microwave power of 2 mW, and a gain of 1.10^5 were used for these spectra. The temperature was 20°C. Data from Benga et al. (1978). *Eur. J. Biochem. 84*, 625–633.

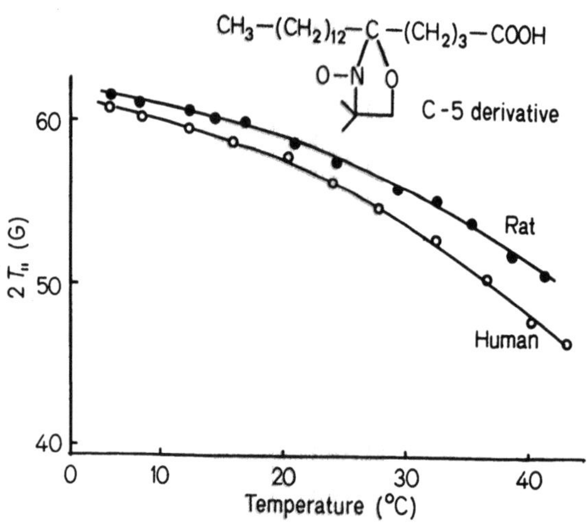

Figure 4.4. Temperature dependence of the hyperfine splitting (2 $T_{\|}$) obtained from the spectra of 5 NS in human and rat liver mitochondria. Preparation and labeling of membranes and ESR measurements were performed as previously described (Benga et al. 1978).

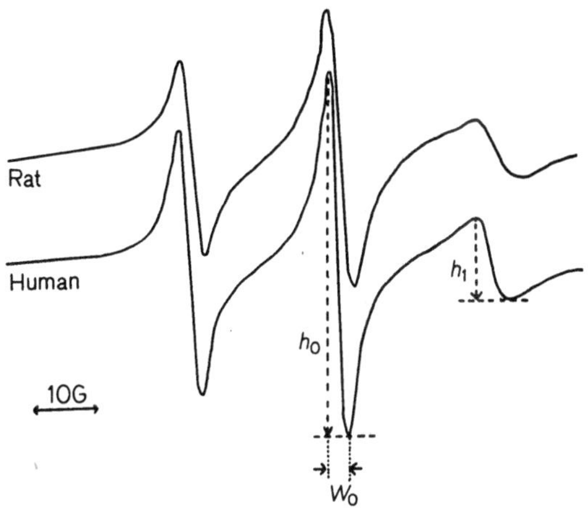

Figure 4.5. The ESR spectra of human and rat liver mitochondria labeled with 16 NS. The microwave power was 2 mW, the gain 1.10^5, and the modulation amplitude 1.6 G. The temperature was 31°C.

dependence of motion of 16 NS showed a higher mobility of the label in human liver mitochondria as compared with rat liver mitochondria (Fig. 4.6).

With 5 NS there are no obvious discontinuities in the plots of 2 $T_{\|}$ against temperature for both human and rat liver mitochondria (Fig. 4.4). With 16 NS for both types of mitochondria, a discontinuity was observed in the plot of τ_c against temperature. The break was at about 22°C for rat and 13°C for human liver mitochondria (Fig. 4.6).

Our data indicate that the fluidity of membrane lipid in human mitochondria, as inferred by the spin-labeling technique, is greater than in the rat. This particularity of human liver mitochondria might be of physiological significance since membrane fluidity plays an important role in cellular processes, such as regulation of enzyme activity, permeability, and transport phenomena. There are three

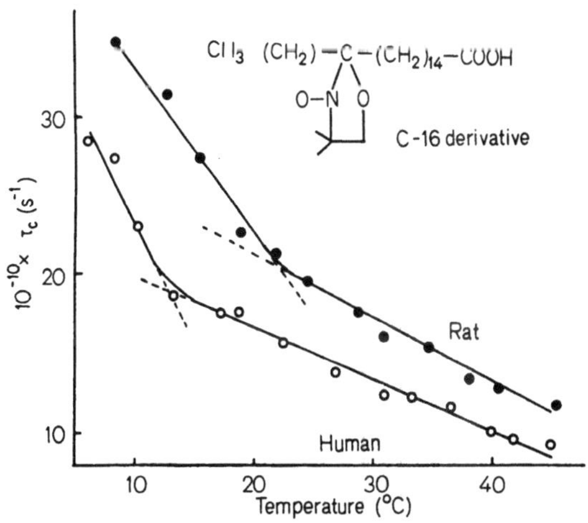

Figure 4.6. Temperature dependence of the rotational correlation time in human and rat liver mitochondria labeled with 16 NS. Preparation and labeling of membranes and ESR measurements were performed as described in previously. Benga et al. (1978). *Eur. J. Biochem. 84*, 625–633.

main factors that control the membrane fluidity: the chain length and unsaturation of fatty acids, the cholesterol, and the proteins. We have seen that the unsaturation of fatty acids in human liver mitochondria is slightly less than in rat liver mitochondria, whereas the cholesterol content is the same. On the other hand, the ratio of lipid to protein in human liver is twice that found in rat liver mitochondria. Purified mitochondrial membranes show the same effects as intact mitochondria (Benga, unpublished data). The higher fluidity of human membranes may be due to the greater immobilization of lipids by proteins compared to that which occurs in rat liver membranes.

Discontinuities in the plots of different ESR parameters as a function of temperature are considered to indicate lipid phase transitions in the membrane and/or conformational changes in the membrane proteins (Mehlhorn and Keith 1972; Raison et al. 1971; Benga and Strach 1975). Liver mitochondria from rats that were fed diets deficient in essential fatty acids contain less linoleic and arachidonic acids than those fed diets supplemented with linoleate and linolenate (Hayashida and Portman 1960; Stancliff et al. 1969). A reduction in the motional freedom of spin-labeled lipids was noticed for mitochondria isolated from rats that were fed diets deficient in essential fatty acids. However, the discontinuities of ESR parameters as a function of temperature occurred at the same temperature for both types of mitochondria (Williams et al. 1972). In our work (Benga et al. 1978) a discontinuity in the plots of the ESR parameter as a function of temperature was noticed at a lower temperature for human liver mitochondria than for those of the rat (Fig. 4.6). This finding is consistent with a higher fluidity of lipids in human membranes.

Enzymic Activities

The composition of human liver mitochondria might explain their unusual stability characteristics (Benga et al. 1972), as well as certain functional peculiarities. An analysis of some enzymic activities of human liver mitochondria has been undertaken, and it was hoped that this might provide a further index of mitochondrial damage in liver and biliary diseases.

The adenosine triphosphatase (ATPase) activity of mitochondria in fact proved to be a useful index of mitochondrial damage (Benga and Mureşan 1974). It is well known from studies of mitochondria from different sources that intact mitochondria exhibit little or no hydrolytic activity toward adenosine triphosphate (ATP) referred to as the "latency" of the ATPase. The classical activator of the "latent" ATPase is 2,4-dinitrophenol (DNP). When mitochondria are damaged, for example, by aging, osmotic shock, or addition of surface active agents, activation by DNP is lost and a simultaneous activation by Mg^{2+} appears.

A marked effect of the presence of albumin on ATPase activity of human liver mitochondria was observed. The addition of albumin to the incubating media led to a decrease of the basal and magnesium-stimulated activity, whereas

Table 4.7. Influence of EDTA on ATPase Activity of Human Liver Mitochondria[a]

	Without EDTA	With 0.5 mM EDTA	With 0.5 mM EDTA and Albumin (1% final concentration)
Basal activity	28.3 ± 5.1[b] (5)[c]	12.6 ± 3.1 (8)	7.7 ± 3.0 (6)
5 mM Mg^{2+}	57.7 ± 11.2 (4)	57.2 ± 10.2 (10)	26.6 ± 8.9 (10)
0.1 mM DNP	57.0 ± 10.5 (3)	53.2 ± 15.1 (10)	69.2 ± 16.5 (10)

Source: Data from Benga and Mureşan (1974). *Biochem Med. 10,* 131–145.

[a]The reaction mixture (final volume 0.2 ml) contained 5 mM ATP, 150 mM sucrose, 50 mM KCl, 50 mM tris-HCl (pH 8.5); ATPase activity was expressed as nanomoles of inorganic phosphate (Pi) per minute per milligram of protein.

[b]Mean and SEM.

[c]Number of observations given in parentheses.

DNP stimulation was considerably greater compared to that in the absence of albumin (Table 4.7). The effect of albumin was more pronounced with lipid-rich mitochondria isolated from fatty livers. This action of albumin could not be replaced by EDTA. The addition of albumin to the incubating media enables the distinction of a "reversible" and an "irreversible" type of damage to the mitochondria. The pattern of the ATPase activity of mitochondria with preserved integrity is characterized by a low basal activity, a small stimulation by Mg^{2+} (about twofold), and a marked stimulation by DNP (up to 20 times). When the basal and Mg-ATPase decreases upon the addition of albumin and the DNP-ATPase rises so that the ratio DNP-ATPase to Mg-ATPase becomes greater than 1, a reversible damage of mitochondria is assumed. This type of damage may occur during the homogenization of the tissue and the isolation of mitochondria possibly due to the deleterious action of agents such as free fatty acids, bilirubin, or bile acids or due to the activation of phospholipases and lipases found in mitochondria, lysosomes and microsomes. In case of an irreversible damage of mitochondria, the normal pattern of the ATPase activity is not restored by adding albumin (Fig. 4.7). This type of damage is probably due to an in vivo alteration of the mitochondria.

Defatted albumin also markedly affects the 3-hydroxy-butyrate dehydrogenase activity of human liver mitochondria, the activity being increased by a factor of 2. No significant effect was noted with the other enzymic activities listed in Table 4.8. Although the specific activities of most of the enzymes are greater in rat than in human liver mitochondria, the activity of isocitrate dehydrogenase is exceptional. As previously mentioned, isocitrate was the best reduced nicotinamide adenine dinucleotide (NADH)-linked respiratory substrate for human liver mitochondria.

One of the most interesting observations was that human liver mitochondria exhibited a much greater cytochrome oxidase activity with human cytochrome

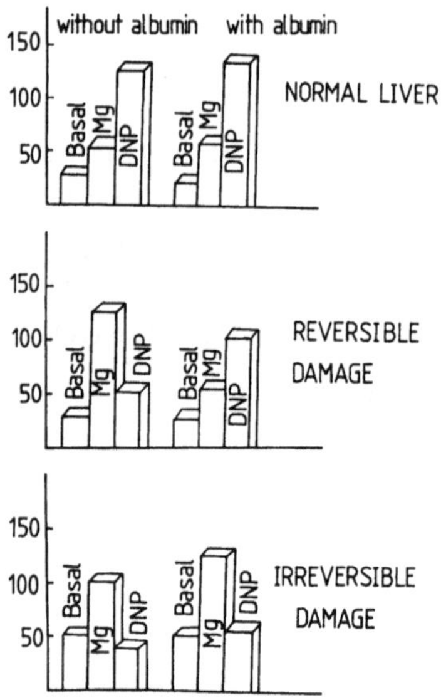

Figure 4.7. The patterns of ATPase activity of human liver mitochondria in cases of normal liver, reversible damage, and irreversible damage. The ATPase activity is expressed in nanomoles per inorganic phosphate per minute per milligram of protein. Data from Benga (1975). *St. Cerc. Biochem. 18*, 83–91.

Table 4.8. Enzymic Activities of Rat and Human Liver Mitochondria[a]

Enzyme	Human Liver	Rat Liver
3-hydroxybutrate-dehydrogenase		
Without BSA	33.7 ± 4.1[a](5)[b]	110 ± 25[a](5)[b]
With BSA	70.5 ± 2.2 (5)	115 ± 23 (5)
Malate-dehydrogenase	1692 ± 140 (5)	3200 ± 230 (4)
Glutamate-dehydrogenase	27.5 ± 4.0 (5)	–
Isocitrate-dehydrogenase[c]	112.0 ± 20 (7)	69 ± 12 (6)
Monoamine oxidase	3.9 ± 0.2 (4)	5.2 ± 0.3 (4)
Adenylate-kinase	614 ± 32 (5)	1800 ± 75 (3)
Cytochrome oxidase[d]		
With horse cytochrome *c*	9.1 ± 2.3 (14)	106 ± 21 (5)
With human cytochrome *c*	47.1 ± 7.2 (10)	92 ± 18 (5)

Note: The activities were measured after a previous solubilization of the organelles with the nonionic detergent Lubrol WX (0.5 mg/mg of protein). The reaction medium of 1 ml final volume contained 0.02–0.20 mg of mitochondrial protein. The specific activity is expressed as nanomoles of substrate transformed per minute per milligram of protein. Other details as previously described in Benga et al. (1980). *Rev. Roum. Biol. 25*, 147–154.

[a]Mean and SEM.

[b]Number of experiments given in parentheses.

[c]The specific activity with NADP as acceptor.

[d]The specific activity is calculated from the first-order reaction velocity constant (milliliters per mg protein per minute).

c than with horse cytochrome *c*, whereas rat liver mitochondria oxidized the horse or human cytochrome *c* at the same rate (Benga and Borza 1975). In other words, human liver mitochondria show a specificity for human cytochrome *c* (Table 4.9). The differences in the reactivity of human liver cytochrome oxidase with these cytochromes *c* were also noted using other buffers, although the maximum activity was obtained with phosphate buffer, in agreement with Smith et al. (1974).

The concentrations of cytochromes aa_3 were practically the same in human liver mitochondria (0.27 nmol/mg of protein) and rat liver mitochondria (0.24).

Our findings can be discussed taking into account the small differences in amino acid sequence between horse and human cytochrome *c* (Dickerson 1972). There are only 12 replacements of amino acids out of 104 residues, six of those being with amino acids with similar chemical properties (Dickerson 1972). Only six replacements involve a change in the functional groups of the side chain–at positions 12, 50, 58, 60, 83, 89, and 92. If we look at the three-dimensional structure of horse cytochrome *c* (Dickerson, 1972), we can see that the residue 12 is in the upper part of the heme crevice next to lysine 13. Okunuki (in Dickerson 1972) has shown that blocking the lysine 13 reduced the reactivity of cytochrome *c* by a factor of 2. The other five residues are all on the left-hand side of the cytochrome *c* molecule (the left and right sides of the molecule are as described by Dickerson 1972). Our data suggest the importance of the left

Table 4.9. Cytochrome Oxidase Activity of Mitochondria with Horse and Human Cytochrome *c*[a]

Source of Mitochondria	Phosphate Buffer (mM) 25	50	75
Rat liver			
Horse cytochrome *c*	15.6[b]	36.4	106 ± 21[c] (N = 5)
Human cytochrome *c*	5.9	34.1	92 ± 17 (N = 5)
Human liver			
Horse cytochrome *c*	9.1 ± 2.4	3.1	33.4
Human cytochrome *c*	21.3 ± 5.5 (N = 10)	47.1 ± 7.2 (N = 30)	

[a]The enzymic activity was expressed in terms of the first-order kinetic constant, *k* (milliliters per mg protein per minute). The experimental conditions were as described previously (Benga and Borza, 1975).

[b]Mean of three experiments.

[c]Mean and SEM for the number (N) of experiments that are given in parentheses.

side of the cytochrome *c* for the oxidase reaction. It was considered for some time that the cytochrome *c* is oxidized on one side and is reduced on the other side. However, we noticed later that the same specificity toward the human cytochrome *c* (compared with horse cytochrome *c*) was exhibited by human liver mitochondria also in the reduction reaction, namely by the succinate-cytochrome *c* reductase (Benga unpublished data). This suggested that the same part of the cytochrome *c* molecule is involved in both oxidation and reduction reactions.

On the other hand, we can take into account the hydrophobicity of all of the 12 replacements in the amino acid sequence. The horse cytochrome *c* has a charged residue (Glu), whereas the human cytochrome *c* has an Ala at position 92. There is another charged residue in horse cytochrome *c*, Lys 60, replaced by Gly in human cytochrome *c*; but the human cytochrome *c* has a charged residue (Glu 89), whereas horse cytochrome *c* has Thr at position 89. If we consider the hydrophobicity of the other replacements according to the hydrophobicity scale of Segrest and Feldmann (1974), the human cytochrome *c* has a greater hydrophobicity than the horse cytochrome *c*. To function correctly, cytochrome *c* must interact with the mitochondrial membrane at the right point. As we have shown, human liver mitochondria contain two to three times more lipids than rat liver mitochondria. All of the 12 replacements in the amino acid sequence are found on the outside of the cytochrome *c* molecule and are possibly involved in penetration into the membrane of human liver mitochondria, which is more "hydrophobic" than its counterpart in the rat.

Medical Applications: Fragility of Human Liver Mitochondria and Alterations in Pathological Conditions

The greater fragility of human liver mitochondria (compared with rat liver mitochondria) was documented by the fractionation of mitochondria into the inner and outer membranes (Benga et al. 1979).

The fractionation procedure of rat liver mitochondria, derived from the method of Schnaitman et al. (1967, 1968), is described in Fig. 4.8. It can be seen that treatment of rat liver mitochondria with digitonin selectively removes the outer membrane leaving inner membranes intact with the associated matrix. When this fraction is treated with Lubrol WX the matrix material is released.

Figure 4.8. Fractionation of rat liver mitochondria. Aliquots (1–2 ml) of mitochondrial suspensions in 0.25 M sucrose containing about 40 mg/protein/ml were placed in an ice bath, and identical aliquots of cold digitonin solution (6 mg/ml in 0.25 M sucrose) were added. The suspensions were incubated at 0°C with continuous stirring for 20 min after the addition of digitonin. The digitonin-treated suspensions were centrifuged at 9500 g for 15 min. The

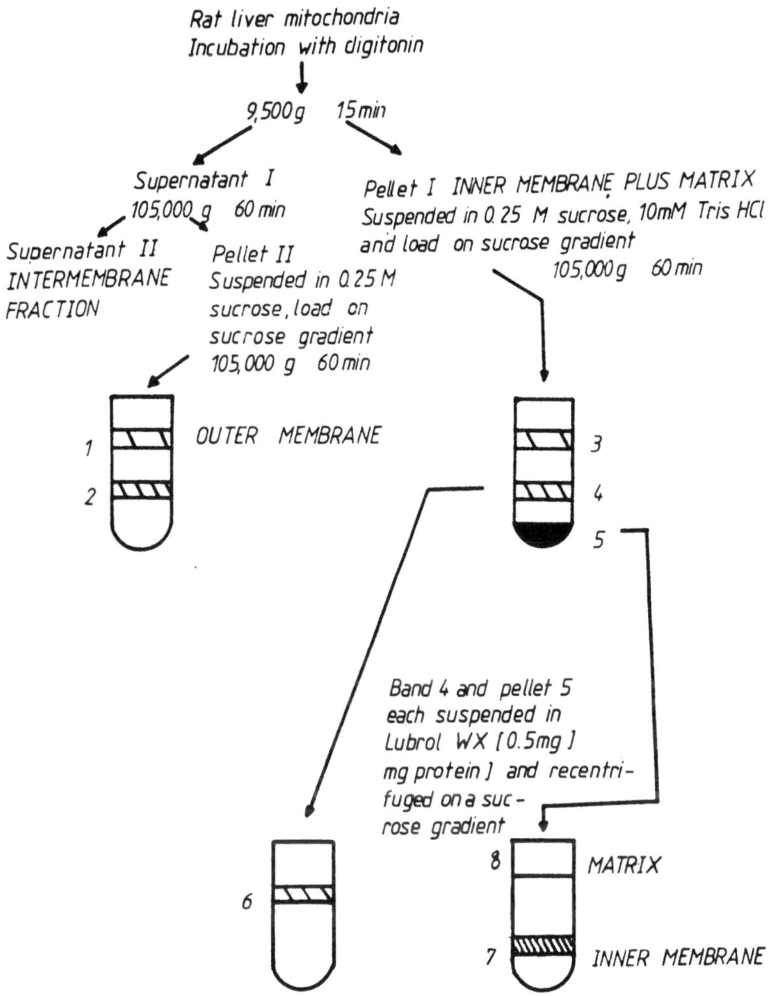

supernatant I contains the crude outer membranes and the intermembrane fraction. The pellet I contains the crude inner membrane plus matrix fraction. Supernatant I was centrifuged at 105,000 g for 60 min. The resulting pellet II contains the crude outer membranes, whereas the supernatant II corresponds to the intermembrane fraction. Both pellets I and II were resuspended in 0.25 M sucrose 10 mM tris-HCl and were purified by centrifugation through a discontinuous sucrose gradient (1.2 ml of 23.2% sucrose, 1.2 ml of 37.4% sucrose, and 1.8 ml of 51.3% sucrose, all in 20 mM phosphate buffer, pH 7.4). From pellet II the purified outer membrane fraction was obtained (layer 1); from pellet I, layer 4 corresponds to inner membrane and pellet 5 to the pure inner membrane plus matrix particles. These two fractions were treated with Lubrol WX (0.5 mg/mg of protein), incubated for 15 min at 0°C, and purified again on similar linear sucrose gradients. The resulting layers, 6 and 7, correspond to the purified inner membranes, whereas layer 8 corresponds to the matrix. Data from Benga et al. (1979). *J. Cell. Sci. 35,* 417–429.

Centrifugation of this material on a linear sucrose gradient separates the inner membranes from the matrix. Thus, three steps are necessary for the purification of rat liver mitochondrial membranes: a digitonin treatment for the removal of the outer membranes, the purification of the inner membrane plus matrix fraction by sucrose gradient centrifugation, and the treatment with Lubrol followed by a second sucrose gradient centrifugation.

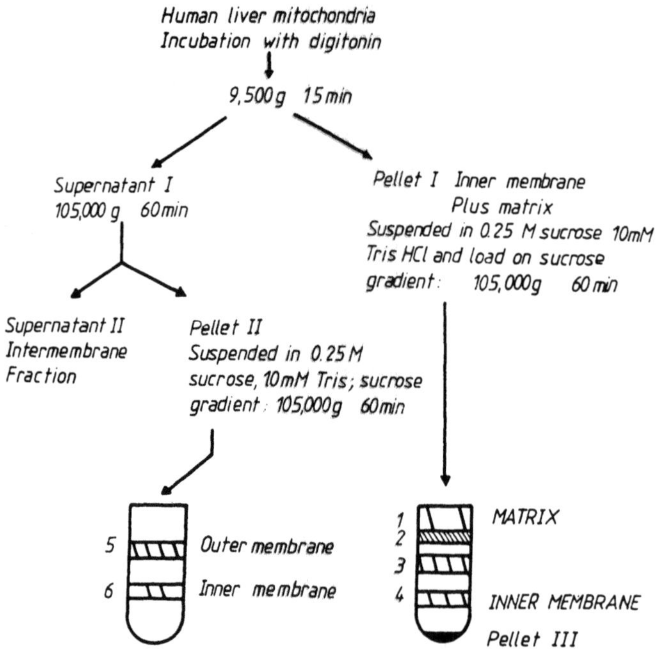

Figure 4.9. Fractionation of human liver mitochondria. Aliquots (0.5–1 ml) of mitochondrial suspensions in 0.25 M sucrose containing about 30 mg protein/ml were placed in an ice bath, and identical aliquots of cold digitonin solutions (6 mg/ml in 0.25 M sucrose) were added. After 20 min of incubation at 0°C with continuous stirring, the digitonin-treated suspensions were centrifuged at 9500 g for 15 min. The supernatant I was centrifuged at 105,000 g for 60 min for separation of intermembrane fraction (supernatant II) from the crude outer membranes (pellet II). From pellet II (resuspended in 0.25 M sucrose, 10 mM tris-HCl, pH 7.4) purified outer membranes (layer 5) and inner membranes (layer 6) are obtained by centrifugation through a discontinuous sucrose gradient (1.2 ml of 23.2% sucrose, 1.2 ml of 37.4% sucrose, and 1.8 ml of 51.3% sucrose, all in 20 mM phosphate buffer, pH 7.4). From pellet I (crude inner membrane plus matrix fraction) several fractions are obtained by centrifugation through a linear sucrose gradient: matrix (layer 1), submitochondrial particles (layer 2), outer membrane (layer 3), inner membrane (layer 4), and inner membrane plus matrix particles (pellet III). Data from Benga et al. (1979). *J. Cell. Sci. 35,* 417–429.

The fractionation of human liver mitochondria is described in Fig. 4.9. It can be seen that after treatment with digitonin (0.5 mg/mg of protein) the outer membrane is removed and can be separated from the intermembrane fraction by a sucrose gradient centrifugation. When the pellet I (crude inner membrane plus matrix) is subjected to centrifugation through the linear sucrose gradient, then purified inner membrane and matrix fractions are obtained (layers 4 and 1, respectively, in Fig. 4.9). This means that the inner membrane plus matrix particles are fragmented during the centrifugation rendering the Lubrol treatment unnecessary. Thus, the fractionation of human liver mitochondria can be achieved in just two steps: the digitonin treatment for the removal of the outer membrane and the centrifugation of the inner membrane plus matrix particles through a linear sucrose gradient yielding the purified inner membranes and matrix.

Our procedure for fractionation of human liver mitochondria, which takes particular advantage of their fragility, resulted in pure inner and outer membranes. Table 4.10 shows the distribution and specific activities of marker enzymes and protein in the fractions of human liver mitochondria. In this case, the inner membranes were purified about six-fold and there was little contamination with outer membrane, despite the fact that only a single centrifugation through a sucrose gradient was used. The outer membranes were purified about 14-fold, as indicated by monoamine oxidase activity. In the case of rat liver mitochondria, the inner membranes were purified about five-fold and the outer membranes about 12-fold.

The purification and identification of the membrane fractions was further examined by electron microscopy. The morphological aspects corroborate the results obtained with marker enzymes. In Fig. 4-10(*a*) the appearance of positively stained human liver mitochondria is presented. The electron micrographs of various fractions obtained after centrifugation confirm the results obtained by the distribution of marker enzymes. The negatively stained fraction 4 (see notation in Fig. 4.9) contains many convoluted membranous structures which, on high magnification [Fig. 4.10(*c*)], show the 9.0 nm knobs characteristic of inner membranes. Taking into account the appearance of the fixed thin section of the same fraction [Fig. 4.10(*b*)], the material appears to consist mainly of inner membranes. Fraction 3 (see notation in Fig. 4.9) is comprised primarily of outer membranes [Fig. 10(*d*)] as is fraction 5 [Fig. 4.10(*e*)]. The latter fraction seems to be purer than fraction 3 and contains smaller vesicles. This might be expected taking into account the effects of digitonin treatment.

In the understanding of the reasons for greater fragility of human liver mitochondria in comparison with rat liver mitochondria, a parallel with other membranes might be relevant. Thus it has been shown that camel erythrocytes have an exceptionally high osmotic stability, much higher than the human erythrocytes, for example (Livne and Kuiper 1973). This has been explained by the particular protein-lipid interactions in the different erythrocyte membranes. Camel erythrocyte membranes have a much higher protein:lipid ratio compared

Table 4.10. Distribution of Proteins and Specific Activities of Marker Enzymes in Human Liver Mitochondrial Fractions[a]

Fraction	Protein (%)	Specific Activity (nanomoles/min/mg/ of protein)				
		Adenylate Kinase	Monoamine Oxidase	3-hydroxy-butyrate Dehydrogenase	ATPase	Glutamate Dehydrogenase
Mitochondria	100	148.0	7.5	43.6	64.5	79.3
Inner membrane						
Layer 4	7.4	0	0.6	289.0	3500	30.2
Layer 6	2.4	0	1.2	284.1	2000	35.7
Outer membrane						
Layer 3	0.7	0	63.0	23.8	46.8	1.2
Layer 5	3.1	0	104.0	27.3	0	0
Intermembrane fraction supernatant II	22.4	486.0	0	0	0	0
Matrix (layer 1)	45.3	0	0	0	0	450
Submitochondrial fragments (layer 2)	8.7	0	14.2	20.2	60.5	75.3
Ratio						
Outer membrane / Mitochondria	–	–	13.4	0.6	–	–
Inner membrane / Mitochondria	–	–	0.1	6.6	–	0.4
Intermediate fraction / Mitochondria	–	3.3	–	–	–	–

Source: Data from Benga et al. (1979). *J. Cell Sci.* *35,* 417–429.

[a]The results are the mean of six to ten experiments.

with human membranes. The membrane proteins of camel erythrocytes are also more basic than those in the human, with a higher proportion of arginine and a lower proportion of glutamic acid. Taking into account the acidic nature of phospholipids in the membranes, stronger electrostatic attractions between lipids and proteins in camel membrane have been postulated as compared to the human erythrocyte membrane to explain their particular stability (Livne and Kuiper 1973). On the basis of lipid and amino acid composition, we might expect that in the case of human liver mitochondria, the role played by the hydrophobic interactions between membrane proteins and the lipid bilayer is structurally more important than in the case of the rat, in which electrostatic interactions could occur to significantly greater degree. The higher fluidity of the human liver mitochondrial membranes as inferred by the spin-label motion is in agreement with weaker protein-lipid interactions for these membranes as compared to the rat liver mitochondrial membranes. The greater fragility of human liver mitochondria could also be important for the alterations of these organelles in liver disease states.

The starting point for the study of human liver mitochondria in a variety of diseases is the establishment of the reference values, or the normal values of the enzymic activities. Patients with ulcers and biliary disease and with "normal" livers have been used as controls (Henley et al. 1964; Pagliaro et al. 1963; Scher-sten et al. 1966). Normal liver means the patients have no symptoms or signs of liver involvement and the routine liver blood tests (bilirubin, thymol test, transaminases) are in the normal ranges. However, we found some differences in oxygen uptake and ACI values among subjects with ulcers, biliary diseases, and those with no digestive diseases (Toader et al. 1976).

There were significant differences between the ulcer and biliary tract disease groups when succinate was the substrate; the oxygen uptake without ADP was higher, and consequently the ACI was lower, for patients with biliary diseases (Table 4.11). Because it was considered that the liver was damaged in any digestive disease, we tried to study mitochondria from patients with no digestive diseases, such as epigastric or umbilical hernia, and uterine fibroma. In comparison with patients with no digestive diseases, for subjects with ulcers or biliary diseases, smaller values of oxygen uptake with ADP and of ACT were noted with glutamate and α-ketoglutarate plus malate as substrate (Table 4.11).

The differences in oxygen uptake and ACI values among subjects with ulcers, biliary diseases, and those with no digestive diseases impose restrictions on selection of subjects with digestive diseases as controls. The liver may be affected by many agents even in the case of subjects with no digestive diseases. We have noted steatosis and fibrosis of liver in some cases with umbilical hernia and obesity. Therefore, we consider that patients undergoing abdominal surgery who are selected as controls in studies on isolated fractions of human liver should fulfill the following conditions: absence of any hepatic antecedents, no clinical evidence of liver involvement; no abnormality in routine liver function tests, a histologic aspect free of pathological conditions, and a normal

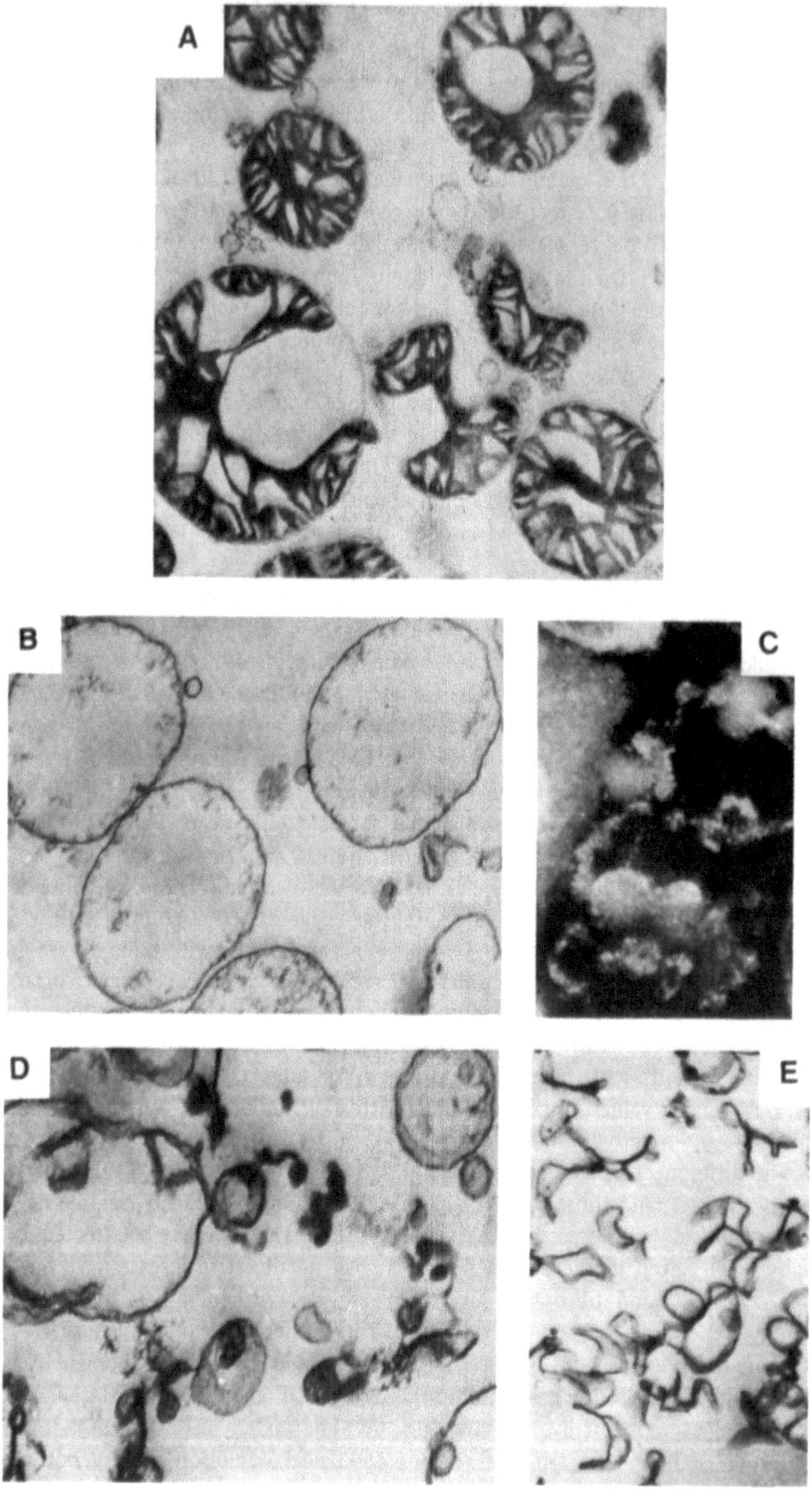
A
B
C
D
E

Table 4.11. Respiratory Activity of Liver Mitochondria Isolated from Patients with Digestive Diseases and Normal Livers[a]

Substrates	Diagnosis	No. of Cases	Oxygen Uptake Without ADP	Oxygen Uptake With 0.2 mM ADP	ACI
Succinate	U[b]	14	22.3 ± 2.0[c]	116.2 ± 26.0	5.3 ± 0.1
	B	7	27.0 ± 2.0	117.5 ± 3.8	4.1 ± 0.2
	N	9	29.1 ± 4.2	120.0 ± 7.4	4.5 ± 0.5
Pyruvate plus malate	U	12	9.2 ± 0.7	28.9 ± 7.9	3.2 ± 0.9
	B	7	9.5 ± 2.3	32.5 ± 4.4	3.4 ± 0.2
	N	7	10.9 ± 1.9	39.0 ± 6.3	3.6 ± 0.4
Isocitrate plus malate	U	13	10.7 ± 1.4	54.1 ± 10.2	5.5 ± 0.1
	B	5	12.6 ± 7.7	51.9 ± 5.9	4.6 ± 0.7
	N	7	14.7 ± 1.5	47.9 ± 3.5	3.3 ± 0.3
α-Ketoglutarate plus malate	U	6	12.1 ± 2.7	37.2 ± 5.7	3.1 ± 0.4
	B	4	12.4 ± 1.3	31.6 ± 5.1	2.6 ± 0.3
	N	6	11.3 ± 2.1	43.2 ± 7.4	3.9 ± 0.2

Source: Data from Toader et al. (1976). *Enzyme 21*, 232–242.

[a]Experimental conditions as described in the text.
Oxygen uptake was expressed as nanogram atoms per minute per milligram protein. Acceptor control index was calculated as ratio of the respiratory rate in the presence of ADP to the rate before ADP has been added.

[b]U, duodenal or gastric ulcers; B, biliary diseases; N, no digestive diseases.

[c]Mean and SEM.

aspect of the tissue during homogenization and fractionation procedure. A higher content of lipids is indicated by a more whitish color of the tissue and the homogenate and during the centrifugation, by the accumulation of lipid at the top of the tube. A macroscopic fibrosis can be seen during the mincing of the tissue; such areas should be removed.

The liver of patients with biliary diseases has a greater lipid content as compared with that of subjects with ulcers. The lipid content of liver mitochondria was about 0.25 mg lipids/1mg protein in patients with ulcers and 0.40 mg lipids/1mg protein in patients with biliary diseases (Toader et al. 1976). In such patients the pattern of ATPase activity typical for a "reversible damage" was found.

In patients with jaundice and especially in patients with prolonged jaundice an "irreversible" damage of mitochondria was noted. In such cases the oxygen

Figure 4.10. Electron micrographs of human liver mitochondria before and after fractionation. (For fraction notations see Fig. 4.9.) (A) Positive staining of intact mitochondria (X 14,000). (B) Positive-stained appearance of fraction 4 (inner membrane) (X 40,000). (C) Negative-stained appearance of fraction 4 (X 80,000). (D) Positive-stained appearance of fraction 3 (outer membrane) (X 40,000). (E) Positive-stained appearance of fraction 5 (outer membrane) (X 36,000). Data from Benga et al. (1979). *J. Cell. Sci. 35*, 417–429.

uptake with the nicotinamide adenine dinucleotide (NAD)-linked substrates and the ACI were lowered. The damage of mitochondria in such patients is also documented by electron microscopy. In the tissue mitochondria are swollen or condensed with bizarre forms. Spin-label studies have indicated a higher fluidity of the mitochondrial membranes in patients with jaundice.

If we look at biological membranes not only in terms of permeability barriers, but also as catalytic surfaces, the membranes constitute a uniquely ordered system for metabolic reactions. Human liver subcellular membranes with their peculiar composition [higher lipid content, a particular fatty acid composition, more hydrophobic amino acid residues in membrane proteins (Benga and Ferdinand 1977) in human compared to rat liver mitochondria], particularities of enzymic activities, and higher fluidity as inferred by spin-label motion provide an interesting system for further studies of protein–lipid interactions in liver membranes in normal and pathological conditions.

Acknowledgments

The results described above have been obtained in collaboration with Adriana Hodârnău, Rozalia Tilinca, Victoria Borza, Letitia Mureşan (Department of Biochemistry, Faculty of Medicine, Medical and Pharmaceutical Institute Cluj-Napoca, Roumania), Silvia Dancea, D. Poruţia (Laboratory of Electron Microscopy, Medical, and Pharmaceutical Institute Cluj-Napoca, Roumania), I. Petrescu (Department of Biochemistry, Faculty of Biology, "Babeş-Bolyai" University, Cluj-Napoca, Roumania), William Ferdinand (Department of Biochemistry, University of Sheffield, UK), and John Wrigglesworth (Department of Biochemistry, Chelsea College University of London, UK).

The discussions with I. Manta and O. Bârzu (Department of Biochemistry, Medical and Pharmaceutical Institute Cluj-Napoca, Roumania), H. Baum (Department of Biochemistry, Chelsea College University of London, UK), D. Chapman (Department of Physical Biochemistry, Royal Free Hospital School of Medicine, University of London, UK) as well as the financial support of The Wellcome Trust (London, UK) and of the Academy of Medical Sciences of Roumania, are gratefully acknowledged.

The author thanks P. Quinn (Department of Biochemistry, Chelsea College, University of London) for revising the English version of the manuscript.

References

Bârzu, O. (1978). Spectrophotometric assay of oxygen consumption. In *Methods in Enzymology,* Fleischer, S. and Packer, L., eds. Academic Press, New York, vol. 14, pp. 485–498.

Bârzu, O., Mureşan, L., and Benga, Gh. (1972). Spectrophotometric method for assay of mitochondrial oxygen uptake. IV. *Anal. Biochem. 46,* 374–387.

Benga, Gh. (1974). Fracţionarea subcelulară a ţesutului hepatic unman. I. Izolarea fracţiunii mitocondriale. *St. Cerc. Biochim. 17,* 123–135.

Benga, Gh. (1975). Fracţionarea subcelulară a ficatului unman. II. Caracterizarea mitocondriilor în condiţii patologice. *St. Cerc. Biochim. 18,* 83–91.

Benga, Gh. (1979). *Biologia Moleculara: A Membranelor cu Aplicaţii Medicale.* Editura Dacia, Cluj-Napoca.

Benga, Gh. and Borza, V. (1975). Differences in reactivity of cytochrome oxidase from human liver mitochondria with horse and human cytochrome c. *Arch. Biochem. Biophys. 169,* 354–357.

Benga, Gh. and Ferdinand, W. (1977). Increased content of hydrophobic amino acid residues in lipid-rich mitochondrial membranes: A comparison of rat and human liver mitochondria. *Int. J. Biochem. 8,* 17–20.

Benga, Gh. and Mureşan, L. (1974). Human liver mitochondria. III. ATPase activity as an index of mitochondrial damage. *Biochem. Med. 10,* 131–145.

Benga, Gh. and Strach, S.J.S. (1975). Interpretation of the electron spin resonance spectra for nitroxide-maloimide-labeled proteins and the use of this technique in the study of albumin and biomembranes. *Biochim. Biophys. Acta 400,* 69–79.

Benga, Gh., Mureşan, L., Hodârnău, A., and Dancea, S. (1972). Conditions for isolation and study of enzymic properties of human liver mitochondria. *Biochem. Med. 6,* 508–521.

Benga, Gh., Petruscu, I., Tărmure, C., and Pop, V. (1980). Species related functional properties of mitochondria: Comparison between rat and human liver mitochondria. *Rev. Roum. Biol. 25,* 147–154.

Benga, Gh., Hodârnău, A., Tilinca, R., Poruţiu, D., Dancea, S., Pop, V., and Wrigglesworth, J. (1979). Fractionation of human liver mitochondria: Enzymic and morphological characterization of the inner and outer membranes as compared to rat liver mitochondria. *J. Cell Sci. 35,* 417–429.

Benga, Gh., Hodârnău, A., Böhm, B., Borza, V., Tilinca, R., Dancca, S., Petruscu, I., and Ferdinand, W. (1978). Human liver mitochondria: Relation of a particular lipid composition to the mobility of spin-labeled lipids. *Eur. J. Biochem. 84,* 625–633.

Benga, Gh., Tărmure, C., and Hodârnău, A. (1971). The determination of proteins in lipid-rich mitochondrial suspensions. *Enzyme 12,* 574–577.

Blouin, A., Bolender, R.P., and Weibel, E.R. (1977). Distribution of organelles and membranes between hepatocytes and non-hepatocytes in the rat liver parenchyma. *J. Cell Biol. 72,* 441.

Dallner, G., Siekevitz, P., and Palade, G.E. (1966). Biogenesis of endoplasmic reticulum membranes. *J. Cell Biol. 30,* 97.

Dickerson, R.C. (1972). The structure and history of an ancient protein. *Sci. Amer. 226,* 58–72.

Fleischer, S. and Fleischer, B. (1977). Membrane diversity in the rat hepatocyte. In *Membrane Alterations as Basis of Liver Injury,* Popper, H., Bianchi, L., Reuter, W., eds. MTP Press Ltd., Lancaster, pp. 31–48.

Green, D. and Tzagoloff, A. (1966). Role of lipids in the structure and function of biological membranes. *J. Lipid Res. 7,* 587–602.

Griffith, O.H., Cornell, D.W., and McConnell, H.M. (1965). Nitrogen hyperfine tensor and g tensor of nitroxide radicals. *J. Chem. Phys. 43,* 2909–2910.

Hayashida, T. and Portman, O.W. (1960). Swelling of liver mitochondria from rats fed diets deficient in essential fatty acids. *Proc. Soc. Exp. Biol. Med. 103,* 656–659.

Henley, K.S., Napier, E.A., Kreyden, R.W., Corssen, G.C., and Berensohn, S. (1964). Mitochondrial oxidase activity and pyridine nucleotide concentrations of normal human liver. *J. Lab. Clin. Med. 64,* 306–312.

Kivelson, D. (1960). Theory of esr linewidths of free radicals. *J. Chem. Phys. 33,* 1094–1106.

Livne, A. and Kuiper, P.J.C. (1973). Unique properties of the camel erythrocyte membrane. *Biochim. Biophys. Act 318,* 41–49.

McConnell, H.M. and McFarland, B.G. (1970). Physics and chemistry of spin labels. *Q. Rev. Biophys. 3,* 91–136.

Mehlhorn, R.J. and Keith, A.D. (1972). Spin labeling of biological membranes. In *Membrane Molecular Biology,* Fox, C.F. and Keith, A.D., eds. Sinauer Associates, Stamford, Conn., pp. 192–227.

Mureşan, L., Dânşoreanu, M., Ana, A., Bara, A., and Bârzu, O. (1980). Cuvette and flow system for fast and sensitive spectrophotometric assays of oxygen consumption. *Anal. Biochem. 104,* 44–50.

Pagliaro, I., Notarbartolo, A., Mannino, V., and Migneco, C. (1963). Reduced ability of mitochondria isolated from liver tissue of patients with cirrhosis to synthesize ATP. *J. Lab. Clin. Med. 62,* 185–192.

Raison, K., Lyons, J.M., Mehlhorn, R.J., and Keith, A.D. (1971). Temperature-induced phase changes in mitochondrial membranes detected by spin labeling. *J. Biol. Chem. 246,* 4036–4040.

Richardson, T., Tapel, A.L., and Gruger, E.H. (1961). Essential fatty acids in mitochondria. *Arch. Biochem. Biophys. 94,* 1–6.

Richardson, T., Tapel, A.L., Smith, L.M., and Houle, C.R. (1962). Polyunsaturated fatty acids in mitochondria. *J. Lipid Res. 3,* 344–350.

Rouser, E., Nelson, G.J., Fleischer, S., and Simon G. (1968). Lipid composition of animal cell membranes, organelles and organs. In *Biological Membranes. Physical Fact and Function,* Chapman, D., ed. Academic Press, New York, vol. 1, pp. 5–69.

Scherstén, T., Björkerud, G., Jakoi, L., and Björntorp, P. (1966). Oxidation and phosphorylation reactions in isolated liver mitochondria in normal and icteric conditions. *Scand. J. Gastroenterol. 1,* 284–291.

Schmidt, E. and Schmidt, F.W. (1970). Enzyme activities in human liver. *Enzymol. Biol. Clin. 11,* 67–129.

Schnaitman, C. and Greenawalt, J.W. (1968). Enzymic properties of the inner and outer membranes of rat liver mitochondria. *J. Cell Biol. 38,* 158–175.

Schnaitman, C., Erwin, V.G., and Greenawalt, J.W. (1967). The submitochondrial localisation of monoamine oxidase. An enzymic marker for the outer membrane of rat liver mitochondria. *J. Cell Biol. 32,* 719–735.

Segrest, J.P. and Feldmann, P.J. (1974). Membrane proteins: Amino acid sequence and membrane penetration. *J. Mol. Biol. 87,* 853-858.

Smith, L., Davies, H.C., Reichlin, M., and Margoliash, E. (1974). Oxidation and reduction of soluble cytochrome *c* by membrane-bound oxidase and reductase systems. *J. Biol. Chem. 249,* 2904-2910.

Stancliff, R.C., Williams, M.A., Utsumi, K., and Packer, L. (1969). Essential fatty acid deficiency and mitochondrial function. *Arch. Biochem. Biophys. 131,* 629-642.

Toader, C., Acalovshi, I., Toader, I., Manta, I., Hodârnău, A., and Benga, Gh. (1976). Factors influencing the establishment of the normal values of the respiratory activities of human liver mitochondria. *Enzyme, 21,* 232-242.

Whittaker, P.A. and Danks. S.M. (1978). *Mitochondria: Structure, Function and Assembly.* Longman, London.

Williams, M.A., Stancliff, R.C., Packer, L., and Keith, A.S. (1972). Relation of unsaturated fatty acids composition of rat liver mitochondria to oscillation period, spin label motion, permeability and oxidative phosphorylation. *Biochim. Biophys. Acta 267,* 444-456.

5

Some Problems Concerning the Role of Water and Protons in the Function of Biological Membranes

V. Vasilescu, E. Katona, A. Popescu, C. Zaciu, and C. Ganea

Biological membranes, the basic architectural units of all cells, constituting in some cases as much as 80% of the total dry cell weight, are involved in most of the fundamental physiological functions. Their major constituents are known to be lipids, proteins, polysaccharides, and water, but their detailed structure and the structure of water therein, the nature of interactions between membrane constituents, and effects of water and metal ions on the conformation of membrane macromolecules are still unknowns.

Considering these uncertainties about the molecular architecture of membranes, the poor understanding of many mechanisms associated with membrane function is not surprising. Therefore, membrane biophysics continues to be an active field of investigation (Chapman and Dodd 1971; Singer 1971; Chapman 1972; Quinn and Chapman 1980).

In addition, detailed involvement of water–often called the Matrix of Life–in biological functions is not well understood, and thus water biophysics also appears to be one of the most important research fields in present-day molecular biology (Cooke and Kuntz 1974; Packer 1977; Finney 1979; Edsall and McKenzie 1980).

The study of the properties and the role of water and protons in the evolution of some fundamental biological processes, such as nerve impulse generation and propagation or cellular bioenergetics, has been the main concern in our laboratory for the last 15 years (Vasilescu et al. 1968; Vasilescu and Margineanu 1971; Vasilescu and Margineanu 1974; Vasilescu and Katona 1980a; Vasilescu and Zaciu 1980; Vasilescu 1981; Zaciu and Vasilescu 1981). This work presents a synthesis of some recent data and also includes certain relevant earlier results interpreted from a new point of view. Taking into account the complexity of the problem, several methods of investigation were used, most of the data being gathered mainly by using three techniques: the controlled vacuum dehydration technique, nuclear magnetic resonance (NMR) spectrometry, and tissue deuteration.

Controlled Vacuum Dehydration Studies

Recently we developed a controlled vacuum dehydration method (Vasilescu et al. 1978b; Popescu et al. 1980; Vasilescu 1981), enabling us to determine by direct weighing the time course of the water mass expelled from biological tissues under vacuum (0.1 torr) and the total water mass extractable under these conditions (a very tightly bound water fraction cannot be removed by this process).

When plotting the weight of water removed against time, typical multiexponential curves are obtained both for tissues previously immersed in water-Ringer solutions and for those submitted in advance to deuteration, by having kept them 20 h at 4°C in deuterium oxide (D_2O)-Ringer solutions. The experimental kinetic curves are presented in Figs. 5.1(*a*) and (*b*). Using a mathematical transformation, these curves can be split into at least two bioexponential components, which shows the existence of two subpopulations of water molecules extractable by vacuum dehydration. A third subpopulation of water molecules (about 2%) still remains in the tissue (after 4 h of vacuum dehydration), and its extraction is possible only by heating the tissue at 105°C.

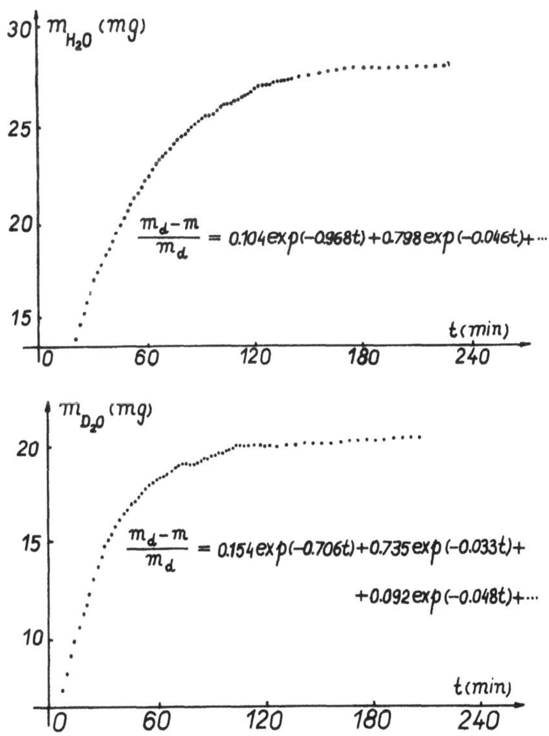

Figure 5.1. The experimental kinetic curve in the case of frog sciatic nerve immersed in H_2O-Ringer (top) and in D_2O-Ringer solutions (bottom), respectively.

The shape of the dehydration kinetic curves is described by the following function:

$$m(t) = m_d[1 - C_1 \exp(-\alpha_1 t) - C_2 \exp(-\alpha_2 t) - \ldots] \qquad (5.1)$$

where $m(t)$ = mass extracted at the time t and m_d = total mass extracted by dehydration. Obviously, the following conditions must be met:

$$m(0) = 0, m(\infty) = m_d$$

so that

$$\sum_i C_i = 1. \qquad (5.2)$$

In the case of a normal sciatic nerve, the dehydration kinetic curve [Fig. 5.1(*a*)] is described by the following expression:

$$m(t) = m_d[1 - 0.104 \exp(-0.968t) - 0.798 \exp(-0.046t) - \ldots] \qquad (5.3)$$

whereas for nerves previously deuterated, the corresponding curve [Fig. 5.1(*b*)] is described by the function

$$\begin{aligned} m(t) = m_d[1 &- 0.154 \exp(-0.706t) \\ &- 0.735 \exp(-0.035t) \\ &- 0.092 \exp(-0.048t) - \ldots] \end{aligned} \qquad (5.4)$$

In terms of the notations in Eq. (5.1), $\alpha_1 \gg \alpha_2$ in both Eqs. (5.3) and in (5.4), which demonstrates that the subpopulation of water molecules characterized by the component $C_1 \exp(-\alpha_1 t)$ is more easily extractable (this may correspond to the so called "free water"), whereas water molecules described by $C_2 \exp(-\alpha_2 t)$ are more difficult to extract (this may correspond to the so called "bound water").

The component $C_3 \exp(-\alpha_3 t)$, namely, $0.092 \exp(-0.048t)$ in Eq. (5.4) is due, in our opinion, to the normal water not exchanged by diffusion after nerve immersion in D_2O-Ringer solutions.

At the same time it is observed from Eqs. (5.3) and (5.4) that $\alpha_1(H_2O) > \alpha_1(D_2O)$ and $\alpha_2(H_2O) > \alpha_2(D_2O)$. These inequalities demonstrate that heavy water occupying the same "compartments" as normal water is more tightly bound than normal water in frog nerve tissue. This result may be explained by the difference that exists between the hydrogen bridge and the deuterium bridge (the former being 15% weaker than the latter), as well as by the mobility difference between protons and deuterons.

Finally, it is interesting to note the equality of the following ratios:

$$\alpha_2(D_2O)/\alpha_1(H_2O) \simeq 1.37$$

and

$$\alpha_2(D_2O)/\alpha_2(H_2O) \simeq 1.39 \qquad (5.5)$$

Nuclear Magnetic Resonance Studies

Nuclear magnetic resonance proved to be a powerful instrument for studying the molecular dynamics in tissue water (Shporer and Civan 1977; Packer 1977; Mathur-De Vré 1979; see also Chapter 7). Some pulsed proton magnetic resonance data on the state of water in peripheral nerves are presented here.

Spin-spin and spin-lattice relaxation behavior of water protons in the isolated frog sciatic nerve and the variation of this behavior with temperature, resonance frequency, deuteration and tissue state were investigated. Data analysis showed that a proper interpretation requires at least a three-state model of tissue water.

Proton spin-spin relaxation behavior was studied by the Carr-Purcell-Meiboom-Gill technique, with a sequence pulse spacing τ varying from 0.04 to 4 ms. At temperatures between 303 and 265°K, spin-spin relaxation decays of nerve water protons always appeared multiexponential. Computer curve-fitting and/or graphical analysis allowed the experimental data to be confidently resolved into three experimentally distinct relaxation components (Fig. 5.2). Assuming that these relaxation components correspond to physically distinct types of water

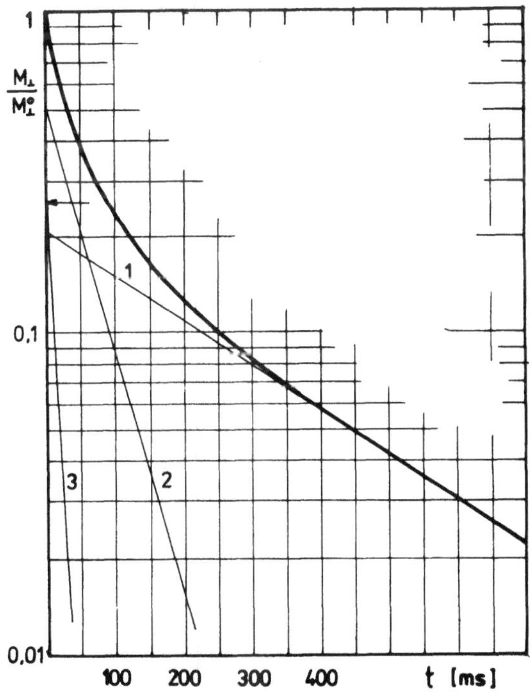

Figure 5.2. Typical spin-spin relaxation decay of water protons in frog peripheral nerve and its graphical analysis revealing the three distinct components.

molecules, we can tentatively consider them to belong to different tissue water compartments: extracellular water (21% of the total water content), axoplasmic water (50%), and the so-called "bound water" representing 0.9 g H_2O/g of dry mass in the case of a normal peripheral nerve. The presence of some temperature-dependent and relatively slow exchange processes between various populations of water molecules was suggested both by the temperature dependence at fixed τ and by the τ-dependence at fixed temperature of the relaxation behavior (Vasilescu et al. 1978a).

Studies of anesthetized and acetylcholine-treated nerves revealed that the size and properties of the three water components depend on the state of the nerve. A slowing of the whole relaxation process (Fig. 5.3) and a marked decrease in the amount of the bound water (from 0.9 g H_2O/g dry mass to 0.6 g H_2O/1g of dry mass) was observed in anesthetized nerves. A significant increase in the amount of the hypothetical extracellular water was noticed in the case of acetylcholine-treated nerves (Fig. 5.4). All these results prove that a change in the state of excitable membranes involves changes in the size of different subpopulations of water molecules and that it is probable that changes in the state of tissue water may induce changes in membrane properties.

At 265°K the bulk of nerve water is frozen. The "nonfreezable" water frac-

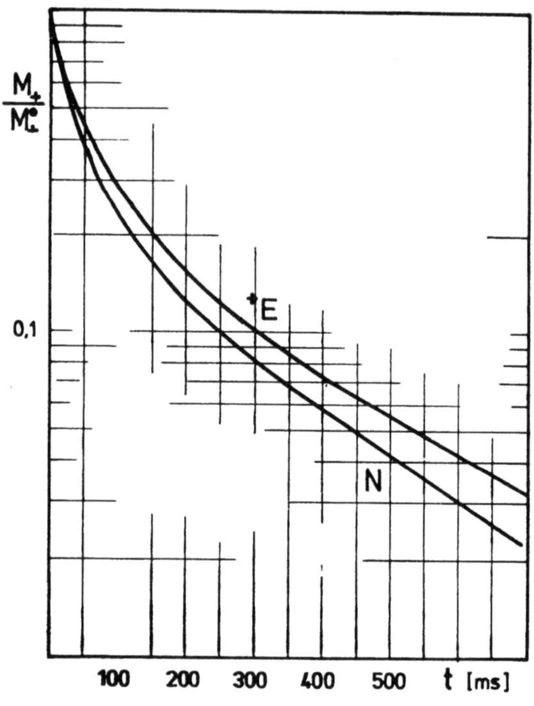

Figure 5.3. Modification of the proton spin-spin relaxation behavior, induced by nerve anesthesia. N, control nerve; E, nerve anesthetized by ethyl ether.

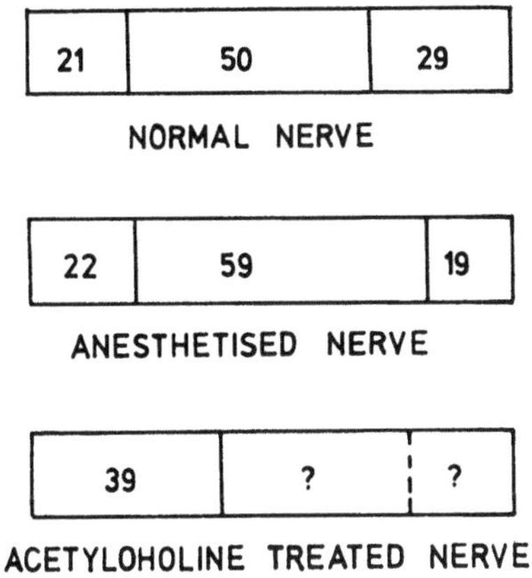

Figure 5.4. Water components revealed by pulsed NMR in frog peripheral nerve. Figures from left to right represent volumes of the extracellular, intracellular, and bound water, respectively, as expressed in percentage of the total water content of the nerve.

tion, identified by the relative amplitude of the mobile part of the water signal, decreases continuously as the temperature is lowered. Even at 203°K the nonfreezable water constitutes 5-9% of the total water content in the fresh nerve, it being equal to 0.2 g H_2O/1g of dry mass (Katona and Vasilescu 1981). We investigated the properties of this nonfreezable water fraction and of the nonexchangeable one in normal and deuterated nerves, respectively. Heterogeneity of these water fractions was proved by the observed (at least) biexponential character of proton spin-spin relaxation decays (Figs. 5.5 and 5.6). The presence of temperature-dependent exchange processes between various components in deuterated nerve was suggested by the temperature dependence of proton spin-spin relaxation decays (Vasilescu and Katona 1980b).

Spin-lattice relaxation time (T_1) data were gathered by the inversion recov-

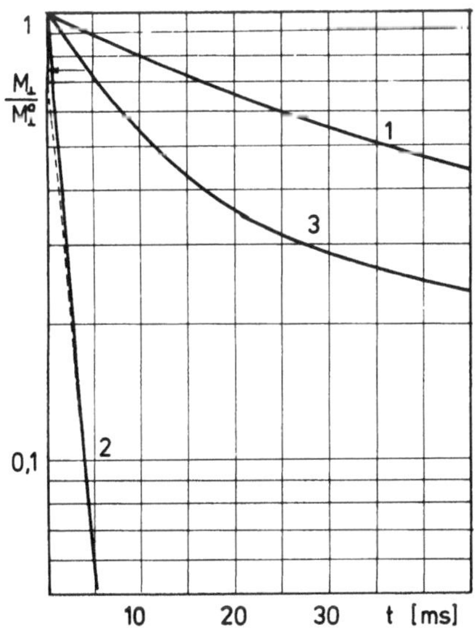

Figure 5.5. Spin-spin relaxation decays of water protons in normal nerves at 300°K (1) and 242°K (2), respectively, and in deuterated nerve at 300°K (3). Biexponentiality of the relaxation decay obtained for normal nerve at 242°K is pointed out by graphical analysis.

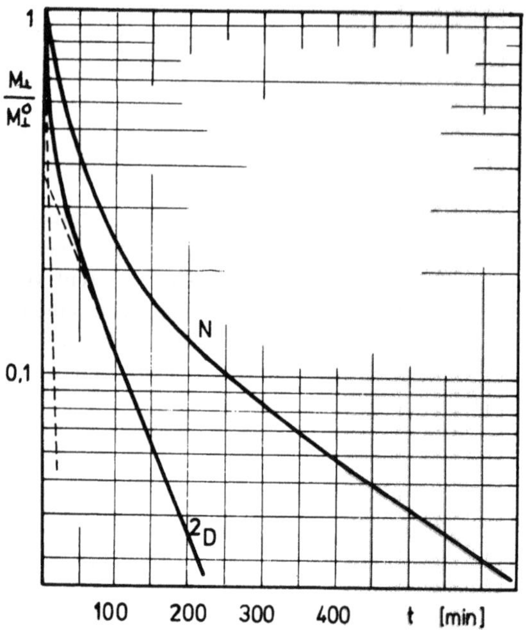

Figure 5.6. Spin-spin relaxation decays of water protons in normal (N) and deuterated (^{2}D) nerves at 300°K. Biexponentiality of the relaxation decay obtained for deuterated nerve is pointed out by graphical analysis.

ery and/or saturation recovery methods. Spin-lattice relaxation decays always appeared simply exponential at all termperatures between 203 and 303°K. The presence of two fairly shallow but distinct minima in T_1 variation as a function of temperature–at each resonance frequency–confirms the heterogeneity of the bound water fraction, as suggested by spin-spin relaxation data. When spin-spin and spin-lattice relaxation data obtained at temperatures below the freezing point of the bulk water were analyzed in terms of a theoretical model assuming a bimodal log-normal distribution of correlation times of molecular motions, the characteristics of the corresponding two distinct fractions of nonfreezable water in nerve were able to be determined (Katona et al. 1982). From the viewpoint of molecular mobility, both components of nonfreezable water in the nerve are more liquidlike than icelike. A likeness between the values obtained for the activation enthalpies (39.3 and 11.3 kJ/mol) and those already known for the hydration water of proteins and lipids, respectively, must be noticed.

Tissue Deuteration Studies

Deuteration proved to be an excellent nondestructive tool for investigating various membrane functions. Studies of the kinetics of H_2O-D_2O exchange in tissues under various conditions, for example, provided information concerning the tissue water distribution and how this distribution might be affected by different tissue components. Furthermore, deuteration modifies all the functions involving water or proton participation, including those depending on the presence of or change in certain water-stabilized conformational states of macromolecules. By following up the time course of D_2O effects on various biosystems, we investigated some aspects of the mechanism of several fundamental biological processes.

Studies of H_2O-D_2O Exchange in Tissues

The kinetics of H_2O-D_2O exchange in tissues was determined both gravimetrically and by infrared photometry. Data analysis in terms of a proper theoretical model enabled us to distinguish in all cases studied three different tissue water compartments characterized by different water accessibiity for diffusion exchange: *fast exchanging* compartment, tentatively ascribed to extracellular water; a *slowly exchanging* compartment, very probably representing the greatest part of intracellular water; and a *nonexchangeable* compartment, "invisible" from the viewpoint of our technique, constituted of water molecules somehow "obstructed," probably including the hydration water of biopolymers and the water molecules trapped inside various macromolecular and supramolecular structures (Vasilescu et al. 1977; Vasilescu and Katona 1980a, b).

The size and properties of these water compartments are dependent on the tissue (Fig. 5.7) and its particular state, and more precisely, on the state of the various subcellular functional units. Experiments performed on muscles (Katona et al. 1979) and nerves show that the state of cell membrane markedly influences the tissue water distribution. Damage to cell membranes in glycerinated muscles leads to a marked decrease in the amount of the nonexchangeable water and to an increase in the fast exchanging water compartment (Fig. 5.8). The

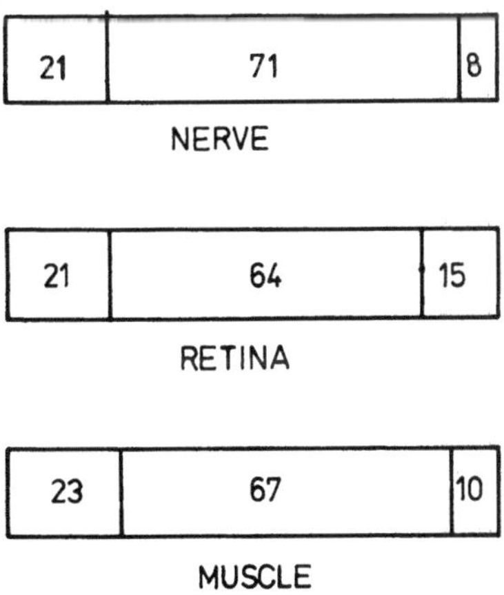

Figure 5.7. Water compartments in different tissues. The significance of the numbers is the same as in Figure 5.4.

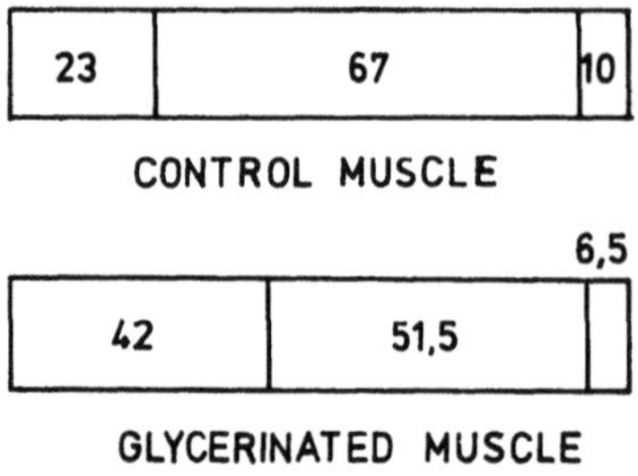

Figure 5.8. Water compartments in frog skeletal muscle. The significance of numbers is the same as in Figure 5.4.

maintenance of membranes in a "resting" state or in continuous function also alters tissue water distribution (Fig. 5.9). We cannot conclude, therefore, that membranes play a unique role in water distribution, but undoubtedly their contribution is essential.

Investigations on the Time Course of D_2O Effects on the Structure and Function of Biosystems

Studies of the effects of D_2O on isolated muscle, heart, retina, nerve trunks, and nerve fibers showed that bioelectrogenic processes and the cellular energy mechanisms were modified. Data analysis suggested that all the effects observed could be explained in terms of the difference between protons and deuterons, that is, between the physical properties of light and heavy water. A synopsis of possible effects of deuteration, as suggested by our results, is presented in Table 5.1.

Studies of the combined action of some general anesthetics and deuteration on sciatic nerve trunks disclosed an increase in the time lag necessary to the settling of anesthesia in deuterated nerves as compared to controls (Vasilescu

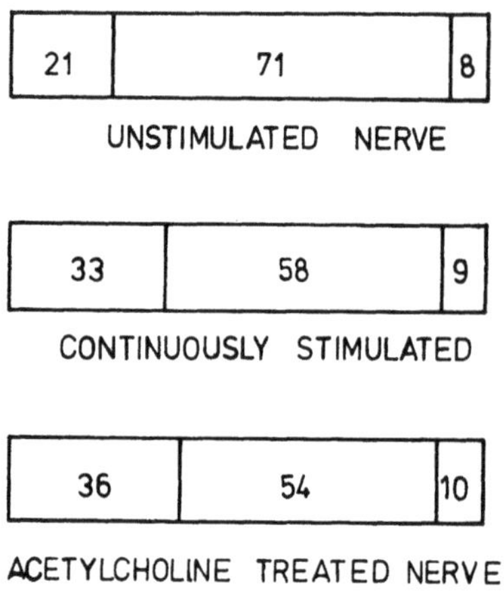

Figure 5.9. Water compartments in frog peripheral nerve. The significance of numbers is the same as in Figure 5.4.

Table 5.1. Possible Effects of Deuteration in Various Biosystems

Substitutions of Deuterons for Protons Leads to Changes in

Mobilities of various species

Strength of intermolecular forces

Stability of various macromolecular conformations | Hydration degree of various species

Modifies	*Interferes with*	*Effects*
Kinetics of various processes Stability of various biosystems Functionality of various supramolecular structures	Cell energy mechanisms affecting ATP $\rightleftharpoons$ ADP equilibrium Early stages of ATP synthesis	Bioelectrogenic processes Early stages of potential generation
	As Indicated by	
Increased settling time of anaesthesia Retarded bioluminescence kinetics Inhibitory effects on isolated heart contractility	Modified dynamics of cellular ATP pools Decoupling of the electrical from the mechanical functions of isolated heart	D_2O effects on excitability properties in dynamic behavior of the Ranvier node membrane Inhibitory effects on isolated heart bioelectrogenesis

and Katona 1980b). This time lag increases under the action of 2.5% ether solutions by approximately 50% at room temperature. A decrease in the ionic mobilities cannot entirely explain these observations. The antagonistic action of deuteration versus that of general anesthetics on the conformation of membrane proteins would perhaps be an explanation. Either the lipid phase or the hydrophobic region of membrane proteins is the site of action of anesthetics. A directly or indirectly induced transition of certain essential proteins to a more unfolded, inactive conformation (Eyring et al. 1973) as a mechanism of action of anesthetic molecules is supported by our results.

In order to check the possible impact of heavy water on the energy metabolism, we followed up the *dynamics of adenosine triphophate (ATP) pool in isolated nerves and retinas* (Vasilescu et al. 1975; Vasilescu et al. 1976; Vasilescu et al. 1978a; Ganea and Vasilescu 1979). A marked decrease of cellular ATP pool in deuterated versus control tissues was found in all cases, the effect being more obvious in continuously stimulated than in unstimulated nerves (35 and 22%, respectively) and much more pronounced in continuously stimulated retina (more than 50%) (Fig. 5.10). These changes in the cellular ATP pool could be accounted for at least partially by a change of the ATP $\rightleftharpoons$ ADP (adenosine diphosphate) equilibrium under the influence of heavy water as shown in Table 5.2. The enhanced hydration of different species under deuteration determines a marked shift of the equilibrium to the right, that is to ATP hydrolysis. However, D_2O affects ATP synthesis and this must also be taken into account.

Experiments performed on the isolated frog heart showed that heavy water

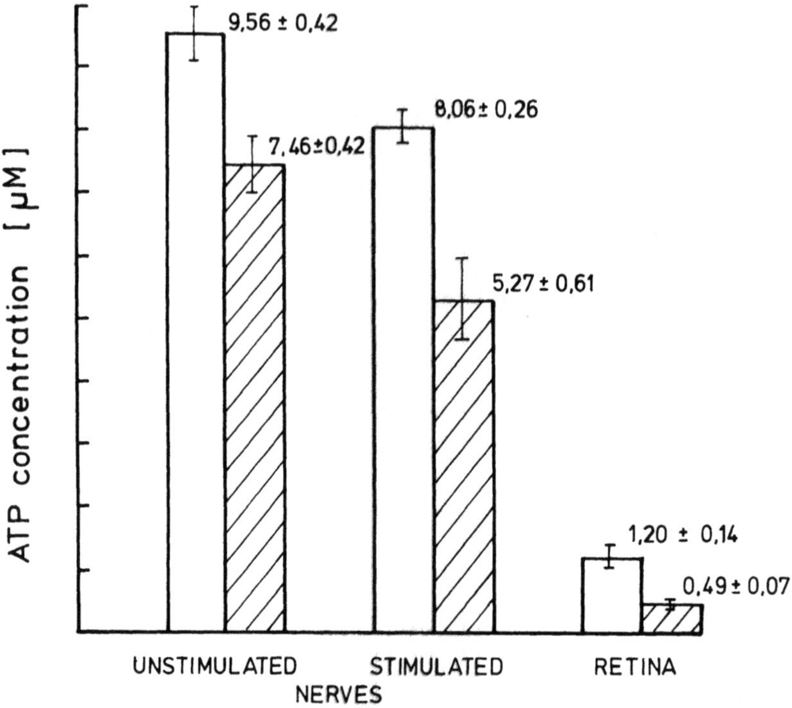

Figure 5.10. Effect of heavy water on cellular ATP content in nerves and retina. The numbers represent the population mean values and their standard errors obtained for 4 to 5 groups of experiments on 5 to 8 pairs of nerves and retina, respectively.

is a strong inhibitory agent both of *heart bioelectrogenesis and of its contractility* (Vasilescu and Ciures 1978). Modified mechanograms and electrocardiograms were recorded after heart deuteration, and a much stronger inhibition of the mechanical activity was observed (Fig. 5.11). Normal water perfusion of the heart after abolishment of the mechanical and electrical activities produced a temporary recovery of both of them. Deuteration-induced conformational

Table 5.2. Half-Time of the Bioluminescence Intensity for Various ATP Concentrations in Water and Heavy Water (Values Calculated from the Kinetic Curves)

ATP concentration (μM)	160	16	1.6	0.16
τ_{D_2O} (min)	0.63	2	8	12
τ_{H_2O} (min)	0.33	1	4	6

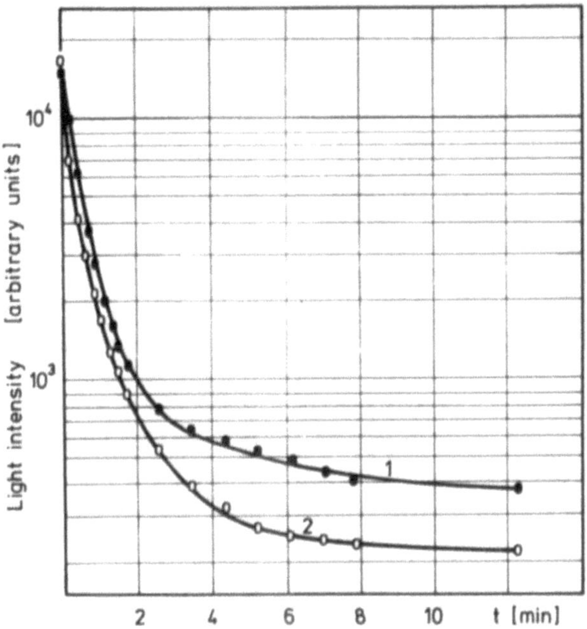

Figure 5.11. Electrical (1) and mechanical (2) activity of isolated frog heart perfused with heavy water. Both electrical and mechanical activities are expressed as percentages of the initial values.

changes in membrane enzymes and in the contractile actomyosin system, as well as the diminution of energy pools under deuteration, were considered the primary factors involved in the inhibitory effects. The higher sensitivity to the action of heavy water is most probably explained by the fact that the energy requirements and complexity of the contraction function are greater than those of bioelectrogenesis.

Progressive increase of rheobase and potential duration and decrease of conduction velocity *D_2O effects* were always observed *in peripheral nerve trunks.* At a constant level of continuous stimulation, the diminution and then disappearance of the nerve action potential were also seen (Vasilescu and Margineanu 1971). The time elapsed until the potential disappearance was found to depend on the stimulus frequency. Because a rise in the stimulus intensity or duration results in the reappearance of the action potential, an increase in the excitation threshold of nerve fibers under deuteration was thought to be the reason for the observed phenomenon (Vasilescu and Katona 1980b). More pronounced effects on higher stimulus frequencies can tentatively be explained by the interference of deuteration with the cellular energy mechanisms.

In order to reach a deeper understanding of the molecular mechanisms involved in the action of heavy water, as well as in bioelectrogenesis, we resumed our investigations concerning *D_2O effects on the nerve,* this time working *on the Ranvier node of the isolated nerve fiber* (Vasilescu and Zaciu 1980; Zaciu and Vasilescu 1981). Experimental and theoretical research based on electric and electrokinetic measurements at the level of the Ranvier node membrane was done in order to clarify the role of water and protons in the genesis of nerve influx.

Our voltage clamp measuring equipment allows sodium and potassium currents through the Ranvier node membrane to be recorded. Experiments that were performed on the Ranvier node membrane, at 12-15°C, revealed that 30 min after its immersion in D_2O, the ionic currents (both Na^+ and K^+) decrease by a factor of 30-40%, whereas the time to peak inward current increases by 20-30%.

We also investigated the effects of deuteration on the excitability properties in dynamic behavior of the Ranvier node membrane (Vasilescu and Zaciu 1980). According to our results, at a constant threshold, the possible maximum frequency of stimulation decreases to half under deuteration conditions, which reveals that heavy water modifies the membrane mechanisms acting during the refractory period. The changes induced have a rapid kinetics (2-3 min). Excitability loss following long stimulation (with $f = 200$ Hz in Ringer-H_2O and $f = 100$ Hz in Ringer-D_2O), as well as recovery at low frequencies (1, 3, and 10 Hz), was observed in both normal and deuterated membranes. The membrane immersed in heavy water needs a higher energy to keep its excitability. Energetic conditions for the stimulus still producing excitation might be expressed by the ratio

$$R = (P_\mu^2 D)_{t_0} / (P_\mu^2 D)_{t_1} ,$$

where t_0 is the moment of stimulation starting (frequency = 100 Hz in D_2O-Ringer and 200 Hz in H_2O-Ringer solution), t_1 is the moment of excitability loss, and P_μ and D are the stimulus amplitude and duration, respectively. Membranes could be kept excitable at different moments of time by varying either the amplitude or the duration of stimulus. In the former case, $R = 30$ and 6.5 in heavy water and water, respectively, whereas in the latter case the corresponding values of R are 10 and 2.5, respectively. Our explanation for the latter case in which the ratios have a smaller value is that, owing to a further condition added (namely that the membrane limits the duration of the rheobasic stimulus to a value equal to τ_{excit}), the excitability is lost before the processes involved in the membrane permeability have been changed significantly.

Conclusions

We have described methodological approaches that can be used to study the role of water in several biological processes. A diverse range of conclusions can be drawn from our studies.

1. At least three distinct tissue water components and temperature-dependent exchange processes between them are revealed by various techniques. The size and properties of these water components are dependent on the state and nature of the tissue.
2. Heterogeneity of nonfreezable and nonexchangeable water fractions in normal and deuterated nerves, respectively, is revealed by NMR measurements.
3. The state of cell membranes markedly influences the tissue water distribution. Any change in the state of the excitable membrane involves changes in the sizes of different subpopulations of water molecules.
4. Settling of anesthesia in nerve is accompanied by a marked decrease in the amount of so-called bound water.
5. Replacement of protons by deuterons leads to a delay in the settling of anesthesia.
6. Substitution of deuterons for water protons affects the hydrolysis and formation of ATP and determines a marked decrease in the cellular ATP content.
7. Heavy water has proved to be a strong inhibitory agent, both of bioelectrogenesis and of contractility, and uncouples electrical and mechanical functions in the isolated frog heart.
8. Deuteration influences the dynamic behavior of the Ranvier node membrane. At constant threshold the possible maximum frequency of stimulation of isolated nerve fibers decreases to half on deuteration.
9. A higher energy is necessary for maintaining the excitability conditions in the deuterated membrane as compared to control.
10. Considering the (sub)molecular level, heavy water effects are the consequences of at least three major phenomena: a change in the hydration degree, a decrease in the mobility of different species, and the stabilization of the native, more compact conformation of large biopolymers.

References

Chapman, D. (1972). Nuclear magnetic resonance spectroscopic studies of biological membranes. *Ann. N.Y. Acad. Sci. 195,* 179–206.

Chapman, D. (1975). Phase transitions and fluidity characteristics of lipids and cell membranes. *Rev. Biophys. 8,* 185–235.

Chapman, D. and Dodd, G.H. (1971). Physiochemical probes of membrane structure (Chapter 2). In *Structure and Function of Biological Membranes,* Rothfield, L.I., ed. Academic Press, New York, pp. 13–77.

Cooke, R. and Kuntz, I.D. (1974). The properties of water in biological systems. *Ann. Rev. Biophys. Bioeng. 3,* 95–126.

Edsall, J.T. and McKenzie, H.A. (1980). Water and proteins: Some history and current perspectives. In *Some Aspects of Water and Metal Cations in Biological Systems,* Pullman, B. and Yagi, K., eds. Japan Sci. Soc. Press, Tokyo, pp. 75-85.

Eyring, H., Woodbury, J.W., and D'Arrigo, J.S. (1973). Molecular mechanism of general anesthesia. *Anesthesiology 38,* 415-424.

Finney, J.L. (1979). Heavy water effects on certain energetic processes in retina. In *Water: A Comprehensive Treatise,* Franks, F., ed. Plenum Press, New York and London, vol. 6, 47-122.

Ganea, C. and Vasilescu, V. (1979). *Rev. Roum. Morphol. Embryol. Physiol. Physiology 16,* 59-62.

Katona, E., Margineanu, D.-G., and Vasilescu, V. (1979). Water compartments in living glycerated and fixed skeletal muscles of the frog. *Cell Tissue Res. 203,* 331-338.

Katona, E. and Vasilescu, V. (1981). Pulsed NMR data on water state in nerve. *Studia Biophys. 84,* 79-80.

Katona, E., Ganea, C., Popescu, A.I., and Vasilescu, V. (1982). In *Biophysics of Water,* Franks, F., ed. John Wiley, England (to be published).

Mathur-De Vré, R. (1979). The NMR studies of water in biological systems. *Prog. Biophys. Molc. Biol. 35,* 103-134.

Packer, K.J. (1977). The dynamics of water in heterogeneous systems. *Phil. Trans. R. Soc. Lond. Biol. B 278,* 59-87.

Popescu, A.I., Ganea, C., Eremia, D., and Vasilescu, V. (1980). International Conference on Water and Ions in Biological Systems, Bucharest, Roumania, p. 171.

Quinn, P.J. and Chapman, D. (1980). The dynamics of water in heterogeneous systems. *CRC Crit. Rev. Biochem. 3,* 1-117.

Shporer, M. and Civan, M.M. (1977). The state of water and alkai-cations within the intracellular fluids: The contribution of NMR spectroscopy. *Current Topics in Membranes and Transport 9,* 1-69.

Singer. S.J. (1971). The molecular organization of biological membranes. In *Structure and Function of Biological Membranes,* Rothfield, L.I., ed. Academic Press, New York, pp. 146-222.

Vasilescu, V. (1981). Water in biosystems. *Studia Biophys. 84,* 3-4.

Vasilescu, V. and Ciures, A. (1978). Influence of heavy water on the cardiac automatism. *Studia Biophys. 71,* 81-84.

Vasilescu, V. and Katona, E. (1980a). In *Frontiers of Bioorganic Chemistry and Molecular Biology,* Ananchenko, S.N., ed. Pergamon Press, Oxford and New York, pp. 445-461.

Vasilescu, V. and Katona, E. (1980b). New data concerning the role of water in biosystems. *Rev. Roum. Morphol. Embryol. Physiol. Physiologie 17,* 3-17.

Vasilescu, V. and Margineanu, D. (1971). Investigations on the effects of deuteration in some functions of biological membranes. *Rev. Roum. Physiol. 8,* 217-226.

Vasilescu, V. and Margineanu, D. (1974). The role of water in biological membrane phenomena as revealed by deuterium isotope effects. *Rev. Roum. Physiol. 11,* 167–171.

Vasilescu, V. and Zaciu, C. (1980). *Studia Biophys. 78,* 107–118.

Vasilescu, V., Margineanu, D.-G., and Katona, E. (1977). Heavy water intake in tissues. II. H_2O–D_2O exchange in the myelinated nerve of the hog. *Experientia 33,* 192–193.

Vasilescu, V., Popescu, A.I. and Eremia, D. (1978b). *IInd Natl. Symp. Freeze-Drying,* Bucharest, p. 109.

Vasilescu, V., Trutia, E., and Zaciu, C. (1975). Contributions to the perfecting of the enzymatic method used for the determination of ATP in biological systems. *Rev. Roum. Morphol. Embryol. Physiol. Physiologie 12,* 87–91.

Vasilescu, V., Zaciu, C., and Trutia, E. (1976). 2 H_2O effects on ATP pool in frog stimulated nerve. *Rev. Roum. Morphol. Embryol. Physiol. Physiologie 13,* 35–38.

Vasilescu, V., Georgescu, D., Scintei, M., and Savopol, I. (1968) Symposium on Biophysical Aspects of Permeability, Jerusalem, p. 4.

Vasilescu, V., Katona, E., Simplaceanu, V., and Demco, D. (1978a). Water compartments in the myelinated nerve. III. Pulsed NMR results.

Zaciu, C. and Vasilescu, V. (1981). Excitability properties of the Ranvier node membrane as investigated by deuteration. *Studia Biophys. 84,* 71–72.

Part II

Factors Affecting Membrane Permeability

6
Proton Flux Across Model and Biological Membranes

D.W. Deamer

A major function of biological membranes is to act as barriers to the free diffusion of ions. This is most clearly seen when one considers the consequences if membranes were freely permeable: Chloroplasts and mitochondria would be unable to synthesize ATP, nerve impulses could not be produced, muscles could not function and blood cells would hemolyse. For these reasons, biophysicists have long been interested in the nature of the barrier to ion flux.

The primary barrier is certainly the lipid bilayer moiety of the membrane. The evidence for this is derived from studies of model membrane systems such as planar lipid membranes and liposomes. When ionic conductance was first measured in such systems, it was found to be orders of magnitude lower than the ionic conductance of biological membranes. It follows that the bilayer is essentially an impermeable barrier to ion flux, and that the relatively high permeability of biological membranes results from conductance through protein channels that penetrate the bilayer.

The most significant ions involved in cell function include sodium, potassium, chloride, calcium, and hydrogen ions or protons. In recent years, increasing interest has focused on proton permeability because of the advent of the chemiosmotic theory (Mitchell 1961, 1966) which showed that electrochemical proton gradients were basic energy sources for coupling membranes. It has been generally assumed that all of the above ions are relatively impermeant to biological membranes. In fact, it was important to prove this for protons in early tests of the chemiosmotic theory. Mitchell and Moyle (1967) showed that proton conductance of mitochondria was 0.45×10^{-6} S cm^{-2} and pointed out that this was a low conductance, even when compared with other ions. However, it is worth taking a closer look at this result.

In the method devised by Mitchell and Moyle (1967), proton conductance was estimated by adding a pulse of an acid or base to a mitochondrial suspension (Fig. 6.1). This produced a pH gradient across the membrane, and the kinetics of decay could be followed by a glass electrode. From the known buffering capacity of the mitochondria, together with an estimate of the membrane area in the system, it was possible to estimate proton conductance. Although proton conductance was in the range of that for sodium or potassium, it should be

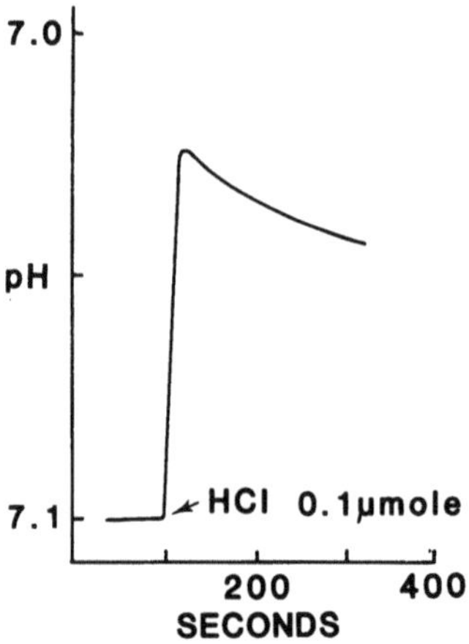

Figure 6.1. Decay of pH gradients in a mitochondrial suspension. When a pulse of hydrochloric acid was added, the external pH first decreased, then increased slowly over a few minutes as proton equivalents passively fluxed inward across the mitochondrial membranes. Redrawn from Mitchell (1966). *Biol. Rev. 41,* 445.

noted that the gradient decays, with a halftime measured in minutes, much more rapidly than expected from knowledge of sodium and potassium permeabilities in other membrane systems.

This apparent paradox brings up a significant point that is frequently overlooked: A low ionic conductance is not necessarily equivalent to a low permeability. Conductance is a measurement of ion flux and depends on the concentration of the ionic species involved. It follows that if protons, sodium, and potassium have similar *conductances* in mitochondria, then the *permeability* of the membrane to protons must be vastly greater, since proton concentration is 10^{-7} M and sodium or potassium concentrations are typically in the range of 10^{-1} M.

From these considerations, it became apparent that the intrinsic permeability of natural membranes to protons appears to be greater than that of other monovalent ions. There are several possible explanations if this is the case. First, it has been known for years that pH gradients rapidly decay across erythrocyte membranes, and this has been accounted for by a protein channel that provides for bicarbonate-chloride exchange, thereby permitting a buffer anion to carry proton equivalents. Mitochondria and chloroplasts may have similar porter proteins that permit such movement (Mitchell 1966), or perhaps nonspecific proton flux occurs through other transmembrane proteins. A second possibility is that cell membranes might contain small amounts of fatty acids or other unknown components that could act to enhance proton permeability. Finally, the lipid bilayer itself could have an unexpectedly high intrinsic permeability to protons.

The third possibility leads to a basic question related to proton flux across membranes: What is the intrinsic proton permeability of the bilayer region of natural membranes and how does it compare with its permeability to other monovalent cations? The simplest way to approach this question experimentally is to use model systems such as liposomes and planar lipid membranes. The first attempt to measure proton permeability coefficients of lipid bilayers was by Nichols et al. (1980) who used the 9-aminoacridine (9AA) method in a liposome system to monitor decay of known pH gradients. In this study, pH gradients were established across liposome membranes composed of egg phosphatidylcholine plus 2 mol% egg phosphatidic acid, which served to prevent aggregation. The liposomes (large unilamellar vesicles–LUV) were prepared by injecting a diethyl-ether solution of lipid into a sodium pyrophosphate buffer at low pH ranges followed by Millipore and gel filtration. The filtration through cellulose nitrate membranes served to size the vesicles and remove aggregates, and the gel filtration removed residual ether and exchanged the external buffer for potassium sulfate. Aliquots of the liposomes were then injected into a potassium phosphate buffer at high pH, and the decay of the gradients (∿3 pH units) was monitored by the resulting decrease in 9AA fluorescence quenching (Deamer et al. 1972). Since the buffering capacity was known and the lipid surface area could be calculated, it was possible to determine net proton-hydroxyl flux and the correlated permeability coefficient. In order to compare proton and sodium permeability, the flux of ^{22}Na was measured in the same system.

A typical experiment is shown in Fig. 6.2. Note that the halftime of decay of the buffered gradients is measured in minutes, comparable with the somewhat faster halftime of decay in mitochondria that was shown earlier. The symbol P_{net} will be used to indicate permeability coefficients in which both proton and hydroxyl flux may have participated. If an attempt has been made to obtain separate measurements, the symbols P_H or P_{OH} will be used. If P_{net} for proton-hydroxyl flux is calculated, the values are in the range of 10^{-4} cm/s.

The result, if correct, reflects a remarkably high permeability of the lipid bilayer to proton equivalents. It was important to test the result in a second system, since the 9AA method required large pH gradients that could have affected the result in some unknown manner. A glass electrode system similar to that described by Mitchell and Moyle (1967) was chosen, since it could be used with pH gradients in the range of a few tenths of a pH unit. In this system,

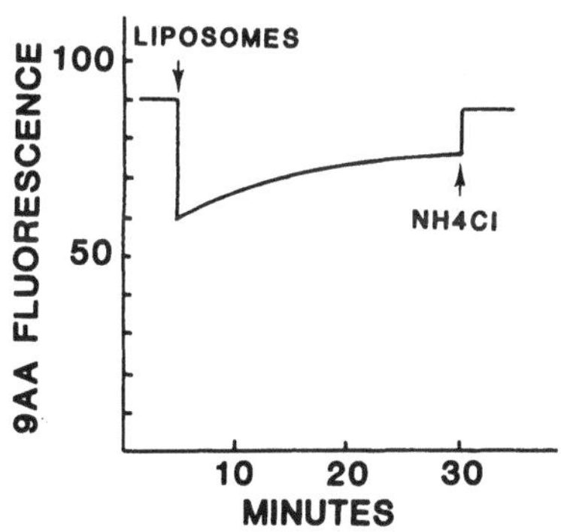

Figure 6.2. Decay of pH gradients in liposomes. Liposomes containing pyrophosphate buffer at pH 5.8 were placed in a second pyrophosphate buffer at pH 8.6 containing 1 μM 9AA. Since the 9AA is a weak base, it becomes concentrated in the acidic volume and undergoes a fluorescence quenching. As the pH gradient decays, 9AA is released and quenching decreases. Ammonium chloride addition causes the gradient to decay completely, and the 9AA fluorescence returns to its original level. Redrawn from Nichols et al. (1980). *Biochim. Biophys. Acta 596,* 393.

impermeant buffers were encapsulated in LUV liposomes as before, and these were placed in a lightly buffered medium. Sucrose was used to balance the internal and external osmotic strength.

A typical result is shown in Fig. 6.3. Note that the buffered gradients again decay with a halftime measured in minutes and that the kinetics of decay are similar in either direction. The permeability coefficient calculated from the proton flux was in the range of 10^{-4} cm/s, and this value was obtained with a variety of lipid mixtures.

Several other recent studies have been directed toward estimating proton permeability, with a nearly unprecedented disagreement about the actual value. Perhaps six orders of magnitude separate the highest and lowest estimates. These studies will be described here and then discussed with a view toward resolving some of the discrepancies.

Clement and Gould (1981) and Biegel and Gould (1981) used an entrapped pH sensitive fluorescent dye to monitor rapid pH changes in the interior of small unilamellar vesicle (SUV) liposome preparations. This method has the advantage of reporting pH shifts in the millisecond range, but is not readily adapted to quantitative measurements of flux. The fluorescent dye pyranine (2.5 mM) was encapsulated during liposome preparation by sonication, the other components being 0.1 M potassium chloride (KCl) and 5 mM Tricine-morpholino-ethanesulfonic acid (MES) buffers. Purified soybean phospholipid was used in these studies because it is commonly used in membrane reconstitution studies. However, the system was also tested with purified lipids such as dimyristoyl phosphatidylcholine. In a typical experiment, a pH gradient was produced in a stop flow apparatus by mixing the liposomes with a second buffer of different pH, and the pH-dependent pyranine fluorescence shift gave an indication of the rate of internal pH change resulting from proton flux. In the absence of valinomycin, the decay had a halftime of a few minutes. If valinomycin was present, the decay halftime was reduced to 300 ms. This result led the authors to conclude that the

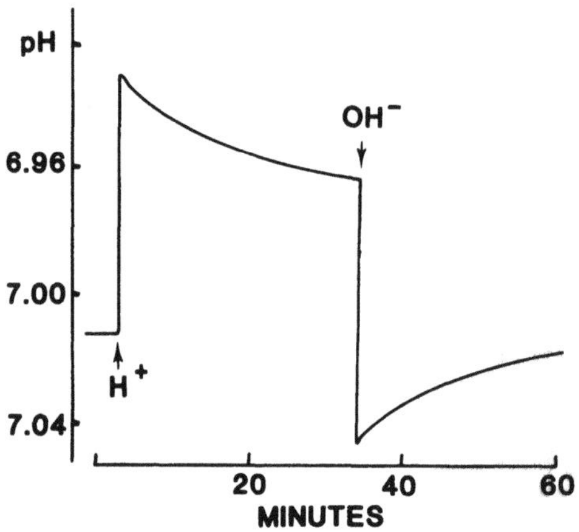

Figure 6.3. Decay of pH gradients in a liposome system. Liposomes containing a concentrated buffer were placed in a relatively unbuffered solution. When the pH was displaced by acid or base addition, the rapid pH shift was followed by a slow return to an intermediate pH as proton equivalents passively crossed the membrane. Redrawn from Nichols and Deamer (1980). *Proc. Natl. Acad. Sci. USA 77,* 2038.

proton permeability was much greater than that of other cations. Potassium was the only counterion available, and in the absence of valinomycin, it was rate limiting. If valinomycin was added so that potassium could move freely, the high proton permeability became apparent, and the pH gradient decayed in less than a second. (The rapid decay relative to the LUV system described previously probably results from the low buffering capability of the encapsulated medium and the small volume of the SUV preparation.)

These results offer independent confirmation that proton permeability is relatively high in liposome systems. Since sonication was involved, rather than ether infusion or detergent dialysis, the possibility of artifacts produced by trace contaminants is reduced. Furthermore, since the same result could be obtained with liposomes prepared from synthetic lipids containing saturated acyl chains, oxidation products could not have contributed to the result.

A considerably lower permeability was reported by Cafiso and Hubbell (1981) who took special care to reduce chloroform and fatty acid traces to minimal levels. In this study, proton flux down a gradient was used to produce a membrane potential that could be detected by a spin-label probe. From the kinetics of the potential buildup, it was possible to calculate proton flux kinetics and permeability, and the values obtained were in the range of 10^{-8}-10^{-9} cm/s. It is significant that these values are still from four to five orders of magnitude greater than the permeability coefficients of sodium and potassium reported in SUV preparations using phosphatidylcholine.

Nozaki and Tanford (1981) studied proton permeability in another LUV system. In this study, liposomes were prepared by octyl glucoside dialysis at pH 4 and 10 in 40 mM aspartate buffer and then titrated to pH 7 to produce a pH gradient across the membrane. The rationale was that aspartate is a good buffer at pH 4 and 10 but buffers poorly at pH 7. A glass electrode could therefore monitor the external pH changes resulting from proton flux down the pH gradients. From the knowledge of the rate of pH change and the aspartate buffer capacity, permeability coefficients could be calculated for protons and hydroxyls at the low- and high-internal pH range, respectively.

In this system, the measured pH change occurred over a period of 40 h, rather than the few minutes reported by other investigators. The apparent P_H was in the range of 10^{-9} cm/s, and P_{OH} was in the range of 10^{-7} cm/s. The flux measured in the presence of nitrate was about 10 times that in the presence of chloride. Since this was in the same order as the relative pK's of nitrate and chloride, the authors concluded that much of the flux, at least in the acid internal pH range, occurred as protonated anions. Using the data of Gutknecht and Walter (1981), the authors were able to subtract the apparent hydrogen chloride contribution for the chloride result and concluded that the true P_H is in the same range (10^{-12} cm/s) as that of other monovalent cations. They account for the high P_{OH} value by suggesting that hydrolysis products may have accumulated in the membranes during the extended dialysis required to remove the octyl glucoside used as a detergent and that these increased the bilayer permeability.

Gutknecht and Walter (1981) used planar lipid membranes to estimate proton permeability. Membranes were formed from several phospholipid species by brushing a decane-lipid solution over 1.8 $(mm)^2$ holes in a polyethylene partition separating two compartments. The phospholipid species included phosphatidylcholine with and without cholesterol and phosphatidylserine. Compartments were either acidified (300 mM HCl) or neutral (50 mM NaCl), and proton flux was measured both as conductance and as H^+ concentration with a glass electrode in the second compartment. This measured the pH shift produced as proton equivalents crossed the membrane. (In some experiments, the first compartment was also made basic, although the membranes were less stable under highly alkaline conditions.) Most of the flux occurred as hydrogen chloride under these conditions, and the calculated hydrogen chloride permeability was 2.9 cm/s. No flux was observed when H_2SO_4 was substituted for hydrogen chloride. The authors concluded that planar lipid sulfuric acid membranes had very low conductance for protons and hydroxyl ions.

Since there are so many variables among the published studies of proton permeability in lipid bilayer systems, it is not possible at this time to decide which represents the "true" value. (See Table 6.1 for summary.) Despite the variation in the results, it is still possible to compare the range of permeability

Table 6.1. Summary of Proton Permeability Coefficients Measured in Lipid Bilayer Systems

System[a]	Permeability Coefficient ($cm\ s^{-1}$)	References
LUV, PC:PA 98:2 9AA method	$P_{net} = 1.4 \pm 1.6 \times 10^{-4}$	Nichols et al. (1980)
LUV, PC:PA 90:10 Glass electrode	$P_{net} = 3.2 \pm 1.1 \times 10^{-4}$	Nichols and Deamer (1980)
SUV, DMPC Pyranine fluorescence	$P_{net} = 3 \times 10^{-3}$ [b]	Biegel and Gould (1981)
SUV, PC Membrane potential	$P_H \sim 10^{-8}$–10^{-9}	Cafiso and Hubbell (1981)
LUV, PC	$P_H = 3 \times 10^{-9}$ [c] $P_{OH} = 1 \times 10^{-7}$	Nozaki and Tanford (1981)
Planar lipid membrane, PC-decane Conductance	$P_H = 3 \times 10^{-9}$ $P_{OH} = 4 \times 10^{-9}$	Gutknecht and Walter (1981)

[a] PC, phosphatidylcholine; PA, phosphatidic acid; DMPC, dimyristoyl phosphatidylcholine; SUV, small unilamellar vesicles; LUV, large unilamellar vesicles.

[b] Calculated from data of Biegel and Gould [(1981). *Biochemistry 20,* 3474] using buffer capacity for Tricine and MES buffers and assuming encapsulated volume of 0.5 L mol DMPC.

[c] This is a measured value in nitrate solution and may not represent an intrinsic value for proton flux. See discussion in text.

Table 6.2. Approximate Proton Permeabilities of Several Biological Membranes

Membrane	Permeability Coefficient ($cm\ s^{-1}$)	References
Mitochondria	$P_{net} = 10^{-3}$ [a]	Mitchell and Moyle (1967) Nichols et al. (1980)
Frog muscle sarcolemma	$P_{net} = 10^{-3}$	Itzuzu (1972)
Sarcoplasmic reticulum	$P_{net} = 10^{-3}$	Meissner (1981)
Erythrocyte membrane	$P_{OH} = 2 \times 10^{-4}$ (pH 9) [b] $P_{OH} = 4 \times 10^{-1}$ (pH 4)	Crandell et al. (1971)

[a] Calculated by Nichols et al. [(1980). *Biochem. Biophys. Acta 596*, 393] from data of Mitchell and Moyle [(1967). *Biochem. J. 104*, 588].
[b] Authors assumed that no proton flux occurred.

values in the model membranes with those of biological membranes. Proton fluxes in the latter have been measured in mitochondria, sarcolemma, and sarcoplasmic reticulum, so it is possible to calculate approximate permeability coefficients in these systems. It is important to note that the net proton flux in the biological membranes may occur by different mechanisms than in the model membrane system, and it is likely that the permeability of the biological membranes will be higher than that of lipid bilayers. However, if any biological membrane is much less permeable to protons than lipid bilayer membranes, it follows that the latter are not good models for studies of the proton barrier properties of natural systems.

Table 6.2 shows some fluxes and computed permeability coefficients of several biological membranes. Significantly, all of the values are greater than even the largest reported permeability coefficient from model systems. We can conclude that the lipid bilayer, even if much more permeable to protons than expected from studies of other ions, nonetheless offers a sufficient barrier to proton flux, since biological membranes are even more permeable and yet are able to maintain transmembrane electrochemical proton gradients.

Although there is no consensus on absolute values of proton permeability in liposome systems, in all of the studies the measured permeability coefficients were orders of magnitude greater than those of sodium or potassium in similar systems. A possible exception is the result of Nozaki and Tanford (1981). The measured permeability coefficient was in the range of that reported by others, but these authors subtracted an assumed flux of hydrogen chloride from the measured flux and calculated that the actual permeability was similar to that of sodium in the same system. One difference in this study is that the authors did not test the effect of valinomycin-potassium on the system, so the proton flux may have been inhibited by the absence of counterion current. Another difference is that the authors did not report any observations in the first few seconds or minutes after establishing the gradients, which is the time interval used by other investigators.

As discussed earlier, Gutknecht and Walter (1981) did not find evidence for relatively high proton flux across planar lipid membranes. The essential point of their paper was that no anomolous conductance could be measured that might have reflected a high proton current, as would be expected if protons were in fact highly permeable. However, electrically neutral proton flux could be measured and was accounted for as hydrogen chloride flux. One difference in this study in relation to the liposome systems is that the measurements were performed at very low pH ranges (300 mM HCl). This is low enough so that the lipid phosphate groups would be associated with protons. The resulting positively charged membrane might indeed be less permeable to a current of positive ions in the form of protons. A second difference is that the planar lipid membranes contain decane or tetradecane as solvent, and the effect of this on proton flux is unknown.

Mechanisms of Proton Flux Across Bilayers

We will close this discussion by speculating on mechanisms by which proton equivalents may cross lipid bilayers. It is a reasonable assumption that in the absence of specific carriers, ions cross lipid bilayers as hydrated species. Therefore, we will first outline some of the concepts that have evolved to account for water flux across bilayers, since water and hydrated ions may share similar pathways.

Two approaches in considering water flux mechanisms have evolved. The first is that transient pores appear in the form of defects in the bilayer structure. Water, as well as solutes, may be able to enter such defects in the bilayer and thereby cross the membrane. This concept is supported by the finding that permeability in synthetic lipid bilayers is greatest near the phase transition temperature where such defects are at a maximum level (Papahadjopoulos et al. 1973). Hauser et al. (1973) have pointed out that if the defect is large enough, the process could not be considered to be equivalent to a true permeability. This remains an unresolved question, particularly for relatively impermeant species such as sodium.

The second approach is that water flux occurs as diffusion of monomers. For example, Traüble (1971) has proposed a molecular mechanism for water flux through bilayer structures in which it is envisaged that gauche-trans "kinks" produce defects in the fluid bilayer structure. Water monomers are able to enter the kinks and thereby diffuse with them across the membrane.

A mechanism involving some form of diffusion is supported by the demonstration of Finkelstein and Cass (1968) that we can calculate a permeability coefficient of water from knowledge of its solubility and diffusion coefficient in a bulk phase hydrocarbon system. The result is 6.4×10^{-3} cm/s, remarkably close to measured permeability coefficients in experimental bilayer systems. This is in part fortuitous, however, since water permeability of bilayers varies over a

10-fold range, depending on lipid composition. Furthermore, the solubility of water in the membrane hydrocarbon phase is unknown and may vary markedly from its solubility in a bulk phase hydrocarbon.

We can now ask whether either of these alternative models helps in understanding the relatively high value of proton permeability obtained in the studies of liposomes that have been described here. Clearly, if water flux is entirely monomeric, then a hydrated proton is no different from any other hydrated cation that may attempt to find its way across the membrane. However, if some fraction of the water within the hydrophobic region of the bilayer is associated with other water molecules in a transient defect, an interesting possibility emerges. It is well known that protons have greater mobilities in aqueous solutions or ice compared with other cations. This is explained by the ability of protons to jump along hydrogen-bonded water molecules, rather than moving as discrete charges. Nagle and Morowitz (1978) have suggested the possibility that associated water may occur in the hydrocarbon region of bilayers, and Nichols and Deamer (1980) have proposed that the relatively high proton permeability of liposomes can be accounted for if protons are able to move through hydrogen-bonded strands or clusters of water. This concept will provide a useful working hypothesis in directing future research.

References

Biegel, C.M. and Gould, J.M. (1981). Kinetics of hydrogen ion diffusion across phospholipid vesicle membranes. *Biochemistry 20,* 3474.

Cafiso, D.S. and Hubbell, W.L. (1981). Spin label detection of electrogenic proton fluxes in phospholipid vesicles. *Biophys. J. 33,* 114a.

Clement, N.R. and Gould, J.M. (1981). Pyranine as a probe of internal aqueous hydrogen ion concentration in phospholipid vesicles. *Biochemistry 20,* 1534.

Crandell, E.D., Klocke, R.A., and Forster, R.E. (1971). Hydroxyl ion movement across the human erythrocyte membrane. *J. Gen. Physiol. 57,* 664.

Deamer, D.W., Prince, R.C., and Crofts, A.R. (1972). The response of fluorescent amines to pH gradients across liposome membranes. *Biochim. Biophys. Acta 274,* 323.

Finkelstein, A. and Cass, A. (1968). Permeability and electrical properties of thin lipid membranes. *J. Gen. Physiol. 52,* 145s.

Gutknecht, J. and Walter, A. (1981). Transport of protons and hydrochloric acid through lipid bilayer membranes. *Biochim. Biophys. Acta 641,* 183.

Hauser, H., Oldani, D., and Phillips, M.C. (1973). Mechanism of ion escape from phosphatidylcholino and phosphatidylservine single bilayer vesicles. *Biochemistry 12,* 4507.

Izutsu, K.T. (1972). Intracellular pH, H ion flux and H ion permeability coefficient in bullfrog toe muscle. *J. Physiol.* (London) *221,* 15.

Meissner, C. and Young, R.C. (1980). Proton permeability of sarcoplasmic reticulum vesicles. *J. Biol. Chem. 255,* 6814.

Mitchell, P. (1961). Coupling of phosphorylation to electron and hydrogen transfer by a chemiosmotic type of mechanism. *Nature 191,* 144–148.

Mitchell, P. (1966). Chemiosmotic coupling in oxidative and photosynthetic phosphorylation. *Biol. Rev. 41,* 445.

Mitchell, P. and Moyle, J. (1967). Acid-base titration across the membrane system of rat liver mitochondria. *Biochem. J. 104,* 588.

Nagle, J.F. and Morowitz, H.J. (1978). Molecular mechanisms for proton transport in membranes. *Proc. Natl. Acad. Sci. USA 75,* 298.

Nichols, J.W. and Deamer, D.W. (1980). Net proton-hydroxyl permeability of large unilamellar liposomes measured by an acid-base titration technique. *Proc. Natl. Acad. Sci. USA 77,* 2038.

Nichols, J.W., Hill, M.W., Bangham, A.D., and Deamer, D.W. (1980). Measurement of net proton-hydroxyl permeability of large unilamellar liposomes with the fluorescent pH probe α-aminoacridine. *Biochim. Biophys. Acta 596,* 393.

Nozaki, Y. and Tanford, C. (1981). Proton and hydroxyl ion permeability of phospholipid vesicles. *Proc. Natl. Acad. Sci. USA 78,* 4324.

Papahadjopoulos, D., Jacobson, K., Nir, S., and Issac, T. (1973). Phase transitions in phospholipid vesicles. *Biochim. Biophys. Acta 311,* 330.

Traüble, H. (1971). The movement of water across lipid membranes: A molecular theory. *J. Membr. Biol. 4,* 193.

7
Water Diffusion Through Erythrocyte Membranes in Normal and Pathological Subjects: Nuclear Magnetic Resonance Investigations

V.V. Morariu and Gh. Benga

Water exchange through erythrocyte membranes occurs rapidly in the order of milliseconds. This relatively high rate of transport requires special methods for investigation among which nuclear magnetic resonance (NMR) techniques have gained increasingly wider use during the last decade. Conlon and Outhred (1972) first reported water exchange measurements through erythrocyte membranes using an NMR method. Studies of the rate of water diffusion through erythrocyte membranes are of fundamental as well as applied interest. At present, the mechanism of water permeation through erythrocyte membranes is not known. However, the effects of various physical or chemical agents on this process, as well as alterations in pathological states, have been investigated. Such studies are relevant not only for understanding the mechanism of water transport but also for determining its medical applications.

The purpose of this review is threefold: (1) to outline the methods for measuring water diffusion through erythrocyte membranes with an emphasis on NMR methods, (2) to review the temperature and pH dependence of the water diffusion through these membranes and discuss the possible mechanism of permeation through the membrane, and (3) to report on some possible medical applications. All this work refers to human erythrocytes.

Methods of Measuring Water Diffusion Through Erythrocyte Membranes

There are two basic strategies for measuring water exchange through the erythrocyte membrane: (1) nonstationary methods and (2) stationary methods.

The nonstationary methods involve subjecting the cells to an osmotic gradient that creates a net flux of water in one direction or the other, depending on whether the cells swell or shrink. The membrane is subjected to stress, and the cell will eventually hemolyze. Much information about water transport through erythrocyte membranes was obtained between 1930 and 1950 by these methods (see the review by Jacobs 1952).

In the case of stationary methods the diffusion movement of water is measured and therefore there is no net flux of water through the membrane. The cells remain in their normal state, which is often considered an advantage over nonstationary methods. The stationary methods can be classified in two groups: (1) radiotracer methods and (2) NMR methods. With radiotracer methods, tritium (^3H) is used to label the water of the suspending solution, and the rate of ^3H_2O entry into the cells is measured. Owing to the high rate of water permeation, this type of experiment requires special techniques that are typically used in the investigation of rapid reactions (Paganelli and Solomon 1957; Barton and Brown 1964; Vieira et al. 1970). These have been adapted for erythrocytes but require relatively large quantities of blood and are time-consuming compared with NMR methods. Nevertheless the use of radiotracer techniques brings an independent estimation of water permeation in red blood cells, to be used in conjunction with NMR methods.

The Principle of Water Exchange Measurements Through Erythrocyte Membranes Using NMR Techniques

We shall briefly explain the principle of the NMR method of measuring water exchange across membranes. This method relies upon the characterization of a system, consisting of two compartments A and B, by two nuclear relaxation times, T_A and T_B, of the same type of nuclei residing in each of the compartments (Fig. 7.1). The nuclear relaxation times are parameters that characterize the return to equilibrium after a suitable radiofrequency (rf) perturbation of

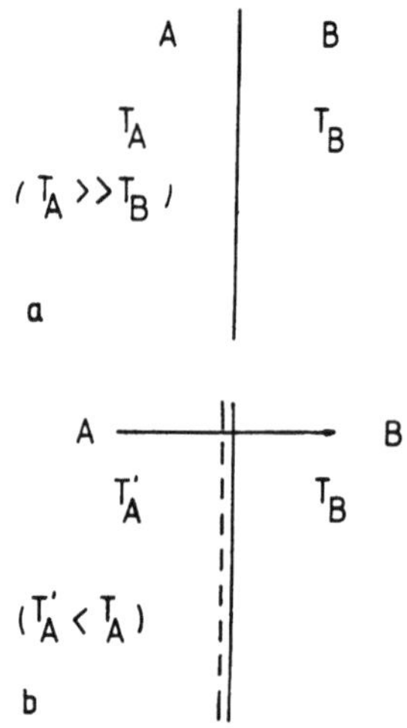

Figure 7.1. Two nuclear compartments having the same type of nucleus; (*a*) No exchange between compartments: The nuclei relax in each compartment with the relaxation time T_A and T_B, respectively (T_A is assumed to be much higher than T_B). (*b*) Fast exchange of nuclei from A to B. The observed relaxation of nuclei in compartment A, which also relax in B due to the exchange process, will be shortened compared to T_A.

the nuclei in an NMR experiment (Farrar and Becker 1971). Long relaxation times are often associated with nuclei that are part of molecules with fast motion. We will next assume that we deal with the same type of molecules distributed in compartments A and B and the corresponding NMR relaxation times differ, for some reason, so that $T_A \gg T_B$. Two cases can be considered: (1) As shown in Fig. 7.1(*a*), there is no exchange of molecules between the two compartments, and (2) as shown in Fig. 7.1(*b*), there is a relatively fast exchange process transferring molecules between compartments. The question is how the exchange process will affect the relaxation times of the two compartments.

In case (1), the two relaxation times, T_A and T_B, will be detected in an actual NMR experiment (Fig. 7.1). (In fact, even if there is a very slow exchange so that the nuclei will have time to relax in each compartment, the result will be much the same.) However, in case (2), the nuclei in A will start relaxing with T_A but will end up in compartment B where its relaxation T_B will be faster. As a result, the observable relaxation time of phase A will be T'_A, which is shortened compared to T_A [Fig. 7.1(*b*)]. The faster the exchange, the shorter T'_A. The equations describing this phenomenon have been derived by Woessner (1963), enabling the calculation of exchange times.

Let us see further how this model can be applied to an erythrocyte suspension (Fig. 7.2). The nucleus of concern in this case is the water proton, which can reside either inside the cell (compartment A) or outside the cell (compartment B). There is an exchange of water molecules between these two compartments. However, if we perform an NMR experiment on such a system we will only detect a single relaxation time. This is due to the fact that the relaxation times of the water protons in both compartments are not very different (of the order of hundreds of milliseconds), and the rapid exchange between compartments makes the distinction between the two compartments impossible [(Fig. 7.2(*a*)]. Obviously some way is needed of making $T_A \gg T_B$ as described above. One way of doing this is the method of paramagnetic doping. If we add a paramagnetic ion, such as manganese, to the cell suspension, then the proton relaxation times, T_B, of water molecules in the suspending solution will become much shorter by a mechanism known as electron-proton interaction (see, e.g., Dwek 1973 for an explanation). Then we will have a system with $T_A \gg T_B$ as described in Fig. 7.1(*b*), and the same type of experimental approach can be applied to erythrocytes. Of course, a prerequisite to this experiment is that the manganese ions do not penetrate the red blood cells. The penetration of manganese ions is hampered in the presence of albumin (Fabry and Eisenstadt 1975). Furthermore, it must be assumed that any manganese bound to the erythrocyte membrane does not alter the permeability of water molecules.

So far we have used the general term "nuclear relaxation time," whereas in reality there are two different relaxation time processes that can be measured: T_1, the so called spin-lattice or longitudinal relaxation time, and T_2, the spin-spin or transversal relaxation time. Both of these relaxation times can be used for the determination of water exchange times.

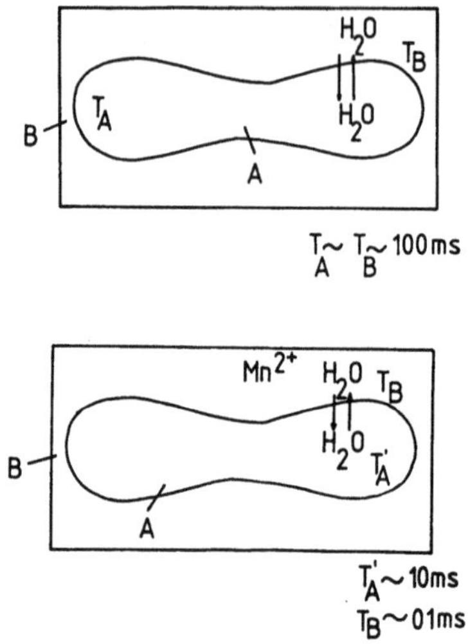

Figure 7.2. Illustration of the NMR paramagnetic doping method for measuring the water exchange through erythrocyte membranes. A is the red blood cell compartment and B is the suspending solution compartment. (*a*) In a suspension of erythrocytes the relaxation times of water protons are similar in the two compartments ($T_A \sim T_B \sim 100$ ms), and fast exchange occurs between them. As a result, a single relaxation time is detected and therefore this experiment cannot be used for the measurement of water exchange. (*b*) If manganese ions are added to the suspending solution, then T_b becomes much shorter (~ 0.1 ms). In the absence of exchange between A and B, T_A should remain unchanged (~ 100 ms). However, fast exchange of water occurs through the membrane, and T_A becomes T'_A (~ 10 ms). The water exchange time can be calculated from T'_A (see the text).

An alternative to the doping technique is the method of diffusion in a magnetic field gradient (see, e.g., Farrar and Becker 1971). This method enables the measurement of self-diffusion coefficients in liquids. However, if this coefficient is measured in very small closed spaces, its measured value will also depend on the size of the space. This is known as restricted diffusion (Chang et al. 1975). Therefore, in an NMR experiment of this type, with a suspension of cells, we will have a method of discriminating between the two compartments by means of two diffusion coefficients: One is the normal bulk diffusion of water and the other is the restricted diffusion inside the cells. This will, in turn, be affected by the rate of water exchange through the membrane. Using this method avoids paramagnetic doping, but there are certain limitations that make this method particularly insensitive at shorter exchange times (Andrasko 1976). This is the case of human erythrocytes.

The principle described above could be adapted to any other type of cell, and it is not necessarily restricted only to protons. If we refer to water, the resonance of ^{17}O in water has also been used to estimate water exchange through erythrocyte membranes (Shporer and Civan 1975). In the case of water ^{17}O resonance, however, it is not necessary to use paramagnetic doping because the relaxation times in the two compartments are sufficiently different to distinguish them, and both are shorter than the exchange times (Shporer and Civan 1975). The disadvantage of this method is its limited sensitivity to exchange rates over only a narrow temperature range.

An NMR method of measuring the transport of lithium and sodium ions across the membrane of phosphatidylcholine vesicles has already been described (Degani and Elgavish 1978). A summary of various NMR methods used for measuring the water exchange through erythrocytes is given in Table 7.1.

Paramagnetic Doping Methods and the Significance of Water Exchange Times

The observed transversal relaxation time T'_{2A} of the erythrocytes is related to the individual relaxation times T_{2A} and T_{2B} and the lifetimes τ_A, τ_B by the following relationship (Woessner 1963):

$$\frac{1}{T'_{2A}} = \frac{1}{2}\left\{\frac{1}{T_{2A}} + \frac{1}{T_{2B}} + \frac{1}{\tau_A} + \frac{1}{\tau_B} - \left[\left(\frac{1}{T_{2A}} - \frac{1}{T_{2B}} + \frac{1}{\tau_A} - \frac{1}{\tau_B}\right) + \frac{4}{\tau_A \tau_B}\right]^{1/2}\right\} \quad (7.1)$$

$\frac{1}{\tau_A}$ and $\frac{1}{\tau_B}$ are related to the populations of the phases by the relationship, $P_A/\tau_A = P_B/\tau_B$.

The significance of the parameters in Eq. 7.1 is as follows: T_{2A} and T_{2B} are the water proton transversal relaxation times of the isolated cells and doped plasma, respectively; τ_A and τ_B are the lifetimes of the water proton in the corresponding compartments; P_A and P_B are the population fractions of water protons in the compartments (or nuclear phases as they are usually termed), therefore $P_A + P_B = 1$.

In order to calculate τ_A, we need to know the following parameters, which can be determined experimentally: T_A, T_B, T'_A, and P_A or P_B.

The population parameter can be estimated by knowing the content of water in erythrocytes. The average fraction of intracellular volume represented by water is 0.71-0.72 and for plasma is 0.95 (Fabry and Eisenstadt 1975; Pirkle et al. 1979). This allows the calculation of P_A and P_B from the hematocrit value. However, a more precise knowledge of P_A should take into account the variations of the hematocrit value and hemoglobin content with age (Vilcu 1977). Our estimations are shown in Fig. 7.3.

Equation (7.1) refers to a T_2 transversal relaxation time experiment. The same expression holds for a T_1 longitudinal relaxation time.

If a relatively low concentration of manganese ions (1-2 mM) is used for doping, Eq. (7.1) should be used for the calculation of τ_A (Pirkle et al. 1979).

Table 7.1. NMR Methods for the Measurements of Water Exchange Time

Method	Nucleus	Advantage	Disadvantage	References
Diffusion in a field gradient	^{1}H	No perturbation of the membrane	The method is more sensitive to larger exchange times (not at physiological temperatures)	Andrasko (1976)
Longitudinal relaxation time (T_1)	^{17}O	No perturbation of the membrane	Requires ^{17}O enriched water; sensitive to longer exchange times only	Fabry and Eisenstadt (1975); Shporer and Civan (1975)
T_1 relaxation time at high concentration of Mn^{2+}	^{1}H	Simple analysis of data	Slower than T_2 measurement	Fabry and Eisenstadt (1975)
T_2 relaxation time at high concentration of Mn^{2+}	^{1}H	Simple analysis of data; fast measurement	Possible effects of Mn^{2+} on membranes; complicated relaxation mechanism by possible chemical shifts induced by Mn^{2+}	Conlon and Outhred (1972)
T_2 relaxation time at low concentration of Mn^{2+}	^{1}H	Little perturbation of membrane; fast measurement	Complex analysis of data; strong discrepancy of data compared to isotonic measurements	Pirkle et al. (1979)

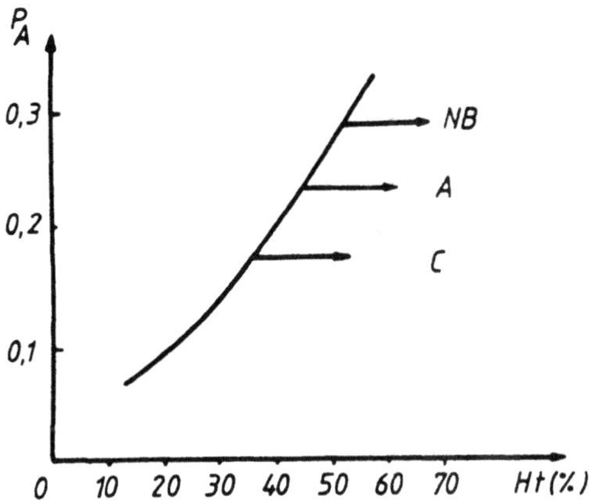

Figure 7.3. The water population fraction P_A (inside the erythrocytes) versus hematocrit (Ht) value. The value of P_A is given for a suspension containing two parts volume of blood and one part doping solution. A, adult; C, children; NB, newborn.

However, if higher concentrations of manganese are used (20-50 mM), then Eq. (7.1) is considerably simplified. This is due to the fact that $T_{2A} \gg T_{2B}$ and Eq. (7.1) can be well approximated by the simple equation (Conlon and Outhred 1972)

$$\frac{1}{T'_{2A}} = \frac{1}{T_{2A}} + \frac{1}{\tau_A} \tag{7.2}$$

or (Morariu and Benga 1977)

$$\frac{1}{T'_{2A}} = \frac{1}{T_{2A}} + \frac{P_B}{P_B \tau_A + T_{2B}} \tag{7.3}$$

which provides a more accurate estimation of τ_A.

So far the majority of workers have used whole blood for their measurements. In their initial work, Conlon and Outhred (1972) diluted 1 ml of blood with 0.5 ml of doping solution containing 20-100 mM of manganese chloride. In later work (Conlon and Outhred 1978), 0.4 ml of whole blood was diluted with 0.9 ml of manganese solution (50 or 100 mM of manganese chloride made isotonic with sodium chloride). This was found necessary to keep the packed cell volume less than 20% in order to eliminate any dependence of relaxation time on the packed cell volume.

We have performed experiments on both whole blood and washed erythrocytes suspended in isotonic saline, buffered solutions (pH 7.4) containing 1% albumin and 1% glucose. The presence of albumin is important to prevent the entry of manganese ions into the cells. These ions are known to bind to albumin forming a manganese-albumin complex (Fabry and Eisenstadt 1975). In our experiments we have used 1 ml of blood or suspension of erythrocytes and 0.5 ml solution of 40 mM manganese chloride made isotonic with sodium chloride.

The lifetime or water exchange time is related to permeability, P, by Eq. (7.4):

$$P = \frac{V}{A \cdot \tau} \tag{7.4}$$

where V is the volume of the cell occupied by water and A is the area of the cell membrane. If V/τ is constant, PA must be independent of cell volume. This has indeed been shown to be the case (Outhred and Conlon 1973). This experiment suggests that water penetrates the membrane through a fixed number of channels.

The Water Exchange Time Through Erythrocyte Membranes at 37°C

The water exchange times through erythrocyte membranes of normal adult subjects at physiological temperatures measured by different NMR methods are listed in Table 7.2. The data available in the literature include measurements at temperatures ranging between 20 and 25°C (Fabry and Eisenstadt 1975; Andrasko 1976; Chien and Macey 1977).

An independent tritium measurement gives a value of 8.9 ms for the water exchange time (Vieira et al. 1970), which may be compared with the values in Table 7.2. The closest value obtained by an NMR method is that obtained by the use of ^{17}O longitudinal relaxation time (Shporer and Civan 1975). However, this method, as mentioned before, is not very sensitive to changes in the exchange time and involves practical difficulties because it requires ^{17}O enriched water.

The next values closest to the tritium value are obtained by the use of doping methods at relatively high concentrations of manganese ions, whereas at lower concentrations the calculated τ_A is much higher. Although the methods of calculation differ among the authors (see entries 1-2 and 3-4 in Table 7.2), the actual long components of the relaxation times determined experimentally are in reasonable agreement.

In conclusion, there is still much debate in the literature concerning the differences among various laboratories and the methods used to estimate τ_A. In our view, the doping method using relatively high concentrations of manganese has practical advantages over other methods, including rapidity in the preparation of the sample and the analysis of the data using standard pulsed NMR equipment. Another advantage is the sensitivity of the doping method compared to the diffusion or ^{17}O methods; it also requires only small volumes of blood.

The Temperature Dependence of Water Exchange of Isolated Erythrocytes at pH 7.4

The temperature dependence of water exchange through erythrocyte membranes has been investigated by several authors (Vieira et al. 1970; Shporer and Civan 1975). None of these investigators reported any significant deviation of the activation energy from linearity at any particular temperature. Other investigators

Table 7.2. Water Exchange Times Through Erythrocyte Membranes of Control Adult Subjects as Measured by Different Nuclear Magnetic Resonance Methods at 37°C

No.	Method	Resonant Nucleus	Resonance Frequency (MHz)	Mn^{2+} Concentration in Plasma (mM)	Sample	A (ms)	References
1	T_2 doping	^1H	9	24–48	Whole blood	8.2 (7.7)[a]	Conlon and Outhred (1972)
2	T_2 doping	^1H	8	34.5–69	Refrigerated whole blood (not more than 48 h old)	7.15 (6.8)[a]	Conlon and Outhred (1978)
3	T_2 doping	^1H	90	19.3	Whole blood	6.0 (6.9)[a]	Morariu and Benga (1977)
4	T_2 doping	^1H	90	19.3	Washed erythrocytes suspended in isotonic buffered solution, pH 7.4, 1% glucose 1% albumin	6.4 (7.2)[a]	Morariu et al. (1981)
5	T_2 doping	^1H	60	1.7	Whole blood, pH 7.5	14.8	Pirkle et al. (1979)
6	T_1 doping	^{17}O	8.13	–	Washed erythrocytes suspended in buffered solution, pH 7.4, prepared from outdated blood	9.4	Shporer and Civan (1975)

[a] These values are the corresponding relaxation times that are used for calculating τ_A.

have, however, demonstrated breaks in the temperature dependence of membrane viscosity or microviscosity (Zimmer and Schirmer 1974; Feinstein et al. 1975), osmotic fragility (Aloni et al. 1977), enzymatic incorporation of ^{32}P into polyphosphoinositides (Buckley and Hawthorne, 1972), and exchange transport of glucose (Lacko et al. 1973). These phenomena have been discussed in terms of phase transitions in erythrocyte membranes, occurring at about 19-25°C. A systematic analysis of the temperature dependence of water transport through erythrocyte membranes should reveal the effect of such transitions on this transport process.

Conversely, it has been suggested from Raman spectroscopic studies that the erythrocyte membranes undergo cooperative, pH-sensitive transitions at physiological temperatures (Verma and Wallach 1976). It is important to know if these transitions have any effects on water transport.

We studied the temperature- and pH-dependent changes of water transport in order to gain a better understanding of the molecular mechanisms of erythrocyte water diffusional transport. Water exchange through isolated erythrocytes appeared to be sensitive to both temperature and pH. These findings are discussed in relation to other membrane processes that showed a similar behavior.

The temperature dependence of the water exchange time τ_A was examined for five samples isolated from different donors. An Arrhenius plot of τ_A of these samples is shown in Fig. 7.4. In the higher temperature range, τ_A varies markedly with temperature but is relatively independent of temperature in the lower range. There is an obvious discontinuity in this plot. The break is at about

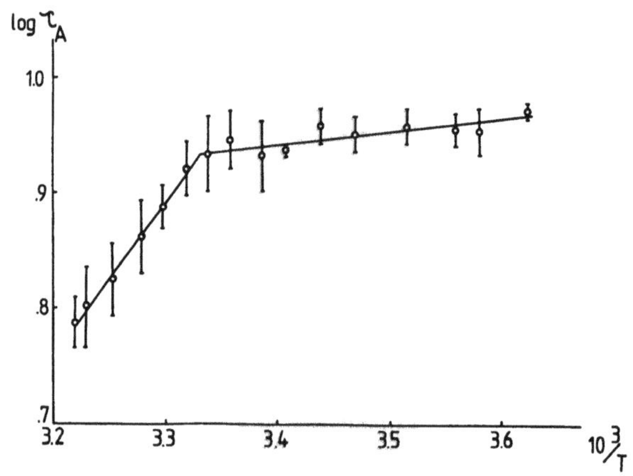

Figure 7.4. Arrhenius plot of water exchange time (τ_A) through the erythrocyte membrane. The measurements have been performed on erythrocytes isolated as described in the text from the blood of five donors. The results represent the mean (o) and the standard deviation (the bars). The plot has been computed using the least-squares method. Data from Morariu et al. (1981). *J. Membr. Biol. 62,* 1–5.

26°C for erythrocytes isolated and suspended in buffered solution at pH 7.4. Two values of the apparent activation energy of the water exchange time through the erythrocyte membranes can be calculated on the basis of this plot. The values of the activation energy, the temperatures at which the discontinuity in the Arrhenius plot occurs, and the water exchange time τ_A at 37°C are given in Table 7.3. The effect of temperature on the water exchange time was fully reversible. The same values of τ_A were obtained when the samples were run from 37 to 2°C and then back from 2 to 37°C. The mean value of the activation energy of the water exchange time at temperatures higher than that of the break is 23.8 kJ/mol. This corresponds to the lower limit of the same parameter reported previously (Morariu and Benga 1977), and it is somewhat lower than the corresponding value given by Conlon and Outhred (1978). The values of the activation energy of the τ_A at temperatures lower than that of the break vary between 0 and 5.9 kJ/mol. There was no observable correlation with the age of donor.

The effects of pH and temperature on water diffusion through erythrocyte membranes have to be discussed within the framework of similar changes described for other transport processes in erythrocytes and with regard to the molecular mechanisms proposed for water transport. As discussed by Sha'afi (1977), two possible mechanisms for water transport in human red blood cell membranes can be considered. One model is based on the idea that thermal fluctuations in membrane lipids can cause conformational changes in the hydrocarbon chains that lead to the generation of mobile structural defects, or "kinks" (Traüble 1972), which may act as intrinsic carriers for water and non-

Table 7.3. Parameters Characterizing the Temperature Dependence of the Water Exchange Time τ_A Through Erythrocyte Membranes at pH 7.4

Donor	Age (years)	$E_t > t_b$ (kJ/mol)	$E_t < t_b$ (kJ/mol)	t_b (°C)	τ_A (ms)
1	22	22.2	0	26	6.8
2	27	23	5.9	26	6.0
3	30	24.3	2.5	27	5.9
4	35	26	0	26	5.9
5	46	23	2.9	26.5	6.0
		23.8 ± 1.7[a]	2.1 ± 2.1[a]		

Note: t_b is the temperature at which a break in the Arrhenius plot of the water exchange time τ_A through erythrocyte membranes was observed; $E_t > t_b$ is the apparent activation energy at temperatures higher than t_b; $E_t < t_b$ is the apparent activation energy at temperatures lower than t_b. The measurements were performed on erythrocytes isolated in buffered solutions at pH 7.4.

[a] Mean ± standard deviation.

electrolytes. The second model assumes the presence of hydrophilic pathways for water transport, or "pores," assembled from membrane integral proteins. Sha'afi (1977) presented evidence that suggested that either band 3 glycoprotein or glycophorin (nomenclature of Fairbanks et al. 1971 and Marchesi et al. 1972, respectively) or both may be involved in the formation of hydrophylic pathways for water transport.

De Gier (1979) has pointed out that the activation energy can be an indication of whether the permeant solute penetrates by the kink mechanism or whether a facilitated diffusion mechanism is available. The apparent activation energy of water diffusion through erythrocyte membranes determined in this report is in agreement with previously reported values (Sha'afi 1977) and is much lower than the activation energy for water permeation through liposomes (50 kJ/mol) (De Gier 1979). This suggests that mechanisms of water permeation in erythrocytes are different from those in liposomes.

Our findings can be explained in terms of a protein channel facilitating water diffusion based on our observation of the break at about 26°C in the Arrhenius plot of the water exchange time. A number of other erythrocyte membrane processes has been reported to show a break in the temperature dependence at about 19-26°C. Zimmer and Schirmer (1974) and Feinstein et al. (1975) attributed the nonlinearities in the temperature dependence of membrane viscosity and microviscosity to phase transitions in membrane lipids. Other authors have also mentioned the involvement of proteins in the discontinuities that occur in the temperature dependence of erythrocyte membrane processes as discussed by Lacko et al. (1973) for the transport of glucose or by Aloni et al. (1977) for the break at 25°C in the osmotic fragility of human erythrocytes. Other membrane properties reported to show breaks or discontinuities as a function of temperature include membrane-bound enzymatic activities (Raison et al. 1971; Lenaz et al. 1975) and various electron-spin resonance parameters of spin-labeled membranes (Inesi et al. 1973; Benga and Strach 1975). Such discontinuities are often interpreted as indicating conformational changes in membrane proteins induced by a temperature-dependent phase change in the membrane (see Lenaz 1973; Van Deenen et al. 1975 for reviews). However, because of the heterogeneity of the membrane system, the physical interpretation of the temperature discontinuities is rather complex. Breaks in Arrhenius plot can have more trivial origins depending on the particular process taking place (see also Chapter 4). It is not clear whether a phase transition of erythrocyte membrane lipids actually occurs between 0 and 37°C. Although some investigators have found evidence for a phase transition occurring at about 20°C (Johnson 1975; Zimmer and Schirmer 1975; Bieri and Wallach 1976), in recent studies using deuterium NMR, a phase transition of the acyl chains of the erythrocyte lipids was not detected (Davis et al. 1979). Bond and Baumann (1978) recently emphasized that it is not likely that an entire membrane will exhibit a phase transition analogous to bulk lipid phase transitions. An interaction between the lipids surrounding a protein channel and the protein complex per se could lead to rather abrupt changes in the conformation of the protein channel itself. Although the findings described in the paper by Bond and Baumann (1978) do not enable us to discriminate between such molecular mechanisms, it is clear that a dependence of the apparent activation energy on the temperature reflects a cooperative process (Volkenstein 1977).

The pH Dependence of the Water Exchange Time of Isolated Erythrocytes

The particular shape of the pH dependence at 24°C of samples of erythrocytes prepared from two donors is shown in Fig. 7.5. Obviously, in the acid range there is a marked increase in water exchange times. A much smaller variation of the same parameter occurs between pH 7.0 and 8.0.

The temperature dependence of τ_A was measured on isolated erythrocytes at pH = 6.4, 7.0, and 7.4 (Fig. 7.6). At pH 6.4 the break in the Arrhenius plot is less evident than at pH 7.4. However, a value of t_b[1] could be estimated by intersecting the tangents at the two extremes of the Arrhenius plot. The activation energies were estimated from the same tangents. The break in the Arrhenius plot is evident at pH 7.0, but the t_b is shifted to lower temperatures. The values of the parameters characterizing the water exchange time at various pH values are listed in Table 7.4. It is obvious from Fig. 7.6 and Table 7.4 that the activation

[1] t_b is the temperature at which a break in the Arrhenius plot of the water exchange time τ_A through erythrocyte membranes was observed.

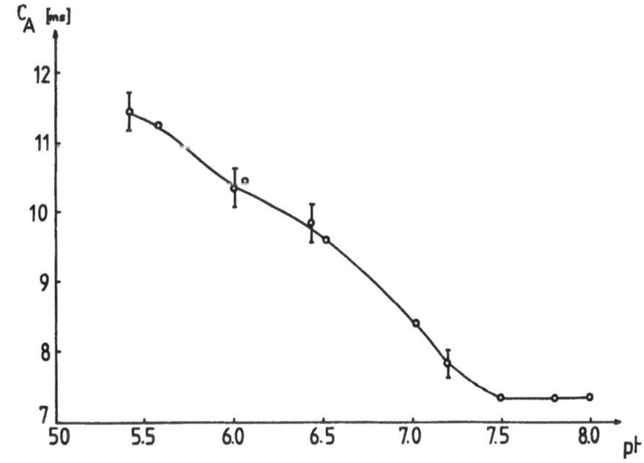

Figure 7.5. The pH dependence of the water exchange time (τ_A) through erythrocyte membranes at 24°C. The erythrocytes from the same batch of blood were isolated and washed in buffered solutions of various pH values as described previously (Morariu et al. 1981). The results are the means (o) and the standard deviations (the bars) of measurements performed in duplicate or triplicate on blood from the two donors.

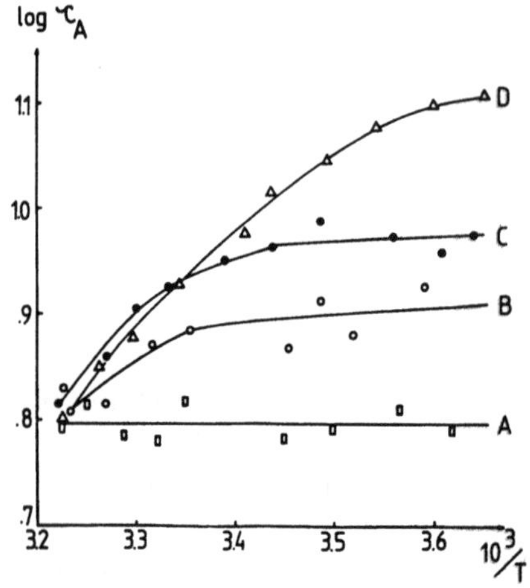

Figure 7.6. Arrhenius plots of water exchange time (τ_A) through erythrocyte membranes at pH 6.4 (D), 7.0 (C), and 7.4 (B). For each pH value the erythrocytes were isolated and suspended in a buffered solution as previously described (Morariu et al. 1981).

energy and t_b are markedly influenced by the pH of the erythrocyte suspension. It also appears that there is a critical pH value at about 7.0 where both the activation energy and t_b show a minimum.

Verma and Wallach (1976) described pH-sensitive cooperative state transitions of erythrocyte membranes between pH 5.5 and 7.0. Based on Raman spectroscopic evidence, these authors suggested that such transitions represent concerted rearrangements of acyl-chain configurations and protein residue orientations in response to the pH variation. The idea of pH-dependent alterations of erythrocyte membranes is further supported by the observation of Goekoop et al. (1978) using freeze-fracture electron microscopy. They showed a decrease in the number of red blood cell membrane elevations by decreasing the pH from 7.5 to 6.5.

It was therefore interesting to observe (Figs. 7.5 and 7.6) that significant changes in water diffusion occurred around the pH values that induce state tran-

Table 7.4. Parameters Characterizing the Water Exchange Time Through Erythrocyte Membranes at Various pH Values[a]

pH	$E_t > t_b$ (kJ/mol)	$E_t < t_b$ (kJ/mol)	t_b (°C)
6.4	20.1	10.9	22
7.0	13	0	17
7.4	24.7	2.1	27

[a] The measurements were performed on erythrocytes isolated in buffered solutions at various pH values. The notations are the same as those in Table 7.3.

sitions in the erythrocyte membrane. This can be explained in terms of a protein channel mediating water diffusion. It is well known that facilitated diffusion is strongly pH dependent. For example, the rate of dissipation of a pH gradient across the human red blood cell is a facilitated transport process (Jennings 1978). The plot of pH dependence of the red blood cell pH equilibrium, mediated by the membrane protein that catalyzes anion exchange (band 3 glycoprotein), is similar to the pH dependence of water diffusion (Fig. 7.5 in this chapter compared to Fig. 3 in Jennings 1978).

As far as the anion transport across the red blood cell membrane is concerned, Jennings and Passow (1979) have pointed out that the transport protein does not simply form an aqueous pore nor does it act like a diffusable carrier. Instead, the transport process is likely to occur by a sequence of conformational changes.

Very little is known about the water channel in human red blood cells. Brown et al. (1975) suggested the involvement of band 3 glycoprotein, whereas other authors (Macey and Farmer 1970; Brahm and Wieth 1977) consider that other water channels exist. The results obtained by us do not contradict a role for band 3 in water diffusion. The similarity in pH dependence of water transport and red blood cell pH equilibration has already been discussed. In addition, other transport processes that may involve band 3 exhibit breaks in the Arrhenius plots at about 20°C, including glucose transport (Lacko et al. 1973) and the transport of chloride and bromine (Brahm 1977). However, it is possible that these transport processes are mediated by different polypeptides migrating in the same regions during electrophoresis or by different segments of the band 3 polypeptide chain. Jennings and Passow (1979) have shown that specific regions of the band 3 molecule are involved in the control of anion transport.

Our studies on the effects of temperature and pH on water diffusion (Morariu et al. 1981) suggest that conformational changes and cooperative effects are involved in the mechanism of water diffusion through erythrocyte membranes. The NMR measurements combined with other physical and biochemical techniques appear to be useful tools with which to investigate this process further.

We have investigated water diffusion through erythrocyte membranes in epileptic children. In preliminary studies we report a reduced permeability of erythrocytes from epileptics compared with controls (Benga and Morariu 1977). Further studies on water permeability of erythrocytes may give clues to the understanding of the molecular mechanisms of this transport process and it is hoped will help identify the defect associated with epilepsy.

Acknowledgments

The authors thank Dr. P. Quinn (Department of Biochemistry, Chelsea College, University of London, UK) and Dr. R. Holmes (Department of Food Science, University of Illinois, Urbana, U.S.) for revising the English version of the manuscript. The financial support of the Academy of Medical Sciences of Roumania is gratefully acknowledged.

References

Aloni, B., Eitan, A., and Livne, A. (1977). The erythrocyte membrane site for the effect of temperature on osmotic fragility. *Biochim. Biophys. Acta 465,* 46-53.

Andrasko, J. (1976). Water diffusion permeability of human erythrocytes studied by a pulsed gradient NMR technique. *Biochim. Biophys. Acta 428,* 304-311.

Barton, T.C. and Brown, D.A.J. (1964). Water permeability of the fetal erythrocyte. *J. Gen. Physiol. 47,* 839-849.

Benga, Gh. and Morariu, V.V. (1977). Membrane defect affecting water permeability in human epilepsy. *Nature 265,* 636-638.

Benga, Gh. and Strach, S.J. (1975). Interpretation of the electron spin resonance spectra of nitroxide-maleimide-labelled proteins and the use of this technique in the study of albumin and biomembranes. *Biochim. Biophys. Acta 400,* 69-79.

Bieri, V.G. and Wallach, D.F.H. (1976). Lipid-protein relationships in erythrocyte membranes revealed by paramagnetic quenching of protein fluorescence. *Biochim. Biophys. Acta 443,* 198-205.

Bond, J.D. and Baumann, G. (1978). Comments on electrostriction and/or cooperativity as viable models of membrane excitability. *Physiol. Chem. Phys. 10,* 381-382.

Brahm, J. (1977). Temperature-dependent changes of chloride transport kinetics in human red cells. *J. Gen. Physiol. 70,* 283-306.

Brahm, J. and Wieth, J.O. (1977). Separate pathways for urea and water, and for chloride in chicken erythrocytes. *J. Physiol.* (London) *266,* 727-749.

Brown, P.A., Feinstein, M.B., and Sha'afi, R.I. (1975). Membrane proteins related to water transport in human erythrocytes. *Nature 254,* 523-525.

Buckley, J.Th. and Hawthorne, J.M. (1972). Erythrocyte membrane phosphoinositide metabolism and the regulation of calcium binding. *J. Biol. Chem. 247,* 7218-7222.

Chang, D.B., Cooper, R.L., Young, A.C., Martin, C.J., and Johnson, B.A. (1975). Restricted diffusion in biophysical systems: Theory. *J. Theor. Biol. 50,* 285-308.

Chien, D.Y. and Macey, R.L. (1977). Diffusional water permeability of red cells. Independence on osmolality. *Biochim. Biophys. Acta 464,* 45-52.

Conlon, T. and Outhred, R. (1972). Water diffusion permeability of erythrocytes using an NMR technique. *Biochim. Biophys. Acta 288.* 354-361.

Conlon, T. and Outhred, R. (1978). The temperature dependence of erythrocyte water diffusion permeability. *Biochim. Biophys. Acta 511,* 408-418.

Davis, J.H., Maraviglia, B., Weeks, G., and Godin, D.V. (1979). Bilayer rigidity of the erythrocyte membrane. ^{2}H NMR of a perdeuterated palmitic acid probe. *Biochim. Biophys. Acta 550,* 362–366.

Degani, H. and Elgavish, G.A. (1978). Ionic permeabilities of membranes ^{23}Na and ^{7}Li NMR studies on ion transport across the membrane of phosphatidylcholine vesicles. *FEBS Lett. 90,* 357–360.

De Gier, H. (1979). Membrane lipids and the general permeability properties of the interface. In *Advances in Inflammation Research,* Weissmann, G., ed. Raven Press, New York, vol. 1, pp. 7–17.

Dwek, R.A. (1973). *Nuclear Magnetic Resonance in Biochemistry: Applications to Enzyme Systems.* Clarendon Press, Oxford.

Fabry, M.E. and Eisenstadt, M. (1975). Water exchange between red cells and plasma. Measurement by nuclear magnetic relaxation. *Biophys. J. 15,* 1101–1110.

Fairbanks, G., Steck, T.L., and Wallach, D.F.H. (1971). Electrophoretic analysis of the major polypeptides of the human erythrocyte membrane. *Biochemistry 10,* 2606–2617.

Farrar, Th.C. and Becker, E.D. (1971). *Pulse and Fourier Transform NMR.* Academic Press, New York.

Feinstein, M.B., Fernandez, S.M., and Sha'afi, R.I. (1975). Fluidity of natural membranes and phosphatidylserine and ganglioside dispersions. Effects of local anesthetics, cholesterol and protein. *Biochim. Biophys. Acta 413,* 354–370.

Goekoop, J.G., Spies, F., Wisse, D.M., Vrielink, R., De Vries, E., and Van Kempen, G.M. (1978). pH-dependent behaviour of erythrocyte membrane elevations. 2. Occurence in moderate acidemic and alkalemic conditions. *Cell Biol. Int. Rep. 2,* 397–402.

Inesi, G., Millman, M., and Eletr, S. (1973). Temperature-induced transitions of function and structure in sarcoplasmic reticulum membranes. *J. Mol. Biol. 81,* 483–504.

Jacobs, M.H. (1952). The measurement of cell permeability with particular reference to the erythrocyte. In *Modern Trends in Physiology and Biochemistry,* Guzman-Barron, ed. Academic Press, New York, pp. 149–172.

Jennings, M.L. (1978). Characteristics of CO_2-independent pH equilibration in human red blood cells. *J. Membr. Biol. 40,* 365–391.

Jennings, M. and Passow, H. (1979). Anion transport across the erythrocyte membrane, in situ proteolysis of band 3, and cross-linking of proteolytic fragments by 4,4′-diisothiocyanodihydrostilbene-2,2′-disulphonate. *Biochim. Biophys. Acta 554,* 498–519.

Johnson, R.M. (1975). Effect of temperature in resealing kinetics of washed human erythrocytes. *Fed. Proc. 34,* 536.

Lacko, L., Witke, B., and Geck, P. (1973). The temperature dependence of the exchange transport of glucose in human erythrocytes. *J. Cell Physiol. 82,* 213–218.

Lenaz, G. (1973). The role of lipids in the regulation of membrane-associated activities. *Acta Vitamin. Enzymol.* (Milano) *27,* 62–95.

Lenaz, G., Parenti-Castelli, G., and Sechi, A.M. (1975). Lipid-protein interactions in mitochondria. Changes in mitochondrial adenosine triphosphatase activity induced by *n*-butyl alcohol. *Arch. Biochem. Biophys. 167,* 72-79.

Macey, R.I. and Farmer, R.E.L. (1970). Inhibition of water and solute permeability of human red cells. *Biochim. Biophys. Acta 211,* 104-107.

Marchesi, V.T., Segrest, J.P., and Kahane, I. (1972). Molecular features of major glycoprotein of human erythrocyte membrane. *ABS PAP ACS 164,* 222.

Morariu, V.V. and Benga, Gh. (1977). Evaluation of a nuclear magnetic resonance technique for the study of water exchange through erythrocyte membranes in normal and pathological subjects. *Biochim. Biophys. Acta 469,* 301-310.

Morariu, V.V., Pop, I.V., Popescu, O., and Benga, Gh. (1981). Effects of temperature and pH on the water exchange through erythrocyte membranes: Evidence for state transitions. *J. Membr. Biol. 62,* 1-5.

Outhred, R. and Conlon, T. (1973). The volume dependence of the erythrocyte water diffusion permeability. *Biochim. Biophys. Acta 318,* 446-450.

Paganelli, C.V. and Solomon, A.H. (1957). The rate of exchange of tritiated water across the human red cell membrane. *J. Gen. Physiol. 41,* 259-277.

Pirkle, J.L., Ashley, D.L., and Goldstein, J.H. (1979). Pulse nuclear magnetic resonance measurements of water exchange across the erythrocyte membrane employing a low Mn concentration. *Biophys. J. 25,* 389-406.

Raison, J.K., Lyons, M.J., and Thomson, W.W. (1971). The influence of membranes on the temperature-induced changes in the kinetics of some regulatory enzymes of mitochondria. *Arch. Biochem. Biophys. 142,* 83-90.

Sha'afi, R.I. (1977). Water and small nonelectrolyte permeation in red cells. In *Membrane Transport in Red Cells,* Ellory, C., ed. Academic Press, New York, pp. 221-256.

Shporer, R. and Civan, M.M. (1975). NMR study of ^{17}O from $H_2{}^{17}O$ in human erythrocytes. *Biochim. Biophys. Acta 385,* 81-87.

Traüble, H. (1972). In *Passive Permeability of Cell Membranes,* Kreuzer, F. and Sledgers, J.F.G., eds. Plenum Press, New York and London, vol. 13, p. 197.

Van Deenen, L.L.M., De Gier, J., Demel, R.A., De Kruiff, B., Blok, M.C., Van der Neut-Kok, B.C.M., Haest, C.W.M., Ververgaert, P.H.J.Th., and Verkleij, A.J. (1975). Lipid-lipid and lipid-protein interaction in model systems and membranes. *Ann. N.Y. Acad. Sci. 264,* 124-141.

Verma, S.P. and Wallach, D.F.H. (1976). Erythrocyte membranes undergo cooperative, pH-sensitive state transitions in the physiological temperature range: Evidence from Raman spectroscopy. *Proc. Natl. Acad. Sci. USA 73,* 3558-3561.

Vieira, F.L., Sha'afi, R.I., and Solomon, A.K. (1970). The state of water in human and dog red blood cell membranes. *J. Gen. Physiol. 55,* 451-466.

Vilcu, Al. (1977). *Eritrocitul.* Editura Medicala, Bucuresti.

Volkenstein, M.V. (1977). *Molecular Biophysics.* Academic Press, New York-San Francisco-London, pp. 26-34.

Woessner, D.E. (1963). Temperature dependence of nuclear-transfer and spin-relaxation phenomena of water adsorbed on silica gel. *J. Chem. Phys. 39*, 2783–2787.

Zimmer, G. and Schirmer, H. (1974). Viscosity changes of erythrocyte membrane and membrane lipids at transition temperature. *Biochim. Biophys. Acta 345*, 314–320.

8
Virally Mediated Changes in Cellular Permeability

C.A. Pasternak

Conferences on membranes are now held as frequently as those on intermediary metabolism or proteins were some years ago; few discussions of membranes, however, deal with disease. And of those that do, the word "virus" hardly ever appears in the index (e.g., Trump and Arstila 1975; Bolis et al. 1976; Miller and Shamoo 1977; Andreoli et al. 1978). Yet it is obvious that viral diseases involve the surface membrane of cells: the initial interaction between viruses and susceptible cells is at the cell surface, membrane damage is a characteristic feature of infected cells (e.g., Thigpen 1971; Fenner et al. 1974), the interplay between infected cells and the immune system takes place at the cell surface, and the action of interferon in protecting against viral attack begins at the cell surface. The features of the cell surface that have been particularly studied to date (e.g., Poste and Nicholson 1977) are the nature of virus receptors (proteins and glycoproteins), the mechanism by which viruses enter cells (endocystosis or, in the case of certain enveloped viruses, virus-cell fusion), the appearance of viral antigens (proteins and glycoproteins) on the surface of infected cells, the mechanism of viral release, and the participation of surface-bound antigens in eliciting an immune (cell-mediated) response. What has been relatively less explored is the way in which the cytopathic effects of viruses come about. The aim of this chapter is to suggest that a *change in cellular permeability* is at the root of some of these effects and that such a change may underlie the pathological consequences, that is, the diseases, that result from viral infection in certain cases.

First, the nature of permeability changes that have been observed during the entry of certain viruses into cells will be described. The information that such changes give with regard to the mechanism of viral entry by the process of virus-cell fusion will be discussed in some detail, as it is relevant to aspects of membrane fusion in general. Next, the nature of the permeability changes that take place in virally infected cells will be described. Finally, the clinical relevance of virally mediated permeability changes will be assessed.

Permeability Changes During Viral Entry

Nature of Change

Some years ago we noticed that when Sendai virus–an enveloped virus of the paramyxovirus group (reviewed by Ishida and Homma 1978)–is added to cells, a change in the permeability characteristics occurs: Low molecular weight intermediates of metabolism, such as phosphoryl choline, sugar phosphates, and nucleotides, that normally do not leak out of cells do so on exposure to virus; the permeability change is temperature dependent and is likely to involve fusion between the viral envelope and cell plasma membrane (Pasternak and Micklem 1973). Although it might be thought that the leakage phenomenon reflects lysis by the virus (paramyxoviruses are known to be hemolytic; Morgan et al. 1948; Kilham 1949; Fukai and Suzuki 1955), this is not so. Cells remain impermeable to trypan blue, and cytoplasmic proteins such as lactate dehydrogenase do not leak out (Poste and Pasternak 1978). On the other hand, other low molecular weight substances such as K^+ leak out (Pasternak and Micklem, 1974a) and Na^+ (Poste and Pasternak 1978) and Ca^{2+} (Poste and Pasternak 1978; Getz et al. 1979; Impraim et al. 1979) leak in. Recently, we have found that cells become permeable to monovalent cations somewhat before they become permeable to phosphorylated intermediates (Bashford et al. 1983). As a result of the ionic changes, water enters. The reason for this is as follows. Normally the colloid osmotic pressure (which would lead to an entry of water) is balanced by the chemical gradient (i.e., the ionic asymmetry) across the plasma membrane. When the membrane becomes leaky to ions as a result of viral action, the chemical gradient collapses [the Na^+ pump, though active (Poste and Pasternak 1978), failing to maintain the gradient], and there is, therefore, nothing to prevent colloid osmotic pressure from exerting its effect. This leads to the lysis of red blood cells (i.e., the hemolytic action of the virus), but in most other cells, the presence of microvilli or other protuberance capable of expanding the surface area prevents lysis and results merely in cell swelling (Knutton et al. 1976).

At first we thought that the Sendai virus affects the cell membrane in another way as well. Because the virus reduces the transport of compounds that are taken up by facilitated diffusion at the same time as it stimulates the transport of compounds taken up by passive (i.e., simple) diffusion (Pasternak and Micklem 1974b; Pasternak et al. 1976), it seemed that the virus has two separate actions on cells. However, this proved not to be the case, and the inhibitory action of virus on facilitated uptake mechanisms can be explained in terms of the ability of the virus to make cells leaky: Failure to retain phosphorylated metabolites such as phosphoryl choline or 2-deoxyglucose 6-phosphate is the reason for the apparent inhibition of choline or 2-deoxyglucose uptake; failure to retain a Na^+ gradient across cells is the reason for the apparent inhibition of glycine or amino isobutyrate uptake (Impraim et al. 1980). Why virus should inhibit the uptake of amino acids that are not Na^+-linked, such as leucine or valine, is not clear; but then it is not clear by what mechanism such amino acids are concentrated. The important point is that the interaction between the Sendai virus and cells prevents the accumulation of nutrients and other low molecular weight compounds at the same time as it causes leakage of intracellular metabolites, cofactors, and (presumably) molecules such as cyclic adenosine monophosphate (cAMP) (Table 8.1).

The permeability changes are inhibited by Ca^{2+} (Pasternak and Micklem 1974a, 1974b; Micklem and Pasternak 1977) and by other divalent cations such

Table 8.1. Permeability Changes Caused by Hemolytic Paramyxoviruses

Change	Virus	References
Leakage of intracellular constituents		
K^+	Newcastle Disease	Klemperer (1960)
K^+	Sendai	Pasternak and Micklem (1974a) Poste and Pasternak (1978)
Phosphorylcholine, phosphorylinositol, inorganic phosphate, sugar phosphates, adenine nucleotides, and so on	Sendai	Pasternak and Micklem (1973) Impraim et al. (1980)
Phosphorylcholine and sugar phosphates	Newcastle Disease	Poste and Pasternak (1978)
Decreased concentrative uptake of nutrients		
Amino acids	Sendai	Pasternak and Micklem (1973) Negreanu et al. (1974) Impraim et al. (1980)
Choline	Sendai	Pasternak and Micklem (1973, 1974b)
Sugars	Sendai	Pasternak and Micklem (1973) Impraim et al. (1980)
Increased passive uptake of nutrients and ions		
Amino acids, choline, sugars	Sendai	Pasternak and Micklem (1974b) Impraim et al. (1980)
Na^+	Sendai	Poste and Pasternak (1978)
Ca^{2+}	Sendai	Poste and Pasternak (1978) Impraim et al. (1979)
Mn^{2+}	Sendai	Getz et al. (1979) Impraim et al. (1979)

as Sr^{2+} and Mn^{2+} (Impraim et al. 1979). The concentration of Ca^{2+} required to cause inhibition, which can vary as widely as from 0.1 to 10 mM external Ca^{2+}, depends on the amount of virus added; the action of Ca^{2+}, however, is not simply to prevent viral attachment to cells (Micklem and Pasternak 1977; Wyke et al. 1980) but is more complex (Micklem and Pasternak 1982). Some other characteristics of virally mediated permeability changes are listed in Table 8.2.

Virally mediated leakage of low molecular weight compounds and ions is presumably through some kind of hydrophilic "pore." By measuring the uptake of a series of peptide derivatives at a time after addition of virus at which leakage of phosphorylated intermediates is maximal, a pore size of approximately 1 nm has been deduced (Wyke et al. 1980), which explains why macromolecules such as proteins do not leak out. Such a pore size is similar to the pore size of intercellular communicating ("gap") junctions (Simpson et al. 1977), and these, too, allow phosphorylated compounds (Subak-Sharpe et al. 1966) and monovalent cations (Kuffler and Potter 1964; Loewenstein and Kanno 1964) but not proteins (Pitts 1976) to cross the plasma membrane; furthermore intercellular

Table 8.2. Characteristics of Permeability Changes Caused by the Hemolytic Sendai Virus

Characteristic	References
Nonlinear temperature dependence	Pasternak and Micklem (1973) Micklem and Pasternak (1977)
Increased permeability to low molecular weight compounds (< $\sim$ 1000 Da) only	Pasternak and Micklem (1973) Poste and Pasternak (1978) Wyke et al. (1980)
Inhibition by Ca^{2+}	Pasternak and Micklem (1974a) Pasternak et al. (1976) Micklem and Pasternak (1977) Impraim et al. (1980)
Inhibition by Mn^{2+} but not by Mg^{2+}	Imprain et al. (1979)
Reversal of Ca^{2+} inhibition by ethylenediaminetetraacetate (EDTA)	Impraim et al. (1980)
Inhibition by concanavalin A	Micklem and Pasternak (1977)
Spontaneous recovery with time	Negreanu et al. (1974) Pasternak et al. (1976) Impraim et al. (1980)
Viral effect not mimicked by treatment with neuraminidase, protease, phospholipase, or local anesthetic	Micklem and Pasternak (1977)
Fusion between viral envelope and cell plasma membrane is a necessary, but not a sufficient, condition	Wyke et al. (1980)

communicating junctions are "closed" by Ca^{2+} (Rose and Loewenstein 1975) though not by Mn^{2+}, so the analogy between junctions and virally mediated pores is reasonably close (Pasternak et al. 1976; Micklem and Pasternak 1977). Structurally, however, they are different: Communicating junctions are characterized by an ordered aggregation of intramembranous particles (McNutt and Weinstein 1970; Gilula et al. 1972; Peracchia 1977), whereas this is not true of virally induced pores (Knutton 1977; Poste and Pasternak 1978).

When the specificity of the leakage phenomenon was investigated, it was found that every cell type tested–normal or transformed cell lines, organ cultures, brain cells or peripheral lymphocytes–is affected in the same way when exposed to the Sendai virus (Table 8.3). The specificity with regard to virus is the opposite: Only Newcastle disease virus (NDV), another paramyxovirus, has been found to elicit the permeability change (Table 8.3). That NDV does so is not unexpected: As long ago as 1960, Klemperer showed that HeLa cells and Krebs 2 ascites tumor cells exposed to NDV lose intracellular K^{+}, just like red blood cells, and he postulated that the lysis of red blood cells (which he showed to be preventable by Ca^{2+}) was due to osmotically induced entry of water (Klemperer 1960). However, he noted that HeLa and Krebs 2 cells also lose protein (Klemperer 1960), and he was therefore unable to distinguish between lysis and the type of permeability change described above; in fact, his conclusions regarding leakage of protein may have been erroneous (Poste and Pasternak 1978).

Not only does no other virus induce the permeability change, but some preparations of Sendai virus do not do so either. For example, "early harvest" Sendai virus (Homma et al. 1976) and virus grown in cultured MDBK cells instead of in hen's eggs are inactive (Table 8.3). What, then, are the essential components required to induce a permeability change? It appears that only the viral envelope, which consists of two glycoproteins, termed F and HN, embedded in a predominantly phospholipid bilayer, is required. Of the glycoproteins, HN (which is responsible for hemagglutination and neuraminidase activity) is not essential, but F (which is required for fusion between viral envelope and cell plasma membrane) is (Wyke et al. 1980). It is because virus grown in MDBK cells contains an inactive precursor of F (Ishida and Homma 1978) that such preparations are inactive. The reason why "early harvest" virus, which contains intact F (Wyke et al. 1980), is inactive is less clear. It is probably related to the fact that the envelope of early harvest virus is impermeant to water-soluble low molecular weight compounds such as uranyl acetate, whereas the envelope of normal virus is permeable to uranyl acetate (Shimizu et al. 1976). In other words, the requirement for eliciting a permeability change is a membrane that contains the fusion factor F glycoprotein and is inherently leaky. That is, a potentially hydrophilic pore is already present in the viral envelope before it ever fuses with a cell plasma membrane. The exact nature of the pore, and the way in which it expands after fusion with the cell plasma membrane, have yet to be established.

Table 8.3. Specificity of Permeability Changes Caused by Addition of Virus to Cells[a]

Virus	Family	Cell Type	Permeability Change	References
Sendai	Paramyxovirus	Lettree cells	+	Pasternak and Micklem (1973)
		Chick red cells	+	Fuchs and Giberman (1973)
		BHK cells	+	Fuchs and Giberman (1973) Foster (1980)
		HeLa cells	+	Negreanu et al. (1974) Pasternak et al. (1976)
		L cells	+	Negreanu et al. (1974)
		P815Y cells	+	Pasternak and Micklem (1974a)
		3T3, SV40-3T3, HTC, monkey kidney, MRC-5 cells, peripheral human lymphocytes	+	Poste and Pasternak (1978)
		Organ cultures of lung and nasal turbinate; cultured and freshly isolated brain cells	+	Foster (1980)
Early-harvest Sendai		Lettree cells	—	Impraim et al. (1979) Wyke et al. (1980)
Sendai grown in MDBK cells		Lettree cells	—	Wyke et al. (1980)
Newcastle Disease	Paramyxovirus	HeLa and Krebs-2 cells	+	Klemperer (1960)
		Lettree cells	+	Poste and Pasternak (1978)
Measles	Paramyxovirus	HeLa cells and peripheral human lymphocytes	—	Pasternak (1980)
Influenza	Orthomyxovirus	Monkey kidney cells; organ cultures of lung and nasal turbinates	—	Poste and Pasternak (1978) Foster (1980)
Rabies and vesicular stomatitis	Rhabdovirus	BHK and Lettree cells	—	Foster (1980)
Cytomegalo	Herpes virus	MRC-5 cells	—	Poste and Pasternak (1978)
Adeno-5	Adenovirus	HeLa cells	—	Poste and Pasternak (1978)

[a] Mammalian red blood cells are omitted from this table: They are lysed by paramyxovirus and it is therefore difficult to discern permeability changes having the characteristics listed in Table 8.2.

Relevance to Membrane Fusion

Paramyxoviruses such as Sendai and NDV promote cell-to-cell fusion, that is, polykaryon or giant cell formation (Okada 1969; Harris 1970), and it might be wondered whether the permeability changes just described are part of that process. We believe they are.

It is important to distinguish between virus-cell fusion and virally mediated cell-cell fusion. The first simply involves fusion between two membranes (that of viral envelope and cell plasma membrane), whereas the second involves three membranes (viral envelope and two cell plasma membranes). Various models for virally mediated cell-cell fusion have been proposed (Hosaka and Shimizu 1977), of which we favor the virus bridge or cell-virus-cell model (Knutton 1978) illustrated in Fig. 8.1. It will be noted that virus-cell fusion (i.e., membrane fusion) is a distinct stage in the process, and that giant cell formation requires osmotic swelling; the way virally induced permeability changes lead to osmotic swelling has been discussed above. The evidence for this three-stage model of giant cell formation is twofold. First, if osmotic swelling is inhibited (by hypertonic media or by the substitution of potassium chloride for sodium chloride in which case cells swell prior to membrane fusion), giant cell formation is prevented (Knutton and Pasternak 1979; Impraim et al. 1980). Second, if early virus harvest is used instead of the usual virus, permeability changes do not occur (Wyke et al. 1980); there is, therefore, no swelling, and no giant cell formation takes place (Knutton and Pasternak 1979); such cells, which can be shown to be fused but arrested at stage 2 (Fig. 8.1), expand into giant cells merely by exposure to hypotonic medium (Knutton and Bachi 1980).

Thus we propose that virally mediated permeability changes are a direct cause of giant cell formation (Table 8.4). A similar mechanism, namely fusion followed by osmotic swelling, is also likely to account for chemically induced giant cell formation (Knutton and Pasternak 1979). But in systems that do not involve a change in shape such as the dumbbell-to-sphere alteration illustrated in Fig. 8.1, osmotic swelling is unnecessary. Cell-cell fusion between adjacent myoblasts

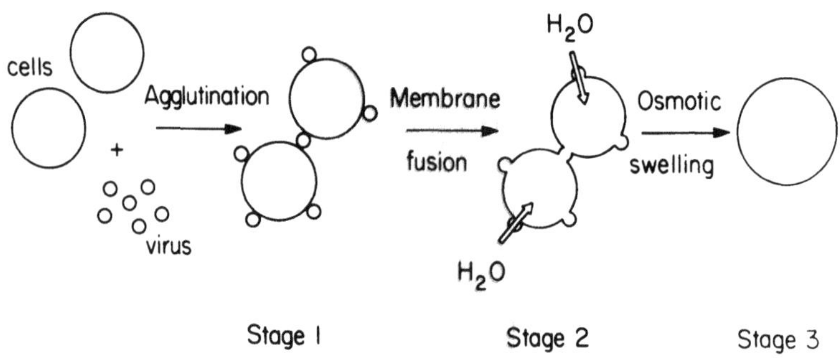

Figure 8.1. A model for virally mediated cell-cell fusion.

Table 8.4. Relevance of Permeability Changes to Systems Involving Membrane Fusion

Virus-cell fusion (or other type of membrane fusion)	Irrelevant: Permeability changes are a consequence (in certain circumstances) not a cause of the fusion event
Giant cell formation (virally mediated or chemically mediated)	Relevant: Permeability changes are a necessary stage in the process

during myogenesis, for example, occurs in the absence of concomitant permeability changes (Curtis et al. 1980).

Implicit in what has been said so far (e.g., Table 8.2) is the assumption that permeability changes result from the fusion of a leaky membrane (the viral envelope) with a nonleaky one (the cell plasma membrane). However, other authors do not view membrane fusion in this way. Instead, they suggest that an increase in intracellular Ca^{2+} (e.g., resulting from a permeability change) is a necessary trigger for fusion (Lucy 1978); the suggestion is based largely on the fact that exocytosis (Gomperts 1976; Rasmussen and Goodman 1977; Hopkins and Duncan 1979), which involves membrane fusion as the terminal event (Henson et al. 1978; Meldolesi et al. 1978; Orci and Perrelet 1978), and fusion between artificial liposomes consisting of negatively charged phospholipids such as phosphatidylserine or phosphatidylglycerol (Papahadjopoulos 1978), each requires Ca^{2+}. Thus, in chemically induced cell-cell fusion, permeability changes, including an influx of Ca^{2+}, are postulated to *precede* the fusion event (Blow et al. 1979), and an entry of Ca^{2+} has been postulated to be necessary for virally induced cell-cell fusion also (Volsky and Loyter 1978). The latter claim receives support from the observation that at sufficiently high concentrations of external Ca^{2+} (40 mM) and OH^- (pH 10.5), spontaneous cell-cell fusion occurs (Toister and Loyter 1973).

It seems likely to us that the Ca^{2+} requirements for exocytosis (approximately 2 mM external Ca^{2+}; 1 μM internal Ca^{2+}) for the fusion of negatively charged phospholipid vesicles (approximately 10 mM Ca^{2+}) and for spontaneous cell-cell fusion (40 mM Ca^{2+}, pH 10.5) are quite unrelated and that the suggestion that membrane fusion in general requires Ca^{2+} is by no means proven. On the contrary, it is clear from our data that virus-cell fusion, which is as good an example of membrane fusion as any other, does *not* require Ca^{2+}. Thus, early harvest Sendai virus does not elicit an increased Ca^{2+} uptake–in contrast to the usual Sendai virus (Impraim et al. 1979)–yet it fuses perfectly well with cells (Knutton 1979), and measurement of virus-cell fusion by a fluorescence enhancement technique, as well as assay by phase contrast microscopy, shows that fusion occurs as effectively in the absence of Ca^{2+} as in its presence (Wyke et al. 1980; Pasternak 1981). In short, there is no reason to suppose that Ca^{2+} entry–or any other permeability change–is a necessary event in membrane fusion

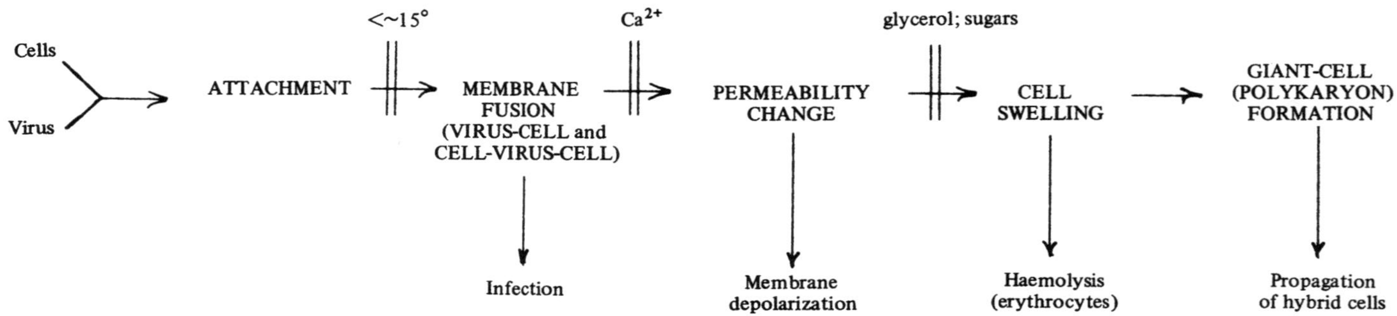

Figure 8.2. Sequence of cellular changes following addition of virus to cells.

(Table 8.4). The stages leading to giant cell formation may then be summarized as shown in Fig. 8.2.

Use as a Tool to Permeabilize Cells

From what has been said so far, it is clear the hemolytic Sendai virus can be used to "permeabilize" cells (Impraim et al. 1980). This has the advantage over other techniques, such as treatment with detergents (Miller et al. 1979) or electric shock (Baker and Knight 1978), in that under appropriate conditions the pores reseal themselves within minutes (e.g., Forda et al. 1982) and the process can be controlled by manipulation of external Ca^{2+} (Impraim et al. 1980).

Cells permeabilized by hemolytic Sendai virus have been used to explore the concentration of intracellular Ca^{2+} required to initiate secretion from endocrine and other cells, by introducing normally impermeant chelators of Ca^{2+} such as ethyleneglycoltetraacetate (EGTA). An example (Gillies et al. 1981) is illustrated in Fig. 8.3. The release of adrenocorticotropin (ACTH) from rat anterior pituitary cells triggered either by hypothalamic extract containing the relevant releasing hormone or by depolarization with 56 mM K^+, requires the presence of >0.5 mM Ca^{2+} in the extracellular medium (Gillies and Lowry 1978; G. Gillies and C.A. Pasternak, unpublished experiments). In the presence of the hemolytic Sendai virus, Ca^{2+} buffered with EGTA at as low a concentration as 0.5 μM causes a release of ACTH [Fig. 8.3(*d*)]; note that in the absence of the virus, release of ACTH (e.g., by K^+) occurs at 0.5 mM Ca^{2+}, but not at 0.1 mM Ca^{2+} [Fig. 8.3(*a*)] or less [Figs. 8.3(*b*)–8.3(*d*)]. The release of ACTH triggered by 0.5 μM–0.1 mM Ca^{2+} in the presence of virus is accompanied by leakage of phosphorylated compounds from the cells (cells were exposed to a mixture of [^{3}H]deoxyglucose and [^{14}C]choline prior to assay; these compounds are efficiently metabolized to [^{3}H]deoxyglucose 6-P and phosphoryl [^{14}C]choline, respectively (Impraim et al. 1980), and the phosphorylated products do not leak out in the absence of virus). It might therefore be argued that the release of ACTH by virus is not by true exocytosis but by some other mechanism. However, Fig. 8.3 shows that when the extracellular Ca^{2+} is restored to 0.5 mM Ca^{2+} [Figs. 8.3(*b*) and 8.3(*d*)] or 2.5 mM Ca^{2+} [Figs. 8.3(*a*) and 8.3(*c*)] so that the virally mediated pore is resealed (as shown by cessation of [^{3}H]deoxyglucose 6-P or phosphoryl [^{14}C]choline leakage), ACTH continues to be released and is susceptible to subsequent stimulation by hypothalamic extract or K^+; note that a second application of virus fails to elicit a significant response because the first application destroyed the receptors necessary for its action (see Forda et al. 1982). One interpretation of these massive secretions of ACTH achieved by stimulation following exposure of cells to 0.5 μM–0.1 mM Ca^{2+} in the presence of virus is that the internal milieu of cells is now *buffered* in this concentration range, and the normal Ca^{2+}-removing mechanisms of cells [Ca^{2+}-adenosine triphosphatase (ATPase) and Ca^{2+}-Na^+ countertransport (Carafoli and Crompton

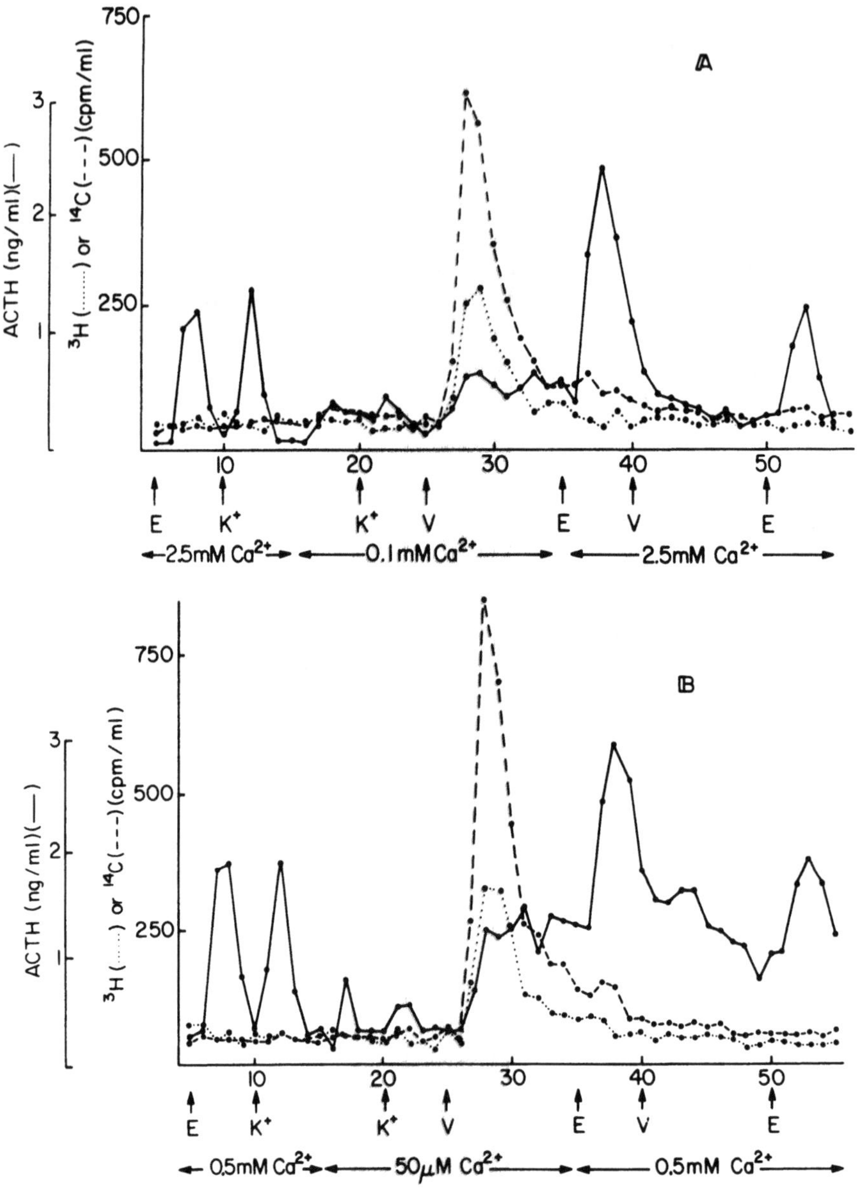

Figure 8.3. Sensitivity of permeabilized anterior pituitary cells to calcium buffers. Rat anterior pituitary cells prelabeled for 30 min at 37°C with [^{3}H]deoxyglucose and [^{14}C]choline were packed in columns and perfused and release of ACTH was measured as described by Gillies and Lowry (1978). Cells were exposed to an extract of rat hypothalamus (stalk median eminance, diluted 30 times) (E), to 56 mM KCl (K^+), or to the hemolytic Sendai virus (approx. 5 HAU/ml) (V) for approx. 2-min periods at the times indicated by the arrows.

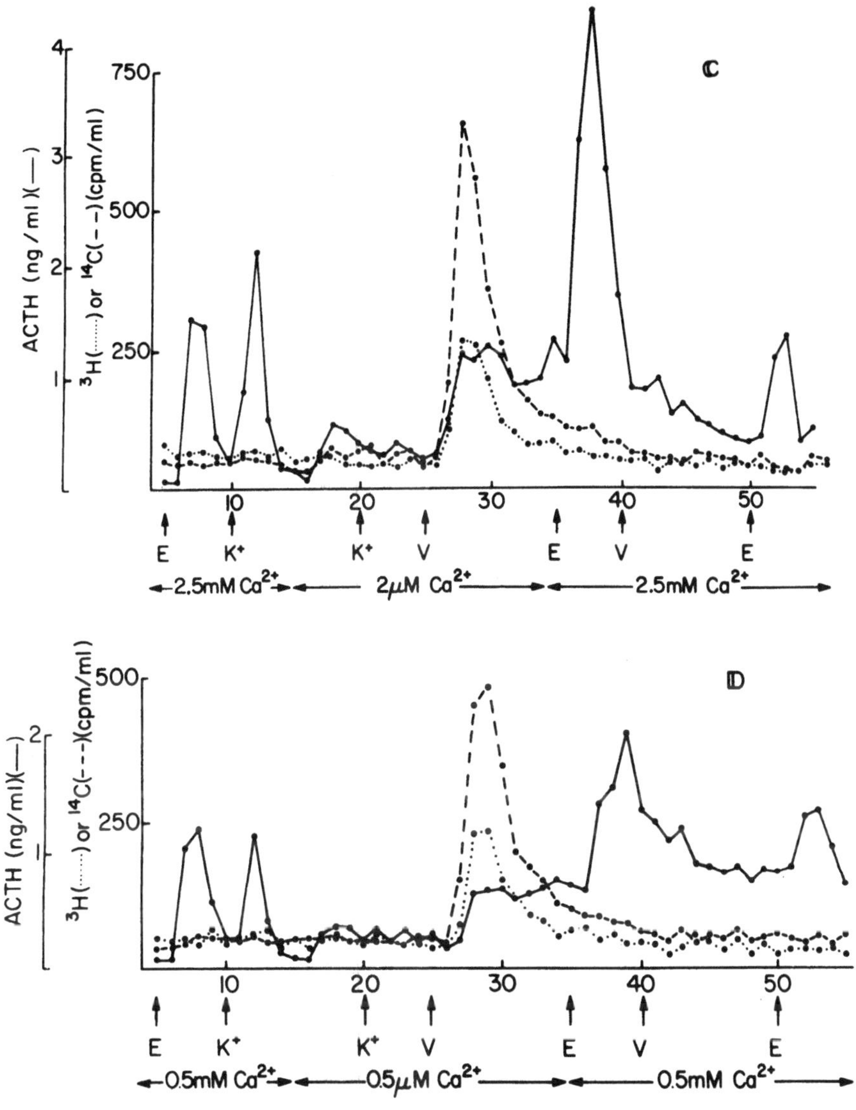

All additions were made in perfusing medium containing 2.5 mM, 0.5 mM, 0.1 mM, 50 μM, 2 μM, or 0.5 μM calcium chloride as indicated; the lower concentrations of calcium chloride were prepared with the following chelating agents (3 mM) to which calcium chloride was added so as to produce (at pH 7.4) the required concentrations of free Ca^{2+}: 0.1 mM Ca^{2+} (nitrilotriacetate–NTA), 50 μM Ca^{2+} (diethylenetriamine pentacetate–DPTA), 2 μM Ca^{2+} (N-hydroxyethylethylenediaminetriacetate–HEDTA), and 0.5 μM Ca^{2+} (EGTA). The column effluents, collected in 55 X 1 ml samples as indicated, were anlayzed for ACTH by radioimmunoassay (Gillies and Lowry 1978) and for radioactivity by conventional techniques. G. Gillies and C.A. Pasternak, unpublished results.

1978)], which normally operate to bring intracellular Ca^{2+} below 0.5 μM, are ineffective at dealing with this situation. Other explanations are possible. What is clear is that this type of experiment confirms the results of other workers (e.g., Hopkins and Walker 1978) who have proposed a role for Ca^{2+} in stimulus-secretion coupling in anterior pituitary cells.

Similar types of experiments have been carried out to determine the intracellular concentration of Ca^{2+} required to elicit secretion of histamine from mast cells (Gomperts et al. 1981, 1983) and of β-glucuronidase from neutrophils (B.D. Gomperts, J.P. Bennett, and J.M. Baldwin, unpublished experiments). Permeabilization of cells with hemolytic Sendai virus could equally well be used to probe the intracellular action of normally impermeant compounds such as cyclic nucleotides, small peptides, or other compounds having a molecular weight of about 300–1000 Da. At very high concentrations of virus, even proteins can be introduced into cells (Tanaka et al. 1975; Yamaizumi et al. 1979).

Permeability Changes in Infected Cells

The permeability changes described in the preceding sections occur during the entry of a virus and are elicited within minutes of adding a virus to cells. They involve only the components of the viral envelope and therefore occur with noninfective virus [e.g., ultraviolet (UV)-inactivated] as well as with infective virus. Do similar membrane alterations take place during the process of viral maturation and release that begins hours after the initial infection has taken place? It is clear that some membrane changes do occur and contribute to the cytopathic effects of virally infected cells; these include rounding up, inability to exclude dyes such as trypan blue, and leakage of intracellular proteins; the presence of lysosomal degradative enzymes among such proteins has led to the suggestion that their release is responsible for the cytolytic action of certain viruses (e.g., Fenner et al. 1974).

The question is whether more subtle alterations, such as those documented above, occur in cells infected with cytolytic viruses prior to leakage of cytoplasmic proteins and inability to exclude trypan blue. One such change, namely an increased permeability to Na^+ leading to an increased intracellular concentration of Na^+, has been suggested to be the very trigger for viral multiplication by favoring viral protein synthesis over host protein synthesis (Carrasco and Smith 1976; Carrasco 1977); experimental evidence in support of this suggestion has not, however, been presented. Since then, some evidence for (Garry et al. 1979; Rhamabhadran and Thach 1981; Garry et al. 1982) and some against (Francoeur and Stanners 1978; Fenwick and Walker 1978; Nair et al. 1979; Norrie et al. 1982) an alteration of intracellular cation concentration at or before the time at which host protein synthesis diminishes and viral protein synthesis begins has appeared. (It may be pointed out that in several instances, for example cells infected with paramyxoviruses, viral proteins are synthesized without significant inhibition of host protein synthesis, so the two events are not necessarily connected.) With regard to possible alterations in membrane permeability, the literature is again not clear. There are certainly several reports of altered ion (Farnham and Epstein 1963; Carrasco and Smith 1976; Egberts et al. 1977; Shrier et al. 1978; Peterhans et al. 1979) and metabolite (Genty 1975; Norkin 1977; Foster et al. 1980; Pasternak 1980) movements, but in some of the earlier work cited, cells appear to have already reached the point at which they were no longer able to exclude trypan blue. In no case has a change in permeability similar to that elicited by the action of hemolytic paramyxoviruses been described.

We decided to carry out a systematic survey of cells infected with enveloped ribonucleic acid (RNA) viruses representative of the following families: paramyxovirus, orthomyxovirus, rhabdovirus, and togavirus. In each case cells were infected with rather low doses of input virus in order to assess the minimal membrane changes that might accompany virus synthesis. Three parameters of membrane function were measured: uptake of radiolabeled compounds, concentration of intracellular cations, and membrane potential. In all cases measurements were made on cells before cytolytic damage occurred; in every case > 70% of cells were able to exclude trypan blue and, where measured, there was no detectable leak of cytoplasmic protein (lactate dehydrogenase) from cells. In each situation measurements were made up to and beyond the time of maximal virus release.

With regard to uptake, the experiments showed that in some cases (monolayer cells infected with paramyxovirus) there were no discernible changes at all, in some cases (suspension cells infected with paramyxovirus or orthomyxovirus) there was a decrease in uptake by concentrative mechanisms and an increase in uptake by simple diffusion, and in some cases (monolayer cells infected with rhabdovirus or togavirus) there was an *increased* capacity to take up sugar (but not other nutrients) by a concentrative mechanism (Table 8.5).

Although the second of these three situations may seem similar to that which occurs in cells treated with hemolytic Sendai virus as described in the first part of this chapter, it differs in several respects (see Table 8.6). Because the changes in infected cells are small, the Na^+ pump, for example, is able to maintain high K^+:Na^+ ratios in cells, unlike the situation in cells treated with hemolytic Sendai virus (Poste and Pasternak 1978); a reflection of this is seen in the fact that in infected cells the dependence of the concentrative uptake of certain amino acids on the Na^+ gradient ("Na^+ linkage") is maintained, whereas in cells treated with hemolytic Sendai virus it is lost (Impraim et al. 1980). Another difference is in the sensitivity to external Ca^{2+}; whereas permeability changes elicited by hemolytic Sendai virus are prevented by Ca^{2+}, those occurring in infected cells are not, or only partially, so (Table 8.6).

With regard to intracellular cation concentration and membrane potential (which was measured by assessing "K^+ null point" with fluorescent dyes as

Table 8.5. Uptake Changes in Virally Infected Cells[a]

Virus	Family	Cell Type	Growth Condition	Change	References
Sendai	Paramyxo	MDBK	Monolayer	No change in uptake of dGlc or amino acids; increased uptake of α-methyl glucoside (due to volume change not to membrane alteration)	Foster (1980) M.A. Gray, unpublished experiments
Measles	Paramyxo	Vero	Monolayer	No change in uptake of dGlc or amino acids	Foster et al. (1983)
Sendai	Paramyxo	L-1210	Suspension	Decreased uptake of dGlc, choline, amino acids at 1 μM; increased uptake of α-methyl glucoside (1 μM), or dGlc, choline, amino acids at > 16 mM	Foster (1980); Foster et al. (1983)
Influenza	Orthomyxo	L-1210	Suspension	Decreased uptake of dGlc, choline, amino acids at 1 μM; increased uptake of α-methyl glucoside (1 μM), or dGlc, choline, amino acids at > 16 mM	Foster (1980); Foster et al. (1983)
Measles	Paramyxo	Peripheral human lymphocytes	Suspension	Decreased uptake of dGlc, choline, amino acids at 1–10 μM	Pasternak (1980) Foster et al. (1983)
VSV	Rhabdo	BHK	Monolayer	Increased uptake of dGlc (10 μM); little or no change in uptake of amino acids (1–10 μM)	Gray et al. (1983a)
SFV	Toga	BHK	Monolayer	Increased uptake of dGlc (10 μM); little or no change in uptake of choline or amino acids (1–10 μM)	Gray et al. (1983a)

[a] Uptake denotes transport across the plasma membrane and subsequent concentrative changes such as phosphorylation that occur with certain compounds (e.g., dGlc or choline at low external concentration). Cells were infected (or mock infected) with virus and changes were measured at time of maximal virus release. Nonhemolytic (early-harvest) Sendai virus (Table 8.3) was used to avoid the "immediate" changes elicited by hemolytic Sendai virus (Tables 8.1 and 8.2).

Table 8.6. Comparison between Permeability Changes in Infected Cells and in Cells Modified by the Addition of Hemolytic Sendai Virus

Suspension Cells Infected with Nonhemolytic Sendai, Measles, or Influenza Virus[a]	Cells Modified by Addition of Hemolytic Sendai Virus[b]
Decreased uptake by concentrative mechanisms	Decreased uptake by concentrative mechanisms
Increased uptake by simple diffusion	Increased uptake by simple diffusion
Changes stable	Changes transient
Magnitude of changes small	Magnitude of changes large
Intracellular K^+:Na^+ ratio retained	Intracellular K^+:Na^+ ratio collapsed
Na^+ linkage of amino acid uptake retained	Na^+ linkage of amino acid uptake lost
Changes insensitive or only partially sensitive (Foster 1980) to external Ca^{2+}	Changes inhibited by external Ca^{2+}

SOURCE: [a]Foster et al. (1983).
[b]See Tables 8.1 and 8.2.

described by Bashford 1981), the following results were obtained (Table 8.7). In no case could an alteration in intracellular K^+ be detected by the time of maximal virus release; at later times, of course, K^+ does begin to leak from cells, which is to be expected since the viruses studied are all to a greater or lesser extent cytolytic. In Semliki Forest virus (SFV)-infected, but not vesicular stomatitis virus (VSV)-infected, BHK cells an increase in intracellular Na^+ (from approximately 30 to 90 mM) was noted. Membrane potential appears to decrease in both SFV- and VSV-infected BHK cells. Since membrane potential also decreases in cells exposed to hemolytic Sendai virus, the effect of Ca^{2+} was studied. Membrane potential of SFV-infected BHK cells was unaffected by external Ca^{2+} (Fig. 8.4), in contrast to the change elicited by the hemolytic Sendai virus (Impraim et al. 1980). It would appear that the reason for the decreased membrane potential of virally infected cells is not the same as that in Sendai-treated cells. Ca^{2+} also failed to reverse the increased uptake of dGlc shown by SFV-infected BHK cells (Fig. 8.4).

We may conclude that in certain situations, dependent on the nature of both cells and viruses, the integrity of the surface membrane of otherwise healthy cells is affected by infection. In no case, however, was a permeability change similar to that elicited by hemolytic Sendai virus (Tables 8.1 and 8.2) or that proposed by Carrasco (1978) and Gray et al. (1983b, c). Nor was there any indication that either the decrease in host protein synthesis or the initiation of viral protein synthesis is dependent on an increased intracellular concentration of Na^+ (Gray et al. 1983b).

Table 8.7. Cation Changes and Membrane Potential in Infected Cells[a]

Virus	Family	Cell Type	Growth Condition	Change	References
Sendai	Paramyxo	MDBK	Monolayer	Loss of K^+; no change in membrane potential	Bashford (1981) Bashford et al. (1981)
Sendai	Paramyxo	BHK	Monolayer	No change in K^+ or membrane potential	M.A. Gray, P. Turek, and C.L. Bashford, unpublished experiments
Measles	Paramyxo	Vero	Monolayer	No change in K^+ or membrane potential	Foster et al. (1983) C.L. Bashford and C.A. Pasternak, unpublished experiments
Measles	Paramyxo	Peripheral human lymphocytes	Suspension	No change in K^+ or membrane potential	Foster et al. (1983) C.L. Bashford, C. Taylor, and C.A. Pasternak, unpublished experiments
VSV	Rhabdo	BHK	Monolayer	No change in K^+ or Na^+; decrease in membrane potential	Gray et al. (1983a) C.L. Bashford, C. Taylor, and C.A. Pasternak, unpublished experiments
SFV	Toga	BHK	Monolayer	No change in K^+; increase in Na^+; decrease in membrane potential	Gray et al. (1983a, b) C.L. Bashford and C. Taylor, unpublished experiments

[a] See legend to Table 8.5.

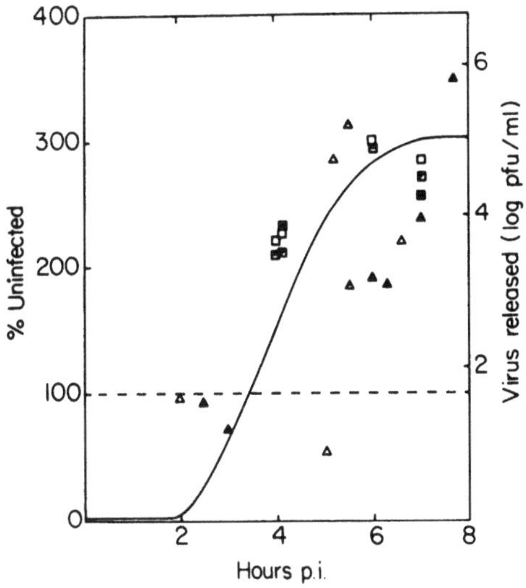

Figure 8.4. Effect of Ca^{2+} on membrane changes in Semliki Forest virus-infected cells. BHK cells were infected with Semliki Forest virus and assessed at various times thereafter (hours post infection). The dGlc uptake (squares) and K^+ null point (Bashford 1981) (triangles) were measured and expressed as a percentage of mock-infected cells assayed in the same medium; symbols of different shape indicate different experiments. Open symbols, measurements carried out in Ca^{2+}-free medium; partially filled-in symbols, in 2 or 4 mM Ca^{2+}; filled-in symbols, in 20 mM Ca^{2+}. Note that an increased null point indicates a decreased membrane potential (intracellular K^+ does not change). Virus release is indicated by the solid line. C.L. Bashford, M.A. Gray, and C.C. Taylor, unpublished results.

Clinical Relevance of Permeability Changes

From the above account of virally mediated permeability changes, it is clear that there are two quite distinct situations (Pasternak et al. 1982): (1) the Ca^{2+}-sensitive permeability change elicited by hemolytic paramyxoviruses when fusing with cells during entry and (2) various Ca^{2+}-insensitive changes that occur during infection of particular cells by particular viruses. Has either situation any clinical relevance?

In regard to the first situation, to date, there have been few reports of human viral disease that can be ascribed to the action of a hemolytic paramyxovirus. However, the correlation between symptoms of disease and viral etiology is often only conjectural, and the fact that hemolytic Sendai virus is able to affect the physiological response of a number of different cell types (Table 8.8) shows that the possibility of such viruses causing a pathophysiological change is a real one (Pasternak and Micklem 1981). In each of the examples in Table 8.8, high extracellular Ca^{2+} was shown to inhibit the response; because most of the

Table 8.8. Physiological Effects Elicited by Hemolytic Sendai Virus[a]

Cell Type	Effect	References
Mast cell	Initiation of histamine release	Sugiyama (1977) Gomperts et al. (1981, 1983)
Neutrophil	Initiation of β-glucuronidase release	B.D. Gomperts, J.P. Bennett, and J.M. Baldwin, unpublished observations
Anterior pituitary cells	Initiation of ACTH, growth hormone, and prolactin release	Gillies et al. (1981) Forda et al. (1982) Pasternak and Micklem (1973)
Embryonic heart	Cessation of spontaneous beating	Forda et al. (1982)
Dorsal root ganglion neurons	Loss of excitability	Forda et al. (1982)

[a] In several cases the effect is transient: Cells return to normal within minutes of the addition of virus.

responses are, however, dependent on *some* Ca^{2+}, the experiments were carried out at subphysiological concentrations of Ca^{2+}. In short, the responses become particularly sensitive to extracellular Ca^{2+} in the presence of virus. Because none of the responses requires the virus to be infective—UV-inactivated virus being fully active—it is clear that infection of cells does not play a role (see Fig. 8.2). That in itself does not detract from possible clinical relevance: The situation would be analogous to diseases caused by bacterial toxins. The symptoms of cholera (Field 1978; Holmgren 1981; Van Heyningen 1982), tetanus (Van Heyningen 1980; Mellanby and Green 1981), or diptheria (Collier 1975; Van Heyningen 1981), for example, can each be ascribed to an initial effect at the plasma membrane of susceptible cells that is quite unrelated to the growth of the invading microbe.

Virally induced changes in membrane permeability may be responsible for the secondary bacterial infections associated with respiratory viruses, since several of these viruses lower the phagocytotic capacity of neutrophils when incubated in vitro. Newcastle Disease virus, which elicits a permeability change (Table 8.3) similar to that induced by the closely related (Poste and Pasternak 1978) Sendai virus, for example, inhibits phagocytosis by neutrophils irrespective of whether it is UV inactivated or not (Faden et al. 1981).

In regard to the second situation, the clinical relevance of permeability changes in infected cells hinges on whether we believe that symptoms of viral disease may be attributed in certain cases to functionally altered yet intact cells rather than merely to cell necrosis caused either by the virus or by an inflammatory and immune consequence of viral infection. Some results indicate that the first possibility, which is likely to be of particular consequence in persistent viral infections, may be a real one. Animals that are generally healthy but persistently infected with lymphocytic choriomeningitis virus show functional defects, such as a reduced secretion of growth hormone from otherwise intact anterior pituitary cells (Oldstone et al. 1984). Rats infected with rabies or pseudorabies virus (a virus of the herpes family) exhibit spontaneous electrical activity in otherwise intact nerves (Dolivo 1980). With isolated cells in vitro, functional alterations that can be ascribed to a permeability change at the plasma membrane have been reported: Embryonic heart cells infected with Herpes simplex virus cease beating (Batra et al. 1976), and dorsal root ganglion neurons infected with Herpes simplex virus lose excitability (Fukada and Kurata 1981), at times *before* the cells succumb to the viral infection. Although the mechanism of such alterations is likely to have more in common with that underlying the changes summarized in Table 8.7 (e.g., decrease in membrane potential) than with changes caused by hemolytic Sendai virus, the outcome bears a striking resemblance to the last two changes listed in Table 8.8. And although this review has dealt only with the effects of cytolytic viruses and has not considered the effects of viral transformation, which clearly do alter cell function without cell destruction, the fact that certain viruses are able to stimulate sugar uptake (Table 8.5) in a manner analogous to the stimulation elicited by transforming viruses (e.g., Pasternak and Knox 1977) and other agents of cell proliferation (e.g., Antoniades and Owen 1982), shows that viruses may affect cell membranes in ways hitherto unrecognized. Much work remains to be carried out in this field of membrane research.

Summary

Certain paramyxoviruses elicit a permeability change on addition to cells: Low molecular weight compounds and ions that normally do not cross the plasma membrane freely do so in virally modified cells. The permeability change is a necessary step in virally mediated giant cell (polykaryon) formation. Several other viruses, including paramyxoviruses, elicit other types of permeability change when growing in susceptible cells. Both changes occur in the absence of cell lysis. Permeability changes may underlie some of the clinical symptoms of viral disease.

Acknowledgments

I am indebted to Kingsley Micklem and other colleagues for many helpful discussions and for permission to cite unpublished experiments. I am grateful to the Cancer Research Campaign, Science Research Council (now Science and Engineering Research Council), Roche Products Limited, and the Cell Surface Research Fund for financial support during the course of the work described in this chapter.

References

Andreoli, T.E., Hoffman, J.F., and Fanestil, D.D. (eds.) (1978). *Physiology of Membrane Disorders.* Plenum Press, New York.

Antoniades, H.N. and Owen, A.J. (1982). Growth factors and regulation of cell growth. *Annu. Rev. Med. 33,* 445–463.

Baker, P.F. and Knight, D.E. (1978). Calcium-dependent exocytosis in bovine adrenal medullary cells with leaky plasma membranes. *Nature 276,* 620–622.

Bashford, C.L. (1981). The measurement of membrane potential using optical indicators. *Biosci. Rep. 1,* 183–196.

Bashford, C.L., Foster, K.A., Micklem, K.J., and Pasternak, C.A. (1981). Optical measurement of the plasma-membrane potential of mammalian cells grown in monolayer culture. *Biochem. Soc. Trans. 9,* 80–81.

Bashford, C.L., Micklem, K.J., and Pasternak, C.A. (1983). Sequential increase of Lettre cell plasma membrane permeability induced by virus. *J. Physiol. 343,* 100–101P.

Batra, G.K., Nahmias, A.J., and De Haan, R.L. (1976). Herpes simplex virus infection stops spontaneous beating of chick heart cells. *Nature 259,* 677–679.

Blow, A.M.J., Bothan, G.M., and Lucy, J.A. (1979). Calcium ions and cell fusion effects of chemical fasogens on the permeability of erythrocytes to calcium and other ions. *Biochem. J. 182,* 525–530.

Bolis, L., Hoffman, J.F., and Leaf, A. (eds.) (1976). *Membranes and Diseases.* Raven Press, New York.

Carafoli, E. and Crompton, M. (1978). The regulation of intracellular calcium. *Current Topics Membr. Transport 10,* 151–216.

Carrasco, L. (1977). The inhibition of cell functions after viral infection: A proposed general mechanism. *FEBS Lett. 76,* 11–15.

Carrasco, L. (1978). Membrane leakiness after viral infection and a new approach to the development of antiviral agents. *Nature 272,* 694–699.

Carrasco, L. and Smith, A.E. (1976). Sodium ions and the shut-off of host cell protein synthesis by picornaviruses. *Nature 264,* 807–809.

Collier, R.J. (1975). Diptheria toxin: Mode of action and structure. *Bacteriol. Rev. 39,* 54–79.

Curtis, D., Micklem, K.J., Gill, K., and Pasternak, C.A. (1980). Changes in the surface membrane during myoblast fusion. *Biochem. J. 190,* 647–652.

Dolivo, M. (1980). A neurobiological approach to neurotropic viruses. *Trends Neuro. Sci. 3,* 149–152.

Egberts, E., Hackett, P.B., and Traub, P. (1977). Alteration of the intracellular energetic and ionic conditions by mengovirus infection of Ehrich ascites tumor cells and its influence on protein synthesis in the midphase of infection. *J. Virol. 22,* 591–597.

Faden, H., Humbert, J., Lee, J., Sutyla, P., and Ogra, P.L. (1981). The in vitro effects of Newcastle disease virus on the metabolic and antibacterial functions of human neutrophils. *Blood 58,* 221-227.

Farnham, A.E. and Epstein, W. (1963). Influence of encephalomyocarditis (EMC) virus infection on potassium transport in L cells. *Virology 21,* 436-447.

Fenner, F., McAuslan, B.R., Mims, C.A., Sambrook, J., and White, D.O. (1974). *The Biology of Animal Viruses.* Academic Press, New York, 2nd ed., p. 338.

Fenwick, M.L. and Walker, M.J. (1978). Suppression of the synthesis of cellular macromolecules by herpes simplex virus. *J. Gen. Virol. 41,* 37-51.

Field, M. (1978). Cholera toxin, adenylate cyclase and the process of active secretion in the small intestine: The pathogenesis of diarrhea in cholera. In *Physiology of Membrane Disorders,* Andreoli, T.E., Hoffman, J.F., and Fanestil, D.D., eds. Plenum Medical Book Co., New York, pp. 877-899.

Forda, S.R., Gillies, G., Kelly, J.S., Micklem, K.J., and Pasternak, C.A. (1982). Acute membrane responses to viral action. *Neurosci. Lett. 29,* 237-242.

Foster, K.A. (1980). The nature of virally-induced permeability changes. Ph.D. Thesis, University of London.

Foster, K.A., Micklem, K.J., Agnarsdottir, G., Lancashire, C.L., Bogomolova, N.N., Boriskin, Y.S., and Pasternak, C.A. (1983). Myxoviruses do not induce non-specific alterations in membrane permeability early on in infection. *Archiv. Virol. 77,* 139-153.

Francoeur, A.M.J. and Stanners, C.P. (1978). Evidence against the role of K^+ in the shut-off of protein synthesis by vesicular stomatitis virus. *J. Gen. Virol. 39,* 551-554.

Fuchs, P. and Giberman, E. (1973). Enhancement of potassium influx, in baby hamster kidney cells and chicken erythrocytes, during the absorption of parinfluenza 1 (Sendai) virus. *FEBS Lett. 31,* 127-130.

Fukada, J. and Kurata, T. (1981). Loss of membrane excitability after herpes simplex virus infection in tissue-cultured nerve cells from adult mammals. *Brain Res. 211,* 235-241.

Fukai, K. and Suzuki, T. (1955). *Med. J. Osaka Univ. 6,* 1-15.

Garry, R.F., Bishop, J.M., Parker, S., Westbrook, K., Lewis, G., and Waite, M.R.F. (1979). Na^+ and K^+ concentrations and the regulation of protein synthesis in sindbis virus-infected chick cells. *Virology 96,* 108-120.

Garry, R.F., Ulug, E.T., and Bose, H.R. (1982). Membrane-mediated alterations of intracellular Na^+ and K^+ in lytic-virus-infected and retrovirus-transformed cells. *Biosci. Rep. 2,* 617-623.

Genty, N. (1975). Analysis of uridine incorporation in chicken embryo cells infected by vesicular stamatitis virus and its temperature-sensitive mutants: Uridine transport. *J. Virol. 15,* 8-15.

Getz, D., Gibson, J.F., Sheppard, R.N., Micklem, K.J., and Pasternak, C.A. (1979). Manganese as a calcium probe: Electron paramagnetic resonance and nuclear magnetic resonance spectroscopy of intact cells. *J. Membr. Biol. 50,* 311-329.

Gillies, G. and Lowry, P.J. (1978). Perfused rat isolated anterior pituitary cell column as bioassay for factor(s) controlling release of adrenocorticotropin: Validation of a technique. *Endocrinology 103,* 521-527.

Gillies, G., Micklem, K.J., and Pasternak, C.A. (1981). Use of Sendai virus to permeabilize anterior pituitary cells to calcium buffers. *Cell Biol. Int. Rep. Suppl. A, 5,* 20.

Gilula, N.B., Reeves, O.R., and Steinbach, A. (1972). Metabolic coupling, ionic coupling, and cell contacts. *Nature 235,* 262-265.

Gomperts, B.D. (1976). In *Receptors and Recognition,* Cuatrecasas, P. and Greaves, M.F., eds. Chapman & Hall, London, series A, vol. 2, pp. 45-92.

Gomperts, B.D., Micklem, K.J., and Pasternak, C.A. (1981). A new method for permeabilising secretory cells which allows direct manipulation of intracellular Ca in the micromolar range. *Cell Biol. Int. Rep. Suppl. A 5,* 20.

Gomperts, B.D., Baldwin, J.M., and Micklem, K.J. (1983). Rat mast cells permeabilized with Sendai virus secrete histamine in response to Ca^{2+} buffered in the micromolar range. *Biochem. J. 210,* 737-745.

Gray, M.A., Micklem, K.J., Brown, F., and Pasternak, C.A. (1983a). Effect of vesicular stomatitis virus on uptake of nutrients and intracellular cation concentration. *J. Gen. Virol. 64,* 1449-1456.

Gray, M.A., Micklem, K.J., and Pasternak, C.A. (1983b). Protein synthesis in cells infected with Semliki Forest virus is not controlled by intracellular cation changes. *Europ. J. Biochem. 135,* 299-302.

Gray, M.A., Austin, S.A., Clemens, M.J., Rodrigues, L., and Pasternak, C.A. (1983c). Protein synthesis in Semliki Forest virus-infected cells is not controlled by permeability changes. *J. Gen. Virol. 64,* 2631-2640.

Harris, H. (1970). *Cell Fusion.* Oxford University Press, Oxford.

Henson, P.M., Ginsberg, M.H., and Morrixon, D.C. (1978). Mechanisms of mediator release by inflammatory cells. *Cell Surface Rev. 5,* 411-508.

Holmgren, J. (1981). Actions of the cholera toxin and the prevention and treatment of cholera. *Nature 292,* 413-417.

Homma, M., Shimizu, K., Shimizu, Y.K., and Ishida, N. (1976). On the study of Sendai virus hemolysis. I. Complete Sendai virus lacking in hemolytic activity. *Virology 71,* 41-47.

Hopkins, C.R. and Walker, A.M. (1978). Calcium as a 2nd messenger in stimulation of lutenizing-hormone secretion. *Mol. Cell. Endocrinol. 12,* 189-208.

Hopkins, C.R. and Dunca, C.J. (1979). *Secretary Mechanisms.* Cambridge University Press.

Hosaka, Y. and Shimizu, K. (1977). Cell fusion by Sendai virus. *Cell Surface Rev. 2,* 129-155.

Impraim, C.C., Micklem, K.J., and Pasternak, C.A. (1979). Calcium, cells and virus: Alterations causes by paramyxoviruses. *Biochem. Pharmacol. 28,* 1963-1969.

Impraim, C.C., Foster, K.A., Micklem, K.J., and Pasternak, C.A. (1980). Nature of virally mediated changes in membrane permeability to small molecules. *Biochem. J. 186,* 847-860.

Ishida, N. and Homma, M. (1978). Sendai viruses. *Adv. Virus Res. 23,* 349–383.

Kilham, L. (1949). A Newcastle disease virus (NDV) hemolysis. *Proc. Soc. Exp. Biol. Med. 71,* 63–66.

Klemperer, H.G. (1960). Hemolysis and the release of potassium from cells by Newcastle disease virus (NDV). *Virology 12,* 540–552.

Knutton, S. (1977). Studies of membrane fusion. II. Fusion of human erythrocytes by Sendai virus. *J. Cell Sci. 28,* 189–210.

Knutton, S. (1978). The mechanism of virus induced cell-fusion. *Micron 9,* 133–154.

Knutton, S. (1979). Studies of membrane fusion. V. Fusion of erythrocytes with non-hemolytic Sendai virus. *J. Cell Sci. 36,* 85–96.

Knutton, S. and Bachi, T. (1980). The role of cell swelling and hemolysis in Sendai virus-induced cell fusion and in the diffusion of incorporated viral antigens. *J. Cell Sci. 42,* 153–167.

Knutton, S., Jackson, D., Graham, J.M., Micklem, K.J., and Pasternak, C.A. (1976). Microvilli and cell swelling. *Nature 262,* 52–54.

Knutton, S. and Pasternak, C.A. (1979). The mechanisms of cell–cell fusion. *Trends Biochem. Sci. 4,* 220–223.

Kuffler, S.W. and Potter, D.D. (1964). Glia in the leech central nervous system: Physiological properties and neuron-glia relationship. *J. Neurophysiol. 27,* 290–320.

Lowenstein, W.R. and Kanno, Y. (1964). Studies on an epithelial (gland) cell junction. I. Modification of surface membrane permeability. *J. Cell Biol. 22,* 565–586.

Lucy, J.A. (1978). Mechanisms of chemically induced cell fusion. *Cell Surface Rev. 5,* 267–304.

McNutt, N.S. and Weinstein, R.S. (1970). The ultrastructure of the nexus: A correlated thin-section and freeze-cleave study. *J. Cell Biol. 47,* 666–688.

Meldolesi, J., Borgese, N., De Cannill, P., and Cessarelli, B. (1978). Cytoplasmic membranes and the secretory process. *Cell Surface Rev. 5,* 510–627.

Mellanby, J. and Green, J. (1981). How does tetanus toxin act. *Neuroscience 6,* 281–300.

Micklem, K.J. and Pasternak, C.A. (1977). Surface components involved in virally mediated membrane changes. *Biochem. J. 162,* 405–410.

Micklem, K.J. and Pasternak, C.A. (1982). Ca^{2+}-sensitive channels induced by certain viruses. *J. Physiol. 326,* p. 11.

Miller, M.W. and Shamoo, A.E., (eds.) (1977). *Membrane Toxicity.* Plenum Press, New York.

Miller, M.R., Castellot, J.J., and Pardee, A.B. (1979). General-method for permeabilizing monolayer and suspension cultured animal cells. *Exp. Cell. Res. 120,* 421–425.

Morgan, H.R., Enders, J.F., and Wagley, P.F. (1948). *J. Exp. Med. 88,* 503–514.

Nair, C.N., Stowers, J.W., and Singfield, B. (1979). Guanidine-sensitive Na^+ accumulation by poliovirus-infected hela cells. *J. Virol. 31,* 184–189.

Negreanu, Y., Reinhertz, Z., and Kohn, A. (1974). Effects of absorption of uv-inactivated parainfluenza (Sendai) virus on the incorporation of amino acids in animal host cells. *J. Gen. Virol. 22,* 265–270.

Norkin, L.C. (1977). Cell killing by Simian virus 40: Impairment of membrane formation and function. *J. Virol. 21,* 872–879.

Norrie, D.H., Wolstenholme, J., Howcroft, H., and Stephen, J. (1982). Vaccinia virus-induced changes in (Na^+) and (K^+) in nela cells. *J. Gen. Virol. 62,* 127–136.

Okada, Y. (1969). Factors in fusion of cells by HVJ, Sendai virus. *Curr. Top. Microbiol. Immunol. 48,* 102–156.

Oldstone, M.B.A., Rodriguez, M., Daughaday, W.H., and Lampert, P.W. (1984). Viral perturbation of endocrine function: Disordered cell function leads to disturbed homeostasis and disease. *Nature 307,* 278–281.

Orci, L. and Perrelet, A. (1978). Ultrastructural aspects of exocytotic membrane fusion. *Cell Surface Rev. 5,* 630–656.

Papahadjopoulos, D. (1978). Calcium-induced phase changes and fusion in natural and model membranes. *Cell Surface Rev. 5,* 766–790.

Pasternak, C.A. (1980). Virally mediated changes in cellular permeability. *Biochem. Soc. Trans. 8,* 700–702.

Pasternak, C.A. (1981). Virus induced cell fusion. *Biochem. Soc. Trans. 9,* 14P.

Pasternak, C.A. and Micklem, K.J. (1973). Permeability changes during cell fusion. *J. Membr. Biol. 14,* 293–303.

Pasternak, C.A. and Micklem, K.J. (1974a). The biochemistry of virus-induced cell fusion changes in membrane integrity. *Biochem. J. 140,* 405–411.

Pasternak, C.A. and Micklem, K.J. (1974b). Virally medicated membrane changes: Inverse effects on transport and diffusion. *Biochem. J. 144,* 593–595.

Pasternak, C.A. and Knox, P. (1977). Normal and transformed cells in vitro. In *Mammalian Cell Membranes,* Jamieson, G.A. and Robinson, D.M., eds. Vol. 3, pp. 89–114. Butterworth, London.

Pasternak, C.A. and Micklem, K.J. (1981). Virally induced alterations in cellular permeability: A basis of cellular and physiological damage. *Biosci. Rep. 1,* 431–448.

Pasternak, C.A. and Micklem, K.J. (1983). Altered function of excitable cells caused by paramyxoviruses. In *Viruses and Demyelinating Disease,* Mims, C.A., Cuzner, M.L., and Kelly, R.E., eds. Academic Press, New York, to be published.

Pasternak, C.A., Strachan, E., Micklem, K.J., and Duncan, J. (1976). New approaches for studying surface membranes. In *Cell Surfaces and Malignancy,* Mora, P., ed. *Fogarty Int. Center Proc. 28,* 147–151. U.S. Government Printing Office, Washington, D.C.

Pasternak, C.A., Gray, M.A., and Micklem, K.J. (1982). Membrane changes during viral infection. *Biosci. Rep. 2,* 609–612.

Peracchia, C. (1977). Gap junction structure and function. *Trends in Biochem. Sci. 2,* 26–31.

Peterhans, E., Haenggeli, E., Wild, P., and Wyler, R. (1979). Mitochondrial calcium uptake during infection of chicken embryo cells with Semliki Forest virus. *J. Virol. 29,* 143–152.

Pitts, J.D. (1976). Junctions as channels of direct communication between cells. In *The Developmental Biology of Plants and Animals,* Graham, C.F. and Wareing, P.F., eds. Blackwell Scientific Publications, Oxford, pp. 96–110.

Poste, G. and Nicolson, G.L., eds. (1977). *Virus Infection and the Cell Surface* (Cell Surface Reviews, Vol. 2). North-Holland, Amsterdam.

Poste, G. and Pasternak, C.A. (1978). Virus-induced cell fusion. *Cell Surface Rev. 5,* 306–349.

Rasmussen, H. and Goodman, D.B.P. (1977). Relationships between calcium and cyclic nucleotides in cell activation. *Physiol. Rev. 57,* 422–507.

Rhamabhadran, T.V. and Thach, R.E. (1981). Translational elongation rate changes in encephalomyocarditis virus-infected and interferon-treated cells. *J. Virol. 39,* 573–583.

Rose, B. and Loewenstein, W.R. (1975). Permeability of cell junction depends on local cytoplasmic calcium activity. *Nature 254,* 250–252.

Shimizu, Y.K., Shimizu, K., Ishida, N., and Homma, M. (1976). On the study of Sendai virus hemolysis. II. Morphological study of envelope fusion and hemolysis. *Virology 71,* 48–60.

Shrier, A., Nahmias, A.J., and De Haan, R.L. (1978). Herpes virus infection alters electrical parameters of heart cell aggregates. *Am. J. Physiol. 234,* C170–C176.

Simpson, I., Rose, R., and Loewenstein, W.R. (1977). Glial maturation factor: Effect on chemical differentiation of glioblasts in culture. *Science 195,* 294–296.

Subak-Sharpe, J.H., Burk, R.R., and Pitts, J.D. (1966). Metabolic cooperation between biochemically marked mammalian cells in tissue culture. *J. Cell Sci. 4,* 353–367.

Tanaka, K., Sekiguchi, M., and Okada, Y. (1975). Restoration of ultraviolet induced unscheduled DNA synthesis of xeroderma pigmentosum cells by the concomitant treatment with bacterophage TY endonuclease V and HVJ (Sendai virus). 95. Viral infections. *Proc. Nat. Acad. Sci. USA 72,* 4071–4075.

Thigpen, J.T. (1971). Viral infections. In *Concepts of Disease,* Brunson, J.G. and Gall, E.A., eds. Macmillan, New York, pp. 644–672.

Toister, Z. and Loyter, A. (1973). The mendranism of cell fusion. II. Formation of chicken erythrocyte polykaryons. *J. Biol. Chem. 248,* 422–432.

Trump, B.F. and Arstila, A.U., eds. (1975). *Pathobiology of Cell Membranes.* Academic Press, New York.

Van Heyningen, S. (1980). Tetanus toxin. *Pharmacol. and Ther. 11,* 141–157.

Van Heyningen, S. (1981). Diphtheria toxin: Which route into the cell? *Nature 292,* 193–294.

Van Heyningen, S. (1982). Cholera toxin. *Biosci. Rep. 2,* 135–146.

Volsky, D.J. and Loyter, A. (1978). Rate of Ca^{++} in virus-induced membrane fusion. Ca^{++} accumulation and ultrastructural changes induced by Sendai virus in chicken erythrocytes. *J. Cell Biol. 78,* 465–477.

Wyke, A.M. (1980). The role of Sendai virus components in membrane fusion. Ph.D. Thesis, University of London.

Wyke, A.M., Impraim, C.C., Knutton, S., and Pasternak, C.A. (1980). Components involved in virally medicated membrane fusion and permeability changes. *Biochem. J. 190,* 625–638.

Yamaizumi, M., Uchida, T., and Ukada, Y. (1979). Macromolecules can penetrate the host cell membrane during the early period of incubation with HVJ (Sendai virus). *Virology 95,* 218–211.

9

Transmembrane Calcium Fluxes and Cell Death

H. Baum and R.F.G. Booth

Some years ago, Shen and Jennings (1972a and 1972b) reported that when dog myocardium was reversibly injured by 10 min of occlusion-induced ischemia, followed by 20 min of arterial reflow, there was no excessive uptake of extracellular calcium. However, 40 min of ischemia, followed by 10 min of arterial reflow, led to a massive uptake of calcium into the irreversibly injured tissue, the loading being primarily localized to the mitochondria. These authors also noted that mitochondria of myocardial cells irreversibly injured by ischemia show a greatly reduced capacity to perform integrated functions.

During hypoxia in the perfused rat heart there is a gradual onset of release of cytosolic enzymes (Hearse et al. 1973). Reoxygenation after a short period of hypoxia prevents such release, whereas reoxygenation once release is evident greatly exacerbates the apparent damage. The damaging effect of reoxygenation after anoxia can be lessened by the inclusion of glucose in the perfusion medium, although the extent of such protection is to some degree species specific (Hearse et al 1976). Hearse et al. (1978) subsequently noted that the damaging effect of reoxygenation of hypoxic-perfused rat heart could be mimicked by calcium repletion following a calcium-free perfusion. These observations, taken together, could be rationalized as follows: Depletion of cytosolic adenosine triphosphate (ATP) during extended periods of hypoxia, especially in the absence of adequate substrate for ATP production by glycolysis, permits extracellular calcium to enter the myocardial cell. Perfusion in a calcium-free medium also causes structural damage sufficient to increase calcium permeability. However, irreversible damage only occurs once oxygenation permits the massive uptake of calcium into the myocardial mitochondria. A general model of the events that might be involved has already been considered by Dhalla (1976), but no precise conclusion was reached as to why calcium uptake into mitochondria in the failing heart should lead to cell death.

A role for Ca^{2+} in the irreversible damage resulting from cerebral ischemia has also been suggested by experiments using synaptosomes as model nerve endings (Booth et al. 1982; Harvey et al. 1982, 1983). We have demonstrated that after a period of in vitro hypoxia followed by reoxygenation, the presence of external Ca^{2+} exacerbates cellular damage as assessed by numerous parameters including the rate of acetylcholine synthesis, tricarboxylic acid (TCA) cycle flux and ATP:ADP (adenosine diphosphate) ratio. The presence of a low external pH was found to potentiate this damage.

The involvement of calcium in irreversible cell death induced by toxic agents (rather than by hypoxia or calcium-free perfusion) has been strongly implicated by Schanne et al. (1979). Primary cultures of rat hepatocytes were treated in the presence or absence of extracellular calcium with 10 different membrane-active toxins. In all cases more than half the cells were killed in 1-6 h in the presence, but not the absence of, extracellular calcium even though an effect of calcium on the primary mechanism of membrane injury by any of the toxic agents could not be implicated. Again, it was concluded that the primary event was damage to the integrity of the plasma membrane, followed by an influx of calcium leading to the initiation of a "final common pathway" for toxic cell death.

What might the nature be of such a common pathway, and how might mitochondria be implicated? Cell death is in fact rather difficult to define. In a nonreplicating cell it might be considered as the irreversible loss of homeostasis, in particular the failure to repair or replace macromolecular structures damaged by external agents or broken down endogenously by enzymes concerned with the normal turnover of cellular constituents.

A consideration of the very large number of plasma membrane cytosolic and mitochondrial enzyme reactions modulated by calcium either free or in a complex with calmodulin (Cheung 1980; Means and Dedman 1980; Denton and McCormack 1980; Carafoli 1981) leads to the conclusion that uncontrolled entry of calcium from the extracellular fluid into the cell would have immediate and catastrophic consequences. These would be exacerbated by structural effects, through disaggregation of microtubules, activation of phospholipases, and possibly facilitation of membrane fusion.

However, mitochondria normally have the capability of buffering cytosolic calcium against the challenge imposed by net influx across the plasma membrane. For example, when squid axons were continuously loaded with Ca^{2+}, cytosolic-free [Ca^{2+}] was buffered between 1 and 3 μM, whereas the total axonal [Ca^{2+}] increased from 0.1 to 2 mM (Brinley et al. 1977). This buffering is achieved in excitable tissue such as the heart by a steady-state cycling involving an electrogenic Ca^{2+} uniporter, a Ca^{2+}-Na^{+} antiporter, and a Na^{+}-H^{+} antiporter (Nicholls and Crompton 1980). In liver, uptake is by the uniporter, the alternative efflux pathway being at present ill defined, although an H^{+}-Ca^{2+} exchange has been postulated as has the cotransport of inorganic phosphate and Ca^{2+} (Rugolo et al. 1981). Also, recent reports (Nedergaard and Cannon 1980) have suggested that the efflux pathway in liver mitochondria may be stimulated by Na^{+}, although more recently this has been questioned (Zoccarato and Nicholls 1982).

Inhibition or uncoupling of respiration will clearly lower the membrane potential and inhibit further uptake and might be expected to permit calcium efflux from mitochondria. However, in the absence of exogenous uncoupling agents and under aerobic conditions, mitochondria can maintain calcium homeostasis. Thus, calcium accumulation into mitochondria would at first appear to be a protective rather than a lethal process. Wherein, therefore lies the paradox?

We have shown (Harris et al. 1979) that under some circumstances a calcium load can constitute a severe challenge to the integrity of the mitochondrial inner membrane. However, it should also be recognized that under certain conditions liver mitochondria can accumulate the cation to millimolar concentration without damage (Zoccarato and Nicholls 1982), although at a certain accumulated concentration, in the presence of inorganic phosphate, complete and irreversible depolarization occurs (Lotscher et al. 1980a). A consideration of all of the agents that inhibit or potentiate calcium-dependent changes in membrane permeability led to the following model.

Passive leakiness of the membrane is limited through the binding to it of magnesium-ADP. Free Ca^{2+} in the matrix tends to displace this, causing efflux of Mg^{2+}, Ca^{2+}, and ADP and hence a nonspecific increase in permeability to small solutes. Depending on the osmotic environment, this can lead to swelling and further structural damage. Also, such permeability changes lead to complete uncoupling, so that reuptake pathways are no longer operative. Hence calcium homeostasis fails.

In the intact cell such a sequence of events would be lethal; not only would all Ca^{2+}-regulated processes be thrown into disarray, but the mitochondrial phospholipase A_2 would be activated; indeed, it has recently been demonstrated that in strontium-induced oscillating ion fluxes in liver mitochondria periodic activations and inactivations of phospholipase A_2 are associated with periodic changes in the incorporation rates of labeled polyunsaturated fatty acids with an apparent phase difference of 180 deg (Wiswedel et al. 1982). Thus, phospholipase A_2 activation induced by an increase in mitochondrial Ca^{2+} fluxes will result in an accumulation of lysophospholipids that are known to have a detrimental effect on membrane function (Weltzien 1979), thus accelerating the catastrophic cascade. The Ca^{2+}-induced activation of phospholipase C that is located within the plasma membrane will also generate phosphatidate, which is a Ca^{2+} ionophore and will therefore enhance the rate of cellular Ca^{2+} influx (Serhan et al. 1981). It should also be recognized that agents that interfere with energy-dependent lysophospholipid reacylation pathways will exacerbate any membrane damage (Beatrice et al. 1980). In this context, the loss of mitochondrial activity as the major source of ATP production, and the activation through uncoupling, of the hydrolytic mode of the mitochondrial adenosine triphosphatase (ATPase) would cause a precipitate fall in cytosolic ATP levels, thus minimizing the possibility of repair processes while increasing the net influx of extracellular Ca^{2+}.

Clearly, these are not the inevitable sequelae of calcium uptake by mitochondria, and we have attempted to delineate in our model the factors that determine the extent to which a given Ca^{2+} load will initiate these destructive events.

Exogenous Mg^{2+} and ADP themselves afford protection, and bongkrekic acid (which greatly enhances ADP binding to the membrane) is a potent protective agent. Marginally low pH (below pH 7.4) is also protective to mitochondria, possibly by facilitating the binding of the ADP anion to an ionizable membrane site. Conversely, agents that decrease ADP binding to the adenine nucleotide carrier generally potentiate Ca^{2+}-dependent damage. (Whether this implicates the carrier itself as the binding site or whether the binding site simply has analogous characteristics is still an open question.) Such agents include atractyloside, fatty acids, and their coenzyme A (CoA) esters. Other potentiators of magnesium-ADP displacement include prostaglandins (Kirtland and Baum 1972; Al-Shaikhaly and Baum 1979), nonsteroidal antiinflammatory agents (Baum et al. 1973), general anesthetics (Grist and Baum 1975), and hypoglycemic agents (Kobayashi et al. 1979). It will be of great interest to determine whether any such potentiators exert specific effects on the adenine nucleotide carrier [or possibly on the guanosine diphosphate (GDP)-binding site that controls H^+ permeability in brown adipose mitochondria (Cannon and Johannson 1980)].

An association between cellular Ca^{2+} homeostasis and thiol content has recently become increasingly apparent. At the cellular level, reoxygenation of hypoxic hearts decreases the cellular content of acid-soluble thiol groups, insoluble thiol groups, and reduced thiol groups (Guarnieri et al. 1980), which correlates directly with the disruption of cellular calcium homeostasis (Nayler et al. 1979). An interdependency at the mitochondrial level has been demonstrated in which the addition of thiol agents such as *N*-ethyl maleimide (in excess) and 5,5′-dithiobis-(2-nitrobenzoate) stimulate efflux, whereas dithiothreitol, an agent that chemically reduces thiol groups, has been demonstrated to diminish the efflux (Harris et al. 1979; Harris and Baum 1980). Also mercurials are powerful potentiators of calcium-dependent damage, presumably by acting directly on the thiol groups, although they might also act via a free-radical-mediated thiol oxidation (Harris and Baum 1980).

The identity of the thiol group(s) essential for the calcium-retention properties is not known, although recent studies using various sulfydryl alkylating agents have implicated that certain thiol groups, associated with the maintenance of mitochondrial inner membrane permeability, are located in the hydrophobic core of the membrane (Le Quoc and Le Quoc 1982). The reduction of these thiols appears to be dependent upon exogenously generated reduced nicotinamide adenine dinucleotide phosphate (NADPH) and may be mediated through a glutathione reductase pathway. This in turn depends upon the metabolic generation of reduced nicotinamide adenine dinucleotide (NADH) and the functioning of an energy-dependent transhydrogenase. Irreversible damage is incurred by the addition of *t*-butyl hydroperoxide plus Ca^{2+} to mitochondria, with the sequence of events being an initial oxidation of NAD(P)H to $NAD(P)^+$ and, in the absence of ATP, subsequent hydrolysis to nicotinamide (Lotscher et al. 1980b) and ADP-ribose, with the latter binding to a specific mitochondrial protein (Hoffstetter et al. 1981). It is the loss of intramitochondrial pyridine nucleotides and generation of ADP-ribose that is proposed to enhance Ca^{2+} release. The oxidation of mitochondrial pyridine nucleotides by oxaloacetate resulting in enhanced Ca^{2+} efflux has also been demonstrated (Lehninger et al. 1978), although this effect is now believed to be dependent upon the nature of the accompanying anion (Wolkowicz and McMillin-Wood 1980).

Paradoxically, it would therefore be expected that cellular hypoxia or ischemia would be cytoprotective by inhibiting respiration and elevating this mitochondrial $NADH:NAD^+$ ratio, whereas it is known that these conditions eventually result in cell death (particularly on reoxygenation). The paradox is resolved by the observation of higher rates of oxygen radical formation when a higher degree of reduction of certain mitochondrial redox couples occurs (Loschen et al. 1973; Nohl and Hegner 1978).

More recently we have demonstrated that superoxide radicals (O_2^-) can severely impair the ability of both heart and liver mitochondria to retain Ca^{2+} (Harris et al. 1982), and this damage correlated directly with the capacity of

Table 9.1 Factors and Agents Affecting the Extent to Which a Given Calcium Load Displaces Magnesium–ADP and Increases Membrane Permeability

Potentiating	Protective
Atractylate	Mg^{2+}
Fatty acids	ADP
Fatty acyl CoA	ATP
Prostaglandins	Bongkrekic acid
Aspirin	Dithiothreitol
Indomethacin	NADH
Halothane	Respiratory substrates
Tolbutamide	Respiratory inhibitors
Mercurials	Low pH
N-ethyl maleimide	Reduced ubiquinone
Fe^{2+}	
Thyroxine	
Thenyl trifluoracetone	
Uncouplers	
Phosphate	
High pH	
Oxaloacetate	
t-butyl hydroperoxide	
ADP-ribose	
Superoxide anions	

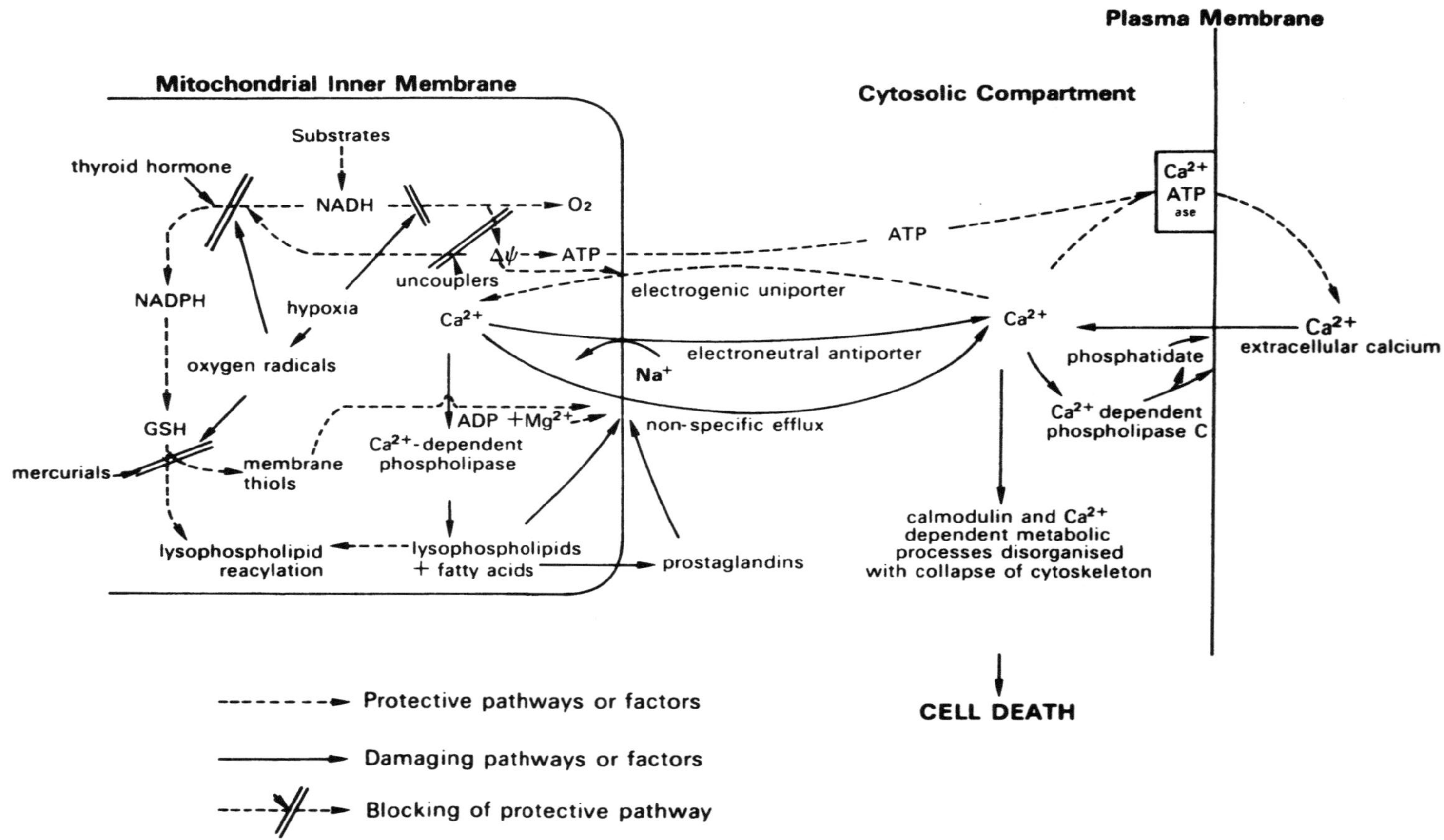

Figure 9.1. Some factors affecting calcium homeostasis in the cell.

the mitochondria to generate reduced thiol groups. Addition of the thiol-reductant dithiothreitol protected against the O_2^- induced damage, as did reduced ubiquinone, which is also known to protect phospholipids from O_2^--induced lipid peroxidation (Booth et al. 1982). Therefore, if cellular peroxidative reactions increase and exceed the capacity of the cellular defence mechanisms (e.g., superoxide dismutase, glutathione peroxidase, α-tocopherol), cellular damage may be mediated by the effects of oxygen radicals on specific reduced thiol groups with a subsequent disruption of mitochondrial Ca^{2+} homeostasis. The reflow period following a hypoxic episode in the isolated perfused rat heart has shown a direct correlation between an increase in lipid peroxidation and loss of cellular thiol groups (Guarnieri et al. 1980).

It should be noted that cellular calcium homeostasis may also be disturbed under certain stressful physiological situations in which a selective adrenergic activation of the Ca^{2+} uniporter of cardiac mitochondria has been reported (Kesser and Crompton, 1981); this is expressed via an α-adrenergic mechanism that can result in an elevation of intramitochondrial-free Ca^{2+} concentrations during Ca^{2+} recycling across the mitochondrial inner membrane. The hormone thyroxine also exerts a potentiating effect on Ca^{2+} efflux, possibly through inhibition of the energy-dependent transhydrogenase (Al-Shaikhaly and Baum 1979).

Some of the factors and agents affecting the extent to which a given calcium load is destructive through increasing mitochondrial membrane permeability are listed in Table 9.1. Many of these are potentially physiological, and presumably the normal balance is such that mitochondria in vivo can support calcium loadings with no damaging effect. However, if for example the calcium challenge becomes too large, or the thioreductase chain becomes inoperative, or fatty acids, prostaglandins, or oxygen free radicals accumulate, the balance may be swung in the other direction, leading to the catastrophic sequence outlined previously (see also Fig. 9.1).

A fuller understanding of the underlying mechanism of these processes and of their modulation is thus very important. It might lead to a strategy to prevent the irreversible damage to vital tissues that otherwise would be the outcome of an ischemic episode.

References

Al-Shaikhaly, M. and Baum, H. (1979). Associated effluxes of calcium and adenine nucleotides from mitochondria: Mediation by thyroxine and other agents and inhibition by bongkrekic acid and adenosine diphosphate. *Biochem. Soc. Trans.* 7, 215–216.

Baum, H., Kirtland, S.J., Crovetti, F., and Carafoli, E. (1973). Prostaglandin E_1, anti-inflammatory agents and calcium: Interactions at the mitochondrial membrane. In *Mechanisms in Bioenergetics*, Azzone, G.F., Ernster, L.,

Papa, S., Quagliariella, E., and Siliprandi, N., eds. Academic Press, New York and London, pp. 365–374.

Beatrice, M.C., Palmer, J.W., and Pfeiffer, D.R. (1980). The relationship between mitochondrial membrane permeability, membrane potential, and the retention of Ca^{2+} by mitochondria. *J. Biol. Chem. 255,* 8663–8671.

Booth, R.F.G., Galanopoulou. G., and Quinn, P. (1982). Protection by ubiquinone and ubiquinol against lipid peroxidation in egg yolk phosphatidylcholine liposomes. *Biochem. Int. 5,* 151–156.

Booth, R.F.G., Harvey, S.A.K., and Clark, J.B. (1983). The effects of in vivo hypoxia on acetylcholine synthesis by rat brain synaptosomes. *J. Neurochem. 40,* 106–110.

Brinley, F.J., Tiffert, T., Scarpa, A., and Mullins, L.J. (1977). Intracellular calcium buffering capacity in isolated squid axons. *J. Gen. Physiol. 70,* 355–384.

Cannon, B. and Johannson, B.W. (1980). Nonshivering thermogenesis in the newborn. In *Molecular Aspects of Medicine,* Baum, H. and Gergely, J., eds. Pergamon, Oxford, Vol. 3, pp. 199–223.

Carafoli, E. (1981). Calmodulin in the membrane transport of Ca^{2+}. *Cell Calcium 2,* 353–363.

Cheung, W.Y. (1980). Calmodulin plays a pivotal role in cellular regulation. *Science 207,* 19–27.

Denton, R.M. and McCormack, J.G. (1980). The role of calcium in the regulation of mitochondrial metabolism. *Biochem. Soc. Trans. 8,* 266–268.

Dhalla, N.S. (1976). Involvement of membrane systems in heart failure due to intracellular overload and deficiency. *J. Mol. Cell. Cardiol. 8,* 661–666.

Grist, E.M. and Baum, H. (1975). The mechanisms of the halothane-dependent efflux of calcium from rat liver mitochondria. *Eur. J. Biochem. 57,* 621–626.

Guarnieri, C., Flamigni, F., and Calderera, C.M. (1980). Role of oxygen in the cellular damage induced by reoxygenation of hypoxic heart. *J. Mol. Cell. Cardiol. 12,* 797–808.

Harris, E.J. and Baum, H. (1980). Production of thiol groups and retention of calcium ions by cardiac mitochondria. *Biochem. J. 186,* 725–732.

Harris, E.J., Booth, R.F.G., and Cooper, M.B. (1982). The effect of superoxide generation on the ability of mitochondria to take up and retain Ca^{2+}. *FEBS Lett. 146,* 267–272.

Harris, E.J., Al-Shaikhaly, M., and Baum, H. (1979). Stimulation of mitochondrial calcium ion efflux by thiol-specific reagents and by thyroxine. The relationship to adenosine diphosphate retention and to mitochondrial permeability. *Biochem. J. 182,* 455–464.

Harvey, S.A.K., Booth, R.F.G., and Clark, J.B. (1982). The effects in vitro of hypoglycaemia and recovery from anoxia on synaptosomal metabolism. *Biochem. J. 206,* 433–439.

Harvey, S.A.K., Booth, R.F.G., and Clark, J.B. (1983). The effect of $[Ca^{2+}]$

and [H^+] on the functional recovery of rat brain synaptosomes from in vitro anoxic insult. *Biochem. J.* To be published.

Hearse, D.J., Humphrey, S.M., and Chain, E.B. (1973). Abrupt reoxygenation of the anoxic potassium-arrested perfused rat heart: A study of myocardial enzyme release. *J. Mol. Cell. Cardiol. 5,* 395–407.

Hearse, D.J., Humphrey, S.M., Feuvray, D., and de Leiris, J. (1976). A biochemical and ultrastructural study of the species variation in myocardial cell damage. *J. Mol. Cell. Cardiol. 8,* 759–778.

Hearse, D.J., Humphrey, S.M., and Bullock, G.R. (1978). The oxygen paradox and the calcium paradox: Two facets of the same problem? *J. Mol. Cell. Cardiol. 10,* 641–668.

Hoffstetter, W., Muhlebach, T., Lotscher, H.R., Winterhalter, K.H., and Richter, C. (1981). ATP prevents both hydroperoxide-induced hydrolysis of pyridine nucleotides and release of calcium in rat liver mitochondria. *Eur. J. Biochem. 117,* 361–367.

Kessar, P. and Crompton, M. (1981). The α-adrenergic mediated activation of Ca^{2+} influx into cardiac mitochondria. *Biochem. J. 200,* 379–388.

Kirtland, S.J. and Baum, H. (1972). Prostaglandin E_1 may act as a "calcium ionophore." *Nature New Biol. 236,* 47–49.

Kobayashi, M., Shimomura, Y., Tanaka, M., Katsumata, K., Baum, H., and Ozawa, T. (1979). On the mechanism of Ca^{2+}-induced swelling of rat liver mitochondria. In *Cation Flux across Biomembranes,* Packer, L. and Mukohata, Y., eds. Academic Press, New York and London, pp. 371–376.

Le Quoc, K. and Le Quoc, D. (1982). Control of the mitochondrial inner membrane permeability by sulfhydryl groups. *Arch. Biochem. Biophys. 216,* 639–651.

Lehninger, A.L., Vercesi, A., and Bababunmi, E.A. (1978). Regulation of Ca^{2+} release from mitochondria by the oxidation-reduction state of the pyridine nucleotides. *Proc. Natl. Acad. Sci. USA 75,* 1690–1694.

Loschen, G., Azzi, A., and Flohe, L. (1973). Mitochondrial H_2O_2 formation: Relationship with energy conservation. *FEBS Lett. 33,* 84–88.

Lotscher, H.R., Winterhalter, K.H., Carafoli, E., and Richter, C. (1980a). The energy state of mitochondria during the transport of Ca^{2+}. *Eur. J. Biochem. 110,* 211–216.

Lotscher, H.R., Winterhalter, K.H., Carafoli, E., and Richter, C. (1980b). Hydroperoxide-induced loss of pyridine nucleotides and release of calcium from rat liver mitochondria. *J. Biol. Chem. 255,* 9325–9330.

Means, A.R. and Dedman, J.R. (1980). Calmodulin–An intracellular calcium receptor. *Nature 285,* 73–77.

Nayler, W.G., Poole-Wilson, P.A., and Williams, A. (1979). Hypoxia and calcium. *J. Mol. Cell. Cardiol. 11,* 683–706.

Nedegaard, J. and Cannon, B. (1980). Effects of monovalent cations on Ca^{2+} transport in mitochondria; a comparison between brown fat and liver mitochondria from the rat. *Acta Chem. Scand. B34,* 149–151.

Nicholls, D.G. and Crompton, M. (1980). Mitochondrial calcium transport. *FEBS Lett. 111,* 261–268.

Nohl, H. and Hegner, D. (1978). Do mitochondria produce oxygen radicals in vivo? *Eur. J. Biochem. 82,* 563–567.

Rugolo, M., Siliprandi, D., Siliprandi, N., and Toninello, A. (1981). Parallel efflux of Ca^{2+} and Pi in energized rat liver mitochondria. *Biochem. J. 200,* 481–486.

Schanne, F.A.X., Kane, A.B., Young, E.E., and Farber, J.L. (1979). Calcium dependence of toxic cell death: A final common pathway. *Science 206,* 700–702.

Serhan, C., Anderson, P., Goodman, E., Dunham, P., and Weissman, G. (1981). Phosphatidate and oxidised fatty acids are calcium ionophores. *J. Biol. Chem. 256,* 2736–2741.

Shen, A.C. and Jennings, R.B. (1972a). Myocardial calcium and magnesium in acute ischaemic injury. *Am. J. Pathol. 67,* 417–440.

Shen, A.C. and Jennings, R.B. (1972b). Kinetics of calcium accumulation in acute myocardial ischaemic injury. *Am. J. Pathol. 67,* 441–451.

Weltzien, H.U. (1979). Cytolytic and membrane-perturbing properties of lysophosphatidylcholine. *Biochim. Biophys. Acta 559,* 259–287.

Wiswedel, I., Barnstorf, U., Augustin, W., Holmuhamedov, E., Medvedeu, B., and Evtodienko, Y. (1982). Involvement of periodic deacylation-acylation cycles of mitochondrial phospholipids during Sr^{2+}-induced oscillatory ion transport in rat liver mitochondria. *Biochim. Biophys. Acta 688,* 597–604.

Wolkowicz, P.E. and McMillin Wood, J. (1980). Dissociation between mitochondrial calcium ion release and pyridine nucleotide oxidation. *J. Biol. Chem. 255,* 10348–10353.

Zoccarato, F. and Nicholls, D.G. (1982). The role of phosphate in the regulation of the independent calcium efflux pathway of liver mitochondria. *Eur. J. Biochem. 127,* 333–338.

Part III

Physiological Changes Affecting Membrane Properties

10 Oxidative Damage in Exercising Muscle and Mitochondria

L. Packer and A.T. Quintanilha

For several years, we have been interested in the role of biological oxidation and mitochondria in the mechanism of oxidative damage and aging. We will illustrate some of our studies by describing the methods, systems, and probes that we are currently developing to assay for free radical and other damage to tissues and bioenergetic membranes.

Although oxidations in biological systems are essential for energy conversion and metabolism, they are also dangerous. They have been shown to result in low levels of accumulative damage during the life span of many animals and are probably responsible for aging and, in many cases, for accelerated aging and pathology (Packer and Walton 1977). When the normal process of aging is enhanced by the intervention of pollutants in the environment, oxidations are probably also involved in the enhanced damage to normal metabolic systems. The data in Fig. 10.1 show that oxygen is dangerous to human cells growing in culture (Packer and Fuehr 1977). This experiment, which took almost 1 yr to complete, showed that normal human diploid cells cultured under 10% oxygen grew better and lived longer than cells cultured under 20% oxygen. All concentrations of oxygen above 20% are extremely dangerous to the cells. In cells grown under 20% oxygen there is an accumulation of damaged products that can be observed by fluorescence microscopy. Although the accumulation of these products may not, in itself, be dangerous, the process leading to their accumulation probably is. Fluorescence-damaged products also accumulate in human cells if the cells are maintained on a diet in which the normal content of serum has been reduced. When the serum concentration was lowered from 10 to 0.1%, the cells maintained themselves but during a period of 7 wk the accumulation of fluorescence-damaged products increased significantly. However, if under the same growth conditions, they were fortified with 10 times the normal serum levels of dl-α-tocopherol (vitamin E), none of the fluorescence-damaged products were observed (Deamer and Gonzales 1974). In addition, Cutler's laboratory has found (Tolmasoff et al. 1980) that the life spans of a variety of species bear a very close correlation to the specific metabolism rate when corrected for by the superoxide dismutase activity.

Life span studies in human cells, however, are difficult because the amount

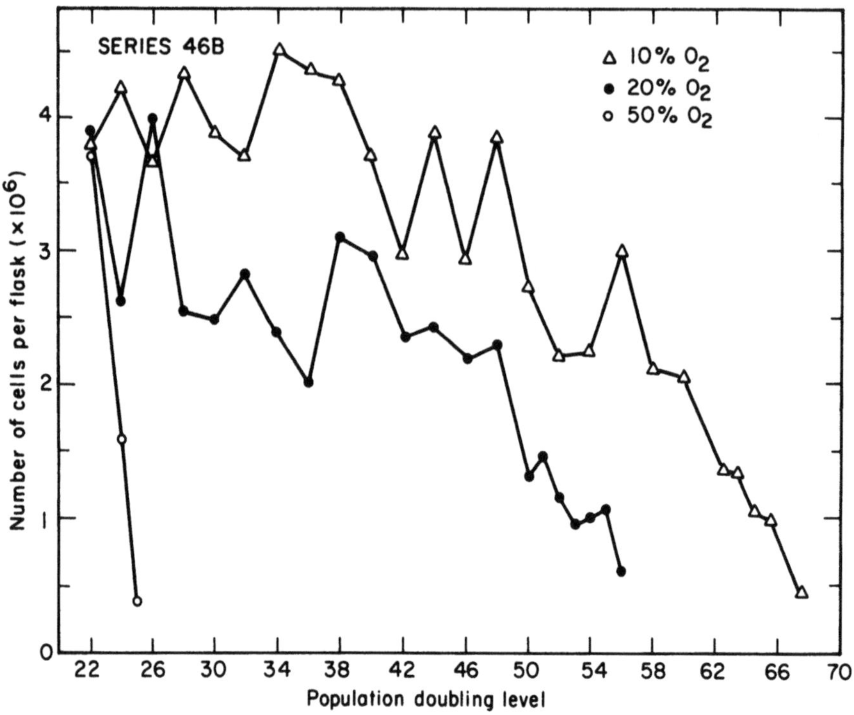

Figure 10.1. Limited in vitro life span of human WI-38 cells in culture as influenced by the ambient oxygen tension.

of available biological material is limited. Hence, we have sought to develop a model system with which we could answer questions at the physiological as well as the tissue level. The test system we chose, in a collaboration between George Brooks and Kelvin Davies of the Exercise Physiology Laboratory at the University of California, Berkeley, and other members from our group, particularly John Maguire and Rolf Mehlhorn, and Peter Dallman of the University of California, San Francisco, was that of exercise in animals. Using this system, we have been studying how metabolism might be involved in oxidative damage.

In this test system animals are exercised by running on a treadmill at an angle of about 15 deg. The animals are induced to move by electrodes situated at the rear of the cage, which, in case they decide to slow down or stop, persuade them to keep moving. By regulating the speed of the treadmill, it is possible to change the workload delivered to the animal. While this is occurring, we also monitor respiratory changes, heart rate, and other physiological parameters and analyze the data on a computer.

We have used endurance training and sprint training. Endurance training involves 10 wk of training for 5 days/wk. The animals have the weekends free, but they are required, over the 10-wk period, to build up the amount of time they have to run to 120 min/day at the relatively slow speed of 1 mph. In the sprint training, they work 7 days/wk for 5 min/day, but at a speed of 3 mph, which is rather fast for rats and near their maximum workload. Control rats are placed in the chambers for short periods of time to learn about the electric shock system.

We observed that as the speed of the treadmill is increased, the amount of oxygen consumption increases with the workload of the animals. To test endurance, both trained and untrained animals ran at 1 mph for as long as they could. The results that are obtained from endurance-trained animals are rather spectacular (Davies et al. 1981). The mitochondrial content of the muscle increased by 100%, indicating a biosynthesis of mitochondrial material as a result of this training regimen. Oxygen consumption during exercise increased by about 15%. The endurance capacity, on the other hand, showed the most dramatic differences: Trained animals could run more than five times longer than untrained animals (Table 10.1). The increase in endurance appears to be correlated to an increase in the amount of mitochondria in the muscles.

Comparable experiments have been carried out with animals that have undergone sprint training. In sprint training, the oxygen consumption increases by about 15%, the same increase, more or less, as was observed in endurance-trained animals. However, the endurance capacity of these animals is unchanged and there is no evidence of increased biogenesis of mitochondria.

The importance of mitochondria and energetics can also be investigated in the exercise system through the intervention of dietary deficiency. With dietary manipulation, we can attempt to understand how events that occur at the cellular level and in cellular energetics might be related to the oxygen delivery system.

We fed grown rats for a period of about 2 mo on a diet in which the iron content was about 2.5 mg/kg of diet, less than the minimal normal iron content of about 50 mg/kg. The resulting iron deficiency caused a very large difference in the ability of these animals to perform. If one imposes a progressively increasing workload, iron deficient animals show a much lower ability to consume greater amounts of oxygen than do control animals (Davies et al. 1982).

Table 10.1 Endurance Training

Group	Mitochondrial Content of Muscle (mg/g)	Muscle Cytochrome Oxidase (Mole/min/g)	VO_2 max ($mg \cdot kg^{-1} \cdot min^{-1}$)	Maximal Endurance (min)
Control	18.2 ± 0.7	37.4 ± 3.1	76.6 ± 1.2	36.3 ± 2.2
Endurance	36.2 ± 1.6[a]	75.9 ± 5.4[a]	87.7 ± 2.0[a]	182.6 ± 10.4[a]

[a] $p < 0.01$ controls versus endurance trained (t test); $N = 10$.

Iron deficiency also results in large changes in mitochondrial function. Mitochondrial pyruvate and malate oxidase, for example, are much reduced. Considering that such activity depends on complex I and complex III of the respiratory chain that contain iron-sulfur centers, this decrease is not surprising. Similarly, the capacity of the deficient animals to consume oxygen is about 50% less than the normal level. Again, the most spectacular finding is that the endurance capacity of the animals has been reduced by more than 90%.

In a subsequent experiment, we reintroduced iron into the diet and followed, over a period of 1 wk, the recovery of the various parameters we have been measuring. Two parameters that are closely correlated are oxygen consumption and hematocrit values. Two other parameters, oxidation of pyruvate by the isolated mitochondria and endurance, are also closely correlated. Again, this suggests that endurance is associated with increased biogenesis of mitochondria, whereas the defects in oxygen consumption have more to do with the oxygen delivery system. In order to prove this unambiguously, we exchange-transfused the iron-deficient animals with blood from normal animals (Davies et al.). In control animals, there are no changes in any of the parameters as a result of blood transfusion. In the case of the iron-deficient animals, there are also no changes in their much-reduced endurance capacity. However, the hemoglobin level is largely restored to normal levels, and oxygen consumption capacity is increased to near the level of the control. Since we had created lesions in the iron-sulfur centers of the respiratory chain, we did not expect to restore the oxygen consumption completely. This experiment showed the relative contribution of the blood in the oxygen delivery system to animal work capacity.

We have also been interested in vitamin E deficiency, which has been known to create nutritional muscular disease. In our experiment, animals were grown on diets deficient in vitamin E and were then tested for parameters of work performance at the whole animal level. Although little change occurred in most of the parameters, the endurance capacity of the deficient animals was reduced by almost 50%. Therefore we would expect that some damage had occurred at the level of the muscle.

Using electron spin resonance (ESR) techniques, whole tissues and homogenates of various tissues have been studied to investigate the stable free radical content of the tissues of control-, exercised-, and vitamin-E-deficient animals. In every case, it was found that exercise-exhausted rats showed an increase in free radical content of up to twofold (Davies et al. 1978). In vitamin-E-deficient animals, we found a larger content of free radicals than in normal animals; this content increased further after exercise. The source of these radicals remains largely unindentified even though more recent studies indicate that they may arise from the ubisemiquinone species in mitochondria (Davies and Hochstein 1982).

During the last few years we have attempted to trace pathways of oxidative damage in relation to the mitochondrial respiratory chain. Figure 10.2 summa-

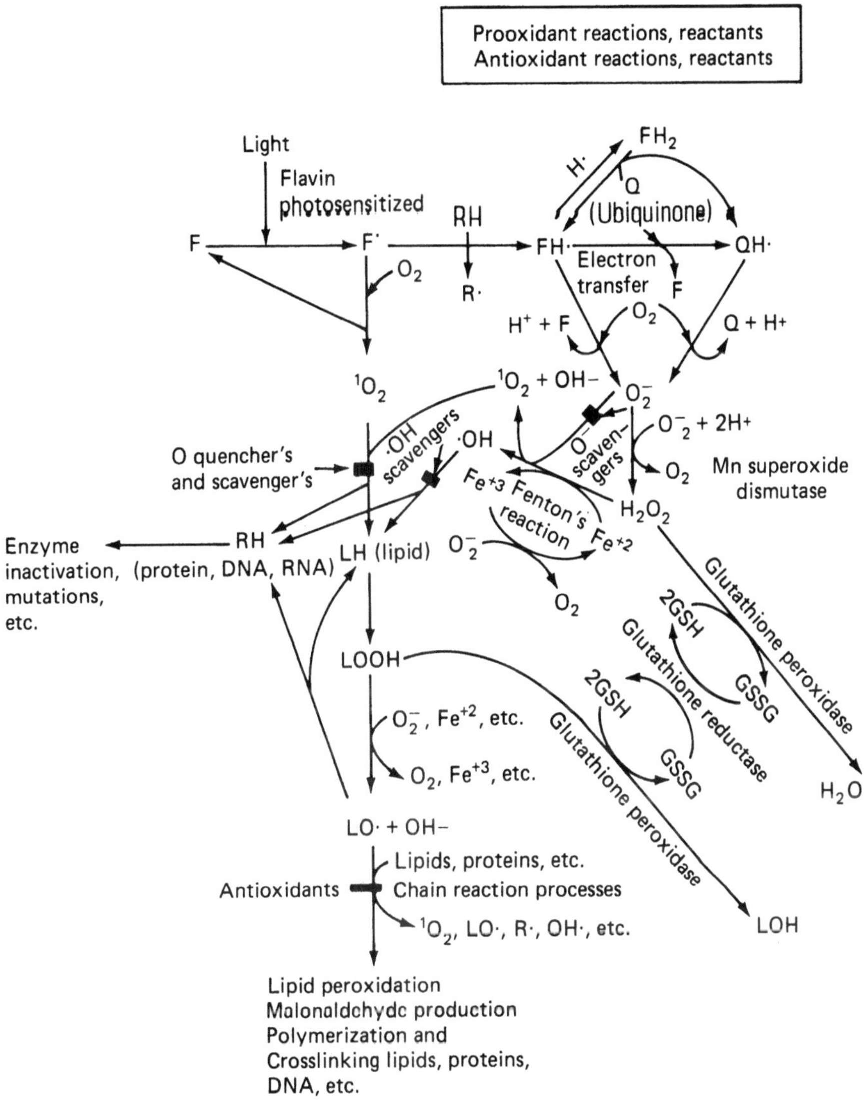

Figure 10.2. Pathways of oxidative damage in mitochondria. Prooxidants, Fe^{+2}, Cu^{+2}, free flavins and hemes, light; antioxidants, α-tocopherol, glutathione, ascorbic acid; superoxide scavengers, ascorbic acid, glutathione, cytochromes, α-tocopherol; singlet oxygen scavengers and/or quenchers, β-carotene, α-tocopherol, histidine, tryptophan, guanine, uric acid; hydroxyl radical scavengers, most organic compounds $k \cong 10^9\ M^{-1}\text{-}s^{-1}$.

rizes the various pathways whereby oxygen radicals might arise as a result of the respiratory chain. Known protective mechanisms are also indicated. Catalase has been omitted, in this case, because of its probable absence in mitochondria. Among the protective substances are glutathione, ascorbic acid and dl-α-tocopherol. These antioxidants prevent the accumulation of aqueous and lipid radicals arising from lipid peroxidation. This pathway shows the formation of superoxide ions (O_2^-) and their dismutation by superoxide dismutase to form hydrogen peroxide. Inorganic hydrogen peroxide can be decomposed by chemical reactions, such as the Fenton reaction, catalyzed by iron or other metals to give rise to dangerous hydroxyl ($H\dot{O}$) radicals. A comparable reaction occurs in organic radicals. Organic hydroperoxides (arising from the addition of oxygen to lipid radicals) are decomposed to form alkoxy radicals, which are suspected to be damaging.

We have investigated the mechanisms of photooxidative damage induced by visible light absorbed by the different chromophores of the mitochondria (Aggarwal et al. 1978). Photosensitized radical mediated damage can be monitored by using some spin-labeled probes to demonstrate the generation of free radicals in this system. Stable nitroxide spin-labels such as TEMPOL can, in principle, interact with other radicals by a single electron exchange (spin reduction or spin oxidation) or by a spin-destruction mechanism in which a radical adduct is formed. In principle, both kinds of processes can occur: spin reduction or spin destruction. Uncharged radicals such as TEMPOL are permeable across membranes. Mehlhorn has shown that if mitochondrial membranes are exposed to light in the presence of TEMPOL (Fig. 10.3), a completely light-dependent, rapid spin destruction occurs (Mehlhorn and Packer 1982). Under these conditions more spin destruction than spin reduction occurs. This can be tested by attempting to reoxidize the spin label. If the spin label has been reduced rather than decomposed, we observe a good measure of recovery of the signal upon addition of ferricyanide, thus demonstrating the extent of spin reduction. The residual signal that is not recovered is evidence of the degree to which spin has been destroyed. The action spectrum for spin destruction coincides with the absorption spectrum of flavin mononucleotide (compare Augusto and Packer 1981; Quintanilha and Davies 1982). It is clear that in this test system, radical generation has probably arisen as a result of light absorption by flavins and flavin-initiated radicals that have been generated in this test system.

Using this test system in vitamin-E-deficient animals, we have investigated some of the bioenergetic parameters in the isolated skeletal-muscle mitochondria. These mitochondria develop transmembrane potentials when energized by adding substrate as pyruvate-malate. We have shown that such potentials are progressively lost as the mitochondria are incubated in vitro over a period of several hours (Quintanilha et al. 1982) and that the process is accelerated in the presence of visible light. Mitochondria from vitamin-E-deficient animals are more suspectible to loss of transmembrane potential than are the controls. Liver mitochondria, interestingly, show similar but slower changes.

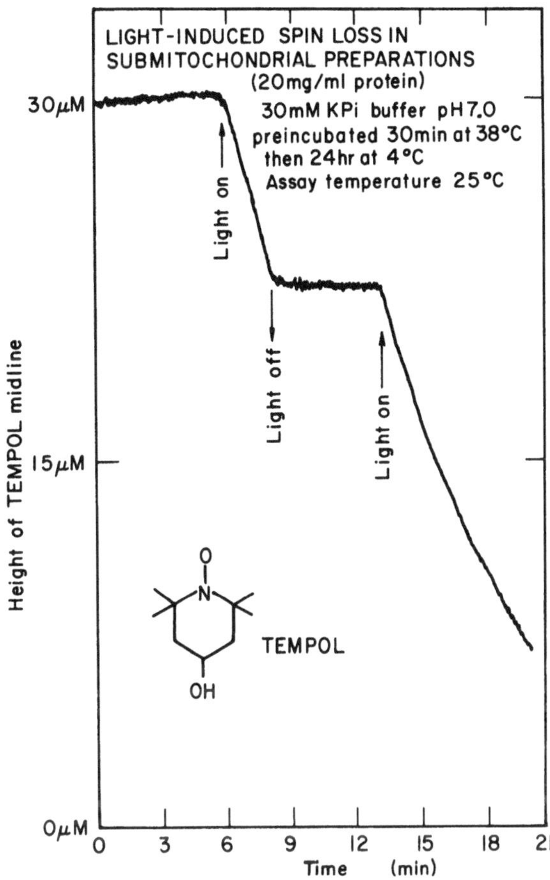

Figure 10.3. Light-induced loss of the ESR signal due to spin destruction and reduction in mitochondria.

Transmembrane potentials have been measured in this experiment by using the permeable cationic tetraphenyl phosphonium (TPP^+) probe and a TPP^+ electrode. This molecule is permeable across membranes because of its lipophilic character, but it senses the membrane potential because it possesses a permanent charge. We measure the transmembrane potential by allowing the probe to equilibrate between the inside and the outside.

Vitamin E deficiency also causes changes in membrane surface charge in mitochondria (Quintanilha et al. 1982). This charge can be investigated by using charged amphipathic spin probes (Mehlhorn and Packer, 1979). In this case, the molecule is impermeable because of its bare positive charge: Because it has a hydrophobic tail 12 carbons in length, it partitions between the outer exposed membrane surface and the aqueous phase. Mehlhorn has made these labels that vary in length from 0 to 22 carbons. By varying the hydrophobic tail length,

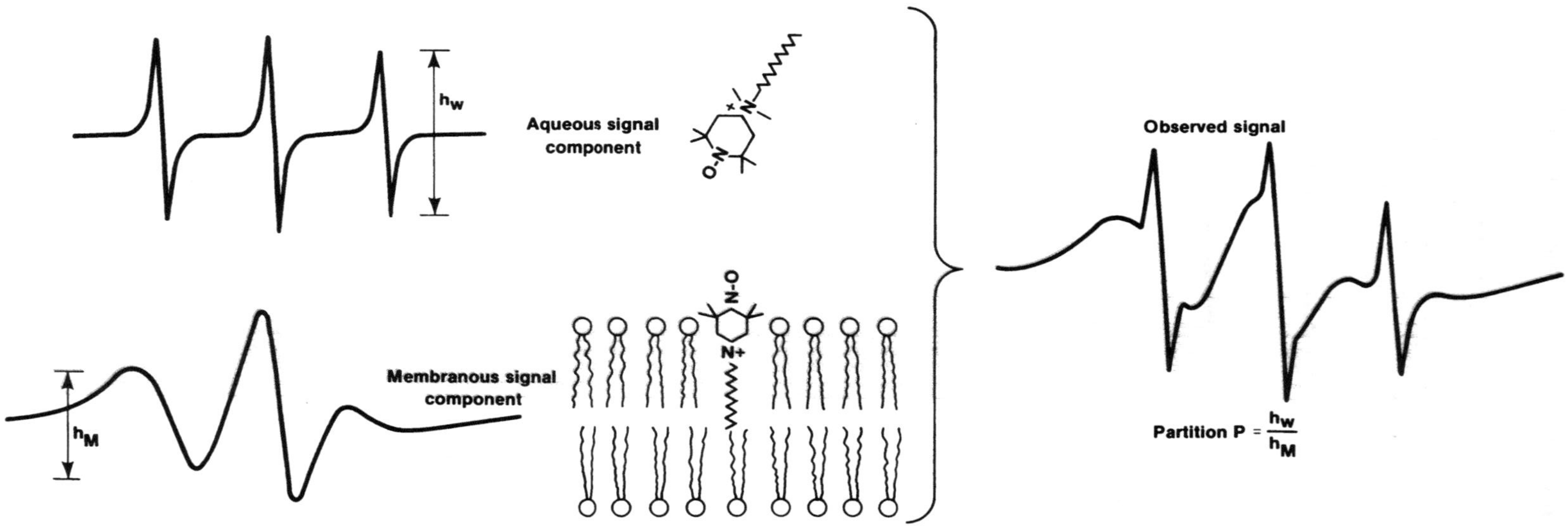

Figure 10.4. Magnetic resonance spectrum of the charged amphipathic (CAT) spin probe when added to an aqueous suspension of membranes. The partition *P*, is a measure of the surface tension charge density; the smaller the value of *P*, the greater the negative charge on the membrane surface.

we can vary the degree of partitioning of the label between the aqueous medium and the membrane. If the label is in the aqueous medium, it shows a narrow sharp ESR spectrum; if it is associated with the membrane, the spin signal is much broader (Fig. 10.4). In the presence of membranes, a composite ESR signal is obtained in which a narrow aqueous component is superimposed on a broader component. If more negative charge appears at the surface of the membrane, the label will be taken up from the aqueous medium onto the membrane to balance the surface charge. The reverse would happen if positive charge were accumulating at the outer surface. We have consistently observed about a 10% increase in negative surface charge density in vitamin-E-deficient mitochondrial membranes. To our knowledge, this is the first time anyone has found a membrane structural affect resulting from vitamin E deficiency. Vitamin E is normally present in membranes in only very small amounts in relation to phospholipids. In mitochondrial membranes, the vitamin E content is perhaps as high as in any natural membrane, about one molecule per 200 phospholipids. It is, therefore, very difficult to detect effects on the membrane structure directly. We find that with in vitro incubation the negative surface charge increases as the result of vitamin E deficiency (Quintanilha et al. 1982). In addition, the muscle mitochondria always have a more negative surface charge density to begin with than liver mitochondria. More recent studies indicate that greater negative surface charge may be correlated with increased lipid peroxidation.

Other changes have also been noted in isolated mitochondria and homogenates from muscle of vitamin-E-deficient animals. Since the transmembrane potential of the mitochondria is decreased as the result of vitamin E deficiency, it is not surprising that the respiratory control ratio also diminishes. One also finds an accumulation of lipid peroxidation products, measured by the formation of malondialdehyde, which agrees with the observation that the tissue stable free radical content is increased as a result of vitamin E deficiency. Lysosomal latency is decreased as a result of vitamin E deficiency. Therefore, it is not surprising, in view of the preceding findings, that vitamin-E-deficient animals have a lowered endurance capacity. It should be noted, however, that such animals can work for short periods of time at high efficiency.

Concluding Statement

There are different kinds of pollutants in our environment, and the ways in which they create damage to the biological system will, therefore, vary considerably. Some environmental substances can be expected to cause damage by means of free radical mediated oxidations at the level of the membrane. We need a variety of tests to understand their mechanism and impact on the biological system. Our group has chosen to use a system that is physiological, reliable, and sensitive and that can give information about the bioenergetic parameters at both the cellular and whole animal levels.

Acknowledgments

We wish to acknowledge financial support from the National Foundation for Cancer Research and Hoffman La Roche.

References

Aggarwal, B.B., Quintanilha, A.T., Cammack, R., and Packer, L. (1978). Damage to mitochondrial electron transport and energy coupling by visible light. *Biochim. Biophys. Acta 502,* 367-382.

Augusto, O. and Packer, L. (1981). Selective inactivation of microsomal drug metabolizing proteins by visible light. *Photochem. Photobiol. 33,* 765-768.

Davies, K.J.A., Donovan, C.M., Refino, C.J., Brooks, G.A., Packer, L., and Dallman, P.R. Iron deficiency and work capacity: Distinguishing effects of anemia and muscle iron deficiency on exercise bioenergetics in the rat. *Am. J. Physiol.,* in press.

Davies, K.J.A. and Hochstein, P. (1982). Ubisemiquinone radicals in liver: Implications for a mitochondrial Q cycle in vivo. *Biochem. Biophys. Res. Commun. 107,* 1292-1299.

Davies, K.J.A., Maguire, J.J., Brooks, G.A., Dallman, P.R., and Packer, L. (1982). Muscle mitochondrial bioenergetics, oxygen supply and work capacity during dietary iron deficiency and repletion. *Am. J. Physiol. 242,* E418-E427.

Davies, K.J.A., Packer, L., and Brooks, G.A. (1981). Biochemical adaptation of mitochondria, muscle, and whole-animal respiration to endurance training. *Arch. Biochem. Biophys. 209,* 538-553.

Davies, K.J.A., Quintanilha, A.T., Brooks, G.A., and Packer, L. (1978). Free radicals and tissue damage produced by exercise. *Biochem. Biophys. Res. Commun. 107,* 1198-1205.

Deamer, D.W. and Gonzales, J. (1974). Autofluorescent structures in cultured W1-38 cells. *Arch. Biochem. Biophys. 165,* 421-426.

Mehlhorn, R.J. and Packer, L. (1979). Membrane surface potential measurements with amphiphilic spin labels. In *Methods in Enzymology,* vol. LVI–Biomembranes; Part G–Bioenergetics: Biogenesis of Mitochondria, Organization and Transport, Fleischer, S. and Packer, L., eds. Academic Press, New York, pp. 613-621.

Mehlhorn, R.J. and Packer, L. (1982). Nitroxide destruction and flavin-photosensitized damage in inner mitochondrial membranes. *Canad. J. Biochem. 60,* 1452-1462.

Packer, L. and Fuehr, K. (1977). Low oxygen concentration extends the lifespan of cultured human diploid cells. *Nature 267,* 432-433.

Packer, L. and Walton, J. (1977). Do free radicals speed and antioxidants slow aging and cancer? *Chemtech 7,* 276-281.

Quintanilha, A.T. and Davies, K.J.A. (1982). Vitamin E deficiency and photosensitization of electron-transport carriers in microsomes. *FEBS Lett. 139,* 241–244.

Quintanilha, A.T., Davies, K.J.A., Szyszlo-Davies, J.M., Racanelli, T., and Packer, L. (1982). Membrane effects of vitamin E deficiency: Bioenergetic and surface charge density studies of skeletal muscle and liver mitochondria. In *Vitamin E: Biochemical, Hematological, and Clinical Aspects,* Lubin, B. and Machlin, L., eds. *Ann. N.Y. Acad. Sci. 393,* 32–47.

Tolmasoff, J.M., Ono, T., and Cutler, R.G. (1980). Superoxide dismutase: Correlation with life-span and specific metabolic rate in primate species. *Proc. Natl. Acad. Sci. USA 77,* 2777–2781.

11

Early Events in Stimulus–Response Coupling Using Neutrophils as a Model System

R.I. Sha'afi and P.H. Naccache

Polymorphonuclear leukocytes (neutrophils), by presenting the first line of defense against foreign and pathogenic elements and by removing endogenous tissue debris from various areas, fulfill extremely important roles in the response of the organism to acute, and possibly chronic, inflammatory stimuli (Murphy 1976). Reduced resistance to infectious disease has, accordingly, been related to impaired leukocytic functions. Neutrophils detect, move toward, and accumulate at sites of injury or infection. This phenomenon, elicited upon the detection of soluble factors, is termed directed locomotion of chemotaxis. Most neutrophil stimuli are chemotactic factors, that is, they cause neutrophils to move up a concentration gradient of the stimulus.

Once at the site of infection, neutrophils ingest or phagocytize the desired particles. This process is followed by the discharge of the neutrophils' various lysosomes and granules into the phagocytic vacuole and the activation of their oxidative metabolism, thereby promoting the degradation and digestion of the engulfed particle or organism.

Chemotaxis, phagocytosis, degranulation, and aggregation with the attendant changes in cell shape, all depend on the mechanical displacement of part of or the whole cell. Mechanochemical coupling is thus an essential component of each of the neutrophil's functions.

Currently there is a great deal of interest in trying to identify and sequence the biophysical and biochemical events that lead to neutrophil activation following the stimulation of the cells by chemotactic factors [the natural chemotactic factor C5a, the small molecular weight fragment of the fifth component of complement, and the synthetic peptide formyl-methionyl-leucyl-phenylalanine (f-Met-Leu-Phe)]. This problem is commonly known in the area of cell physiology as "excitation-response" coupling. In the pursuit of this, it is important to identify and differentiate primary events–events that lie on the main sequence leading to cell activation–from those that are secondary events that are incidental and not absolutely required.

It is generally agreed that chemotactic factors activate neutrophils by interaction with membrane-bound receptors of which two types, high-affinity and low-affinity states, may be present (Aswanikunar et al. 1977; Sha'afi et al. 1978; Chenoweth and Hugli 1979; Showell et al. 1976; Mackin et al. 1982). This conclusion is based on various experimental findings. First, there is a great deal of specificity in the biological responses among closely related synthetic oligo peptides. Second, high- and low-affinity sites for radiolabeled formylated peptide have been detected on neutrophil membranes by equilibrium binding analysis. Third, there is apparent desensitization; that is, preincubation with high concentration of chemotactic factors decreases cell response to subsequent stimulation and also decreases the number of demonstrable binding sites. Fourth, there is evidence for receptor internalization following the binding with the formylated peptide.

Calcium Ion as the Second Messenger in Neutrophil Activation

Because of the need for mechanochemical transduction in neutrophil functions and of the precedent and analogy for skeletal muscle, it has long been assumed that calcium ions were intimately related to neutrophil activation, perhaps as the second messenger. If so, then a minimum of four experimental conditions must be satisfied. First, cell activation should be modulated by the concentration of calcium in the suspending medium. Second, introduction of calcium into the cell by means other than the first messenger should activate the cell. Third, inhibitors that antagonize the activity of the intracellular receptor of calcium (calmodulin) should inhibit cell activation. Fourth, the interaction of the first messenger, with its receptors, should lead to an increased level of the intracellular exchangeable calcium.

Before stating the experimental evidence supporting the view that calcium is the intracellular second messenger in the neutrophil, it is important to discuss the various mechanisms involved in regulating the intracellular concentration of this cation. The regulation of the intracellular concentration of calcium in the neutrophils where mitochondria are relatively scarce is achieved by pump–leak systems at the plasma membrane and by binding of Ca^{2+} by cytoplasmic constituents and plasma membrane. In Fig. 11.1, we depict a schematic representation of the processes possibly involved in the regulation of calcium metabolism in neutrophils.

As shown in this model, two separate channels by which Ca^{2+} could leak into the cells may exist. The first channel, which is denoted by 1, is independent of membrane potential, whereas the second channel, which is denoted by 2, is controlled by the membrane potential. Although the presence of the first channel can be demonstrated experimentally in the neutrophils, there is no evidence to support the presence of the second channel. In addition, there are at least two different energy-dependent mechanisms (3 and 4) that control the rate and extent of Ca^{2+} efflux. The mechanism denoted by 3 is a specific calcium pump driven by the hydrolysis of ATP by "Mg^{2+}, Ca^{2+}"-activated ATPase, the presence

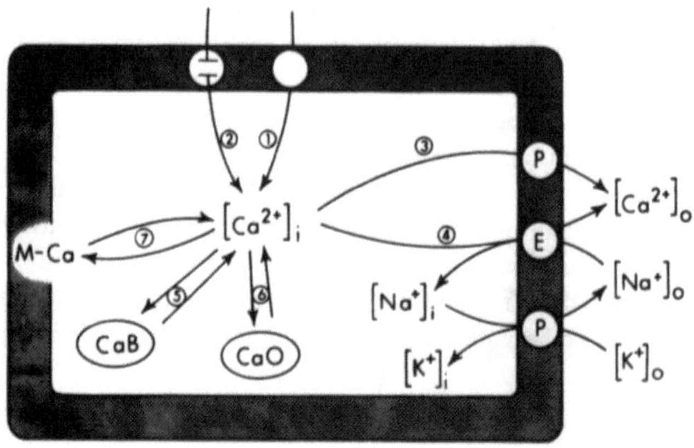

Figure 11.1. A schematic representation of the various possible processes involved in regulating cell calcium in neutrophils. These possible processes are identified as follows: (1) passive calcium leak; (2) passive leak controlled by membrane potential; (3) specific calcium pump that derives its energy from adenosine triphosphate (ATP) by membrane-associated calcium activated adenosine triphosphatase (ATPase); (4) Na^+-Ca^{2+} exchange that is dependent on the activity of the "Na^+, K^+" activated-ATPase for the maintenance of the Na^+ gradient across the cell membrane; (5) calcium binding to intracellular proteins and other cellular compounds (CaB); (6) calcium binding and sequestering by intracellular granules and organelles (CaO); (7) calcium binding to plasma membrane (M-Ca).

of which has been demonstrated in the plasma membrane of neutrophils and inside-out membrane vesicles (Volpi et al. 1982; Jackowski et al. 1979). The second one identified by 4 is a Na^+ influx, Ca^{2+} efflux exchange pump. This component depends on maintaining the Na^+ gradient across the cell membrane by the classical "Na^+, K^+" pump. Using inside-out membrane vesicles prepared from rabbit neutrophils, we were able to demonstrate the presence of a Ca^{2+}-efflux pump but not a Na^+-Ca^{2+} exchange mechanism directly. In addition, there is no evidence of a dependence of Ca^{2+} efflux on the presence of extracellular Na^+ in rabbit neutrophils.

In addition to these membrane events, control of cytosol Ca^{2+} is also dependent on binding to the membrane (M-Ca), buffering by cytosol constituents such as soluble proteins and others (CaB), and the accumulation into intracellular organelles and/or granules (CaO).

It is generally agreed that calcium ion plays the role of the second messenger in neutrophil activation by chemotactic factors. This conclusion is based on the results of functional studies both in the presence and absence of calcium or calmodulin inhibitors in the suspending medium, tracer measurements, fluorescence studies, the use of divalent cation ionophores, electron microscopic visualization, and other techniques and experimental manipulations (for review see

Sha'afi and Naccache 1981). In addition, the time course, concentration dependence, and receptor specificity of the calcium mobilization induced by the stimuli are consistent with its early role in the initiation of the various neutrophil functions. Thus, a model has been developed according to which the activation of the neutrophil is brought about by the mobilization of Ca^{2+}, either from internal stores, by displacement of previously bound calcium from membranous and other internal compartments, or from the extracellular medium by increasing the plasma membrane permeability to Ca^{2+}.

Chemotactic factors such as the synthetic peptide f-Met-Leu-Phe and the small molecular weight fragment of the fifth component of complement (C5a) have been shown to increase rapidly the intracellular level of exchangeable calcium in neutrophils. This increase in the intracellular level of exchangeable calcium comes from two sources. The first is an extracellular source manifested indirectly by increases in the initial rate of ^{45}Ca influx (Fig. 11.2), an increase in the steady-state level of cell-associated ^{45}Ca in the presence of extracellular calcium concentration greater than 50 μM (Fig. 11.3), and finally by the enhancement of the functional responsiveness of neutrophils that is observed in the presence of calcium as compared to that seen in its absence. The second is an

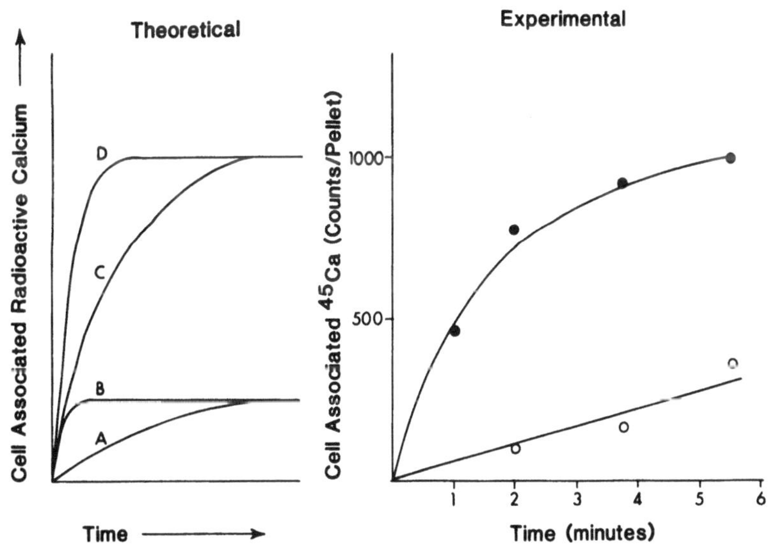

Figure 11.2. Time course of calcium uptake by neutrophils. The letters in the theoretical case denote the following: A, control; B, an increase in membrane permeability; C, an increase in the level of intracellular exchangeable calcium with no change in membrane permeability; D, an increase in both membrane permeability and intracellular level of exchangeable calcium. In the experimental case: O–O, control conditions; ●–●, in the presence of 10^{-9} M f-Met-Leu-Phe. The outside calcium concentration is 0.5 mM and the pellet represents 5×10^6 cells.

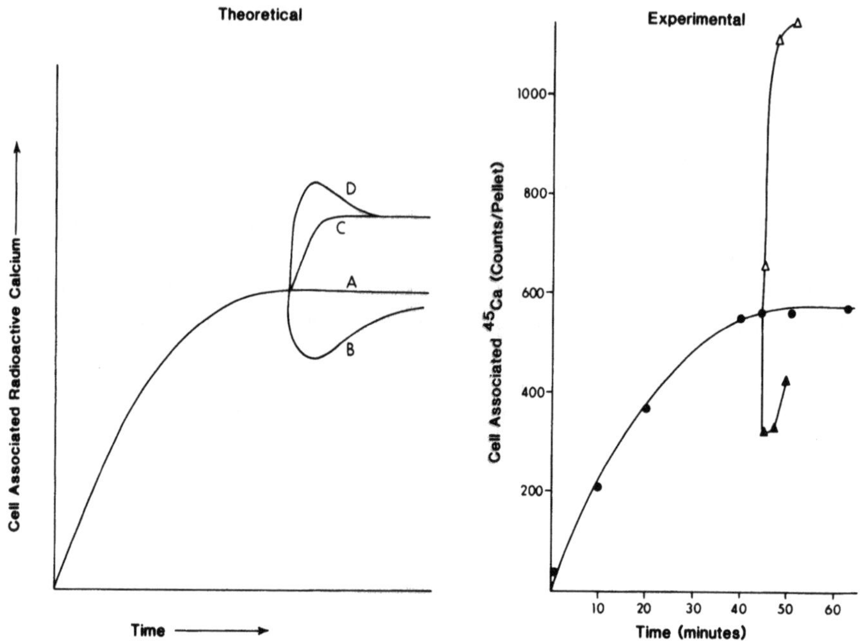

Figure 11.3. Effect of the stimulus on the steady-state level of cell-associated radioactive calcium in the presence of 0.5 mM outside calcium. The cell suspensions are incubated in the presence of ^{45}Ca for 45 min, a time long enough to reach steady state. The stimulus is then added while the cells are still incubated with ^{45}Ca. The letters in the theoretical case denote the following: A, no change in steady-state condition; B, an increase in the level of intracellular exchangeable calcium as the result of release of previously nonexchangeable calcium without an increase in membrane permeability; C, an increase in the level of intracellular calcium as a result of a net uptake of calcium from the bathing medium without any increase in membrane permeability; D, an increase in both membrane permeability and the level of intracellular exchangeable calcium can be due to net uptake or intracellular redistribution. In the experimental case: O–O, control conditions; Δ–Δ in the presence of 10^{-8} M f-Met-Leu-Phe; ▲–▲ in the presence of 10^{-11} M f-Met-Leu-Phe. The pellet represents 5×10^6 cells.

intracellular source experimentally detected by measuring one of three parameters: (1) the steady-state level of cell associated ^{45}Ca in the presence of low (< 50 μM) concentrations of extracellular calcium (Fig. 11.4), (2) the rate of ^{45}Ca efflux from preloaded cells (Fig. 11.5), and (3) the fluorescent characteristics of chlorotetracycline, a fluorescent chelate probe sensitive to the presence or absence of calcium in the environment (Fig. 11.6).

The results in Fig. 11.6 also suggest that the two best described neutrophil chemotactic factors, f-Met-Leu-Phe and C5a, appear to interact with the same pool of membrane calcium and to release it to the cytoplasmic side of the plasma membrane. Intracellular calcium then appears to bind back to the membrane(s) from where it can be displaced by additional stimulation. Recently, we have shown (Molski et al. 1983) that the chemotactic factor-induced changes in calcium fluxes reflect intracellular events and that the pool of calcium mobilized by f-Met-Leu-Phe and the increase in cell-associated calcium upon stimulation by the latter are insensitive to the presence of ethylene glycoltetraacetate (EGTA) (Fig. 11.7).

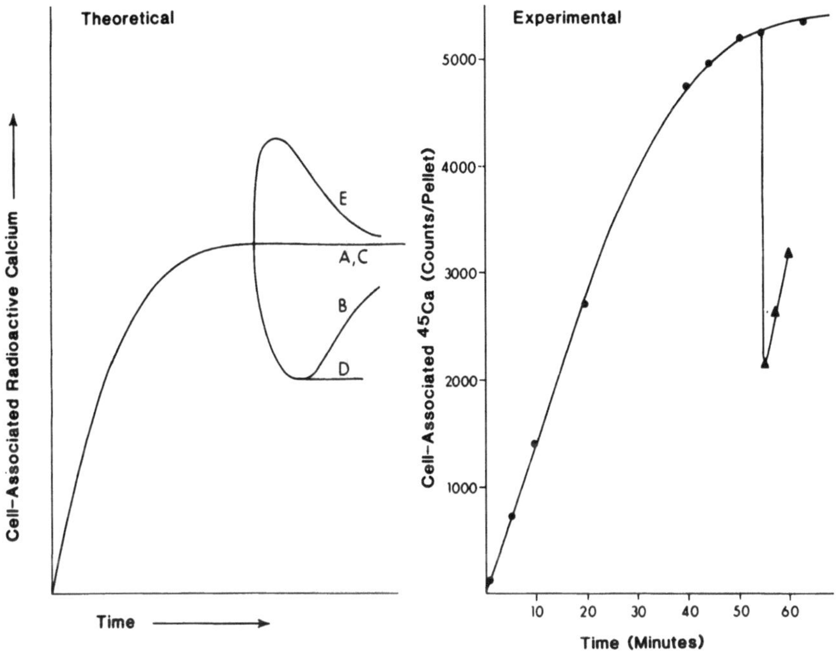

Figure 11.4. Effect of the stimulus on the steady-state level of cell-associated radioactive calcium in the absence of any external calcium ($< 5 \times 10^{-6}$ M). The cell suspensions are incubated in the presence of ^{45}Ca for 45 min, a time long enough to reach steady state. The stimulus is then added while the cells are still incubated with ^{45}Ca. The letters in the theoretical case denote the following: A, no change in steady state; B, an increase in the level of intracellular exchangeable calcium as the result of release of previously nonexchangeable calcium with or without an increase in inward membrane permeability to calcium as long as this increase is not greater than the rate of outward calcium movement; C, an increase in membrane permeability without a change in steady-state level; D, direct activation of calcium pump with or without a redistribution of intracellular calcium—the final steady-state level depends on the degree of pump activation; E, an increase in the level of intracellular exchangeable calcium as the result of release of previously nonexchangeable calcium and an increase in the inward membrane permeability to calcium as long as this increase in membrane permeability is much greater than the rate of outward calcium movement. In the experimental case: ●–●, control conditions; ▲–▲ in the presence of 10^{-8} M f-Met-Leu-Phe.

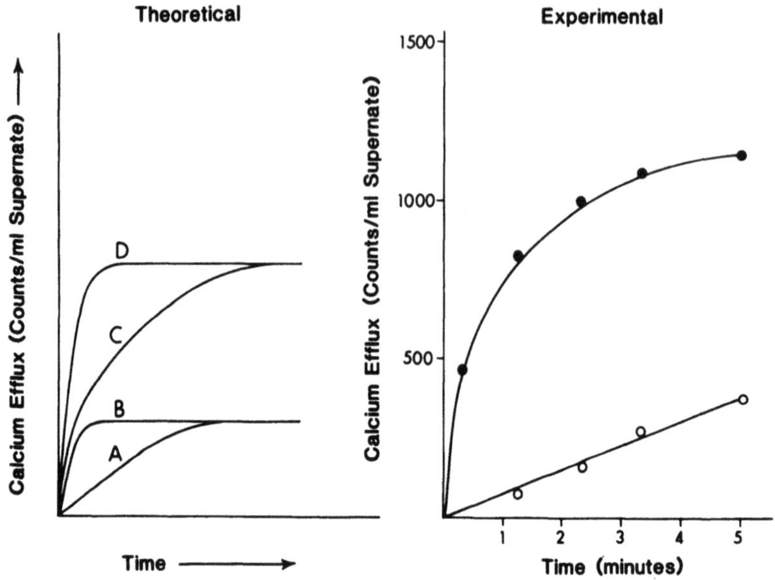

Figure 11.5. Time course of calcium efflux in neutrophils. The letters in the theoretical case denote the following: A, control; B, an increase in membrane permeability; C, an increase in the level of intracellular exchangeable calcium with no change in membrane permeability; D, an increase in both membrane permeability and the level of intracellular exchangeable calcium. In the experimental case: O—O control conditions, ●—● in the presence of 10^{-9} M f-Met-Leu-Phe. The outside calcium concentration is 0.5 mM.

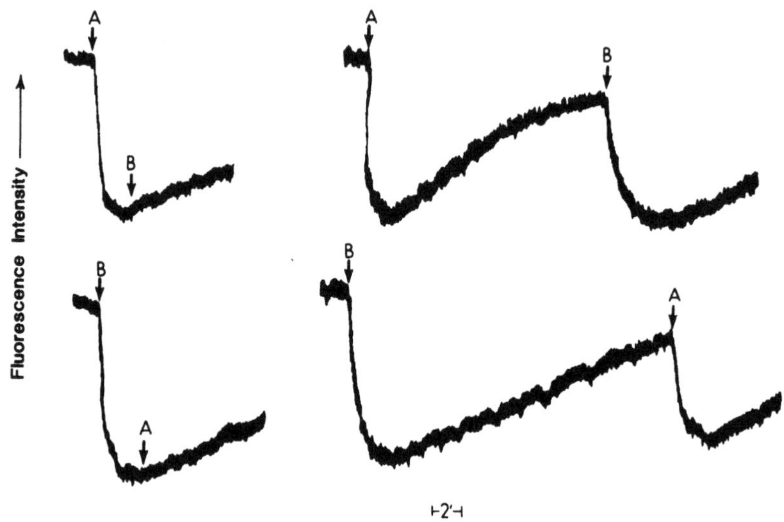

Figure 11.6. Sequential addition of f-Met-Leu-Phe (A) and $C5_a$ (B) to rabbit neutrophils during chlorotetracycline uptake experiments. Chlorotetracycline (10 μM) was added first to equilibrate rabbit cell suspensions thermally (3.0 $\times$ 10^6 cells/ml) followed 1 min later by calcium (0.5 mM). The chemotactic factors f-Met-Leu-Phe (5 $\times$ 10^{-9} M) or $C5_a$ (1/2000) were added at the steady-state point of the chlorotetracycline uptake curve.

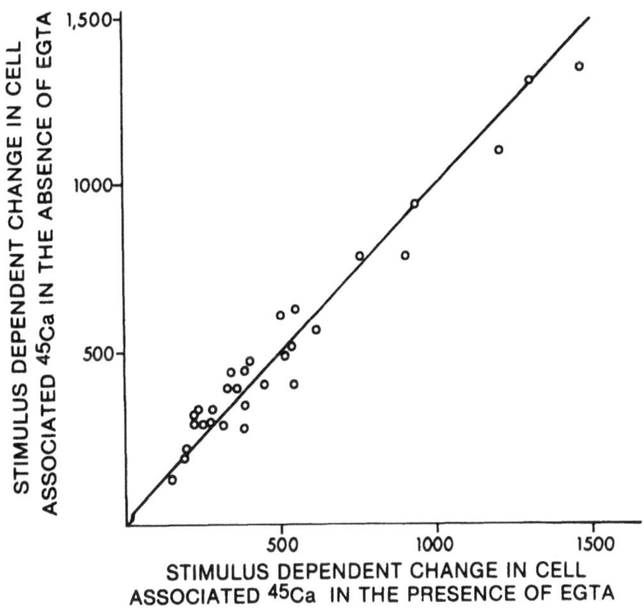

Figure 11.7. The relationship between the stimulus-dependent change in cell-associated calcium in the presence and absence of EGTA. The solid line represents the line of identity.

Nature of the Involvement of Sodium and Hydrogen Ions in Neutrophil Activation

In addition to mobilizing calcium, the chemotactic factor f-Met-Leu-Phe also increases the activity of the Na^+-H^+ exchange mechanism in rabbit neutrophils (Sha'afi et al. 1982). This is manifested by a rapid increase in Na^+ influx and H^+ efflux. However, it is generally found that the removal of sodium ions from the suspending medium does not inhibit cell activation nor does the addition of sodium ionophores activate the cells. Furthermore, the potassium sparing diuretic, amiloride, at concentrations that totally inhibit the Na^+-H^+ exchange mechanism, only marginally inhibit f-Met-Leu-Phe stimulated neutrophil aggregation, degranulation, and oxygen consumption. Based on the available experimental findings it is thus safe to conclude that although maintaining the proper ionic environment is necessary for optimal neutrophil function, the role of Na^+ influx and/or H^+ efflux in neutrophil activation is modulatory in nature and is not intimately related to the mechanism of signal transduction in these cells (Sha'afi et al. 1982; Kurchak and Weissmann 1980; Showell and Becker 1976; Showell et al. 1977; Sha'afi et al. to be published).

Role of Cyclic Nucleotides in Neutrophil Activation

Another important parameter to consider when trying to understand the overall scheme of excitation-response coupling is the intracellular level of cyclic nucleotides. It is generally known that exogenous cyclic nucleotides and agents that elevate intracellular cyclic nucleotide levels modulate the interrelated neutrophil functions of chemotaxis, phagocytosis, and lysosomal enzyme secretion (Zigmond 1978; Zurier et al. 1974; Hill et al. 1975). For example, extracellular cyclic 3′,5′-guanosine monophosphate (3′,5′-GMP) and cholinergic amines that generally elevate cyclic GMP levels promote these neutrophil responses. Chemotaxis, phagocytosis, and lysosomal enzyme secretion are inhibited, on the other hand, by cyclic 3′,5′-adenosine monophosphate (3′,5′-AMP), epinephrine, isoproterenol, cholera toxin, prostaglandins A_1 and E_1, as well as inhibitors of cyclic AMP-dependent phosphodiesterase, all of which are hormonal or pharmacological agents generally capable of raising intracellular cyclic AMP levels. The measured intracellular levels of cyclic GMP in human neutrophils increase in response to both zymosan-induced phagocytosis and the chemotactic synthetic peptide, f-Met-Ala. This peptide is less effective than f-Met-Leu-Phe but nonetheless chemotactic. We and others have also found that addition of the synthetic peptide, f-Met-Leu-Phe, to neutrophils transiently elevates the cellular level of cyclic 3′,5′-AMP (Jackowski and Sha'afi 1979; Hatch et al. 1977; Simghowitz et al. 1980). Stimulation of the cyclic AMP level reaches a maximum within the first minute, and subsequently the cyclic AMP level returns to the basal value within 5 min. Although the exact nature of the role of cyclic AMP in neutrophil activation is not fully understood, it is commonly believed that it is involved in modulating the signal and not in its transduction.

Nature of the Involvement of Arachidonic Acid and its Metabolites in Neutrophil Activation

Two critical features of the handling of arachidonic acid by mammalian cells underlie its modulatory role of various cellular functions: (1) the availability of arachidonic acid as substrate for the enzymes that convert it to biologically active metabolites is normally very limited, and (2) hormones, transmitters, and inflammatory mediators, often acting through highly specific receptors, can induce the rapid mobilization of arachidonic acid from lipid stores. Under resting conditions most animal cells contain little if any free arachidonic acid. This may be due in part to the presence of highly efficient arachidonyl-coenzyme A (CoA) synthetase (Wilson et al. 1982) and acyltransferase activities selective for arachidonic acid, which rapidly incorporate the free fatty acid into phospholipids (23). Thus, phospholipids represent a storage depot for arachidonic acid that becomes available upon specific types of cell activation. Relatively large changes in the concentration of free arachidonic acid can thereby be accomplished by the liberation of small amounts of previously esterified fatty acid. The biochemical mechanisms related to the stimuli-induced liberation of arachidonic acid are still somewhat controversial, although the involvement of one or more lipases is generally accepted. Once released, arachidonic acid is rapidly converted to several sets of biologically highly potent compounds, that is, prostaglandin endoperoxides and prostaglandins, trhomboxanes, and leukotrienes. Thus the liberation of a small amount of fatty acid results in the generation of large amounts of biological activity. Besides funneling free arachidonic acid into this metabolic machinery, the phospholipases involved play a role in the generation of other products that may have significant biological activities, that is, lysophospholipids, diacylglycerol, phosphatidic acid, and platelet activating factor (PAF)-acether.

Arachidonic acid metabolism generates three families of compounds with biological activities. The synthesis of the prostaglandins and the thromboxanes is initiated by the enzyme prostaglandin endoperoxide synthetase and that of the leukotrienes and other hydroxy fatty acids by the lipoxygenase pathways. The relative importance of each of these pathways depends on the cell source. In some cells such as platelets, arachidonic acid is metabolized by the three pathways, whereas in other cells such as the neutrophils, the metabolism of arachidonic acid proceeds predominantly through the lipoxygenase pathway.

Several lipoxygenases with differing positional specificities have been described. Here again the exact mix of these various enzymes depends on the cellular source, the platelet exhibiting predominantly the 12 lipoxygenase and the neutrophils exhibiting both the 5 and 15 lipoxygenases. In each case, the corresponding hydroperoxy derivative of arachidonic acid is formed. The latter can either be spontaneously hydrolyzed to form the corresponding hydroxyeicosatetraenoic acid (HETE) or, in the case of the 5-lipoxygenase, be enzymatically converted to an unstable epoxide, leukotriene A_4, that reacts rapidly and enzymatically with water to form leukotriene B_4 (a dihydroxy derivative of arachidonic acid) or, in the presence of cysteine-containing peptides, to leukotriene C_4, D_4, and E_4 (Samuelsson 1979). Leukotriene B_4 has recently been shown to be ω oxidized to form 20-OH leukotriene B_4, the latter being further metabolized to the dicarboxylic acid 20-COOH-leukotriene B_4 (Jubiz et al. 1982).

The interrelation between the various lipoxygenases and their products is currently under active investigation. Double dioxygenation of arachidonic acid through the 5 and 12 lipoxygenase and the 5 and 15 lipoxygenase have thus been demonstrated (Borgeat et al. 1981). In addition hydroperoxyeicosatetraenoic acid (5-HPETE) has been shown to increase the release of arachidonic acid induced by chemotactic factors in the human promyelocytic leukemia cell line HL60 (Stegel et al. 1982).

As early as 1975, before the description of the lipoxygenase pathway in neutrophils, Turner et al. (1975) reported on the chemotactic activity of various HETEs and on arachidonic acid itself. Recently, arachidonic acid, 5-HETE, and

12-HETE have also been shown to cause neutrophil degranulation. In a complementary approach to these studies, it has been found that known inhibitors of arachidonic acid metabolism such as 5,8,11,14-eicosatetraynoic acid (ETYA) and nordihydroguaiaretic acid (NDGA) and others inhibit the f-Met-Leu-Phe and arachidonic-acid-induced neutrophil activation (for review, see Sha'afi and Naccache 1981). At the present time, it is generally agreed that the agonist properties of arachidonic acid toward neutrophils are due to some of the metabolites by the lipoxygenase pathway, the most important of which is leukotriene B_4 (LTB_4). In addition, arachidonic acid, when added simultaneously with the chemotactic peptide f-Met-Leu-Phe, inhibits the ability of the latter to initiate several but not all of its effects on rabbit peritoneal neutrophils (Naccache et al. 1982). Stimulated neutrophil aggregation, calcium uptake, and increases in the steady-state level of exchangeable calcium are all inhibited by 1–10 μM arachidonic acid. The binding of f-Met-Leu-Phe and the parameters of intracellular calcium redistribution (calcium efflux and changes in the steady-state level of exchangeable calcium in the absence of extracellular calcium) and of stimulated sodium uptake are, on the other hand, unaffected by the same concentrations of arachidonic acid. Arachidonic acid was found not to inhibit f-Met-Leu-Phe stimulated aggregation and calcium uptake. Arachidonic acid, therefore, in addition to its well described agonist properties, also possesses antagonist activities toward rabbit neutrophils. These results add a new level of complexity to the study of the role of arachidonic acid in cell activation. The antagonistic activities of arachidonic acid toward neutrophils is due to direct interaction of the fatty acid with the cell membrane. This conclusion is based on our findings that most if not all unsaturated but not saturated fatty acids have the same antagonistic activities (Naccache et al. to be published).

Among all the arachidonic acid metabolites generated through the lipoxygenase pathway (Fig. 11.8), leukotrienes are the most important biologically. Recently, leukotrienes have been shown to have various biological effects, some of which are summarized in Table 11.1 (for more details, see Samuelsson 1982). The significance of the various leukotrienes rests in their profound effects on both the cells of origin and target cells. Leukotrienes C_4, D_4, E_4 have thus been shown to be the active constituents of slow-reacting substances of anaphylaxis (SRS-A) and as such as potent bronchoconstrictors (leukotrienes are on a molar basis about a 1000-fold more active than histamine in this respect) and vascular permeability increasing factors. Leukotriene B_4 is one of the most active neutrophil stimuli, inducing at nanomolar and subnanomolar concentrations aggregatory, chemotactic, and degranulation responses from neutrophils of several species. The complexity of the in vivo effects of leukotrienes and their modulation has been highlighted by the description of synergistic effects on plasma exudation between leukotrienes and prostaglandins and by their opposing effects on bronchial constriction.

The neutrophils-directed activities of leukotriene B_4 activity have been found to be extremely sensitive to the stereogeometry of the double bonds in the

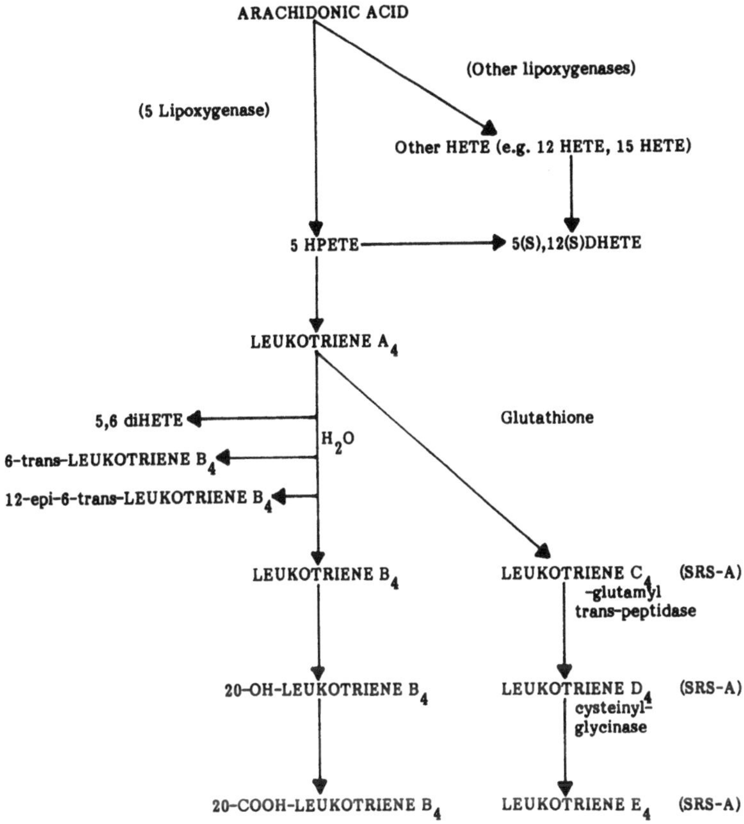

Figure 11.8. Diagram of arachidonic acid metabolism in leukocytes.

Table 11.1. Biological Activities of the Leukotrienes

Leukotrienes C_4, D_4 cause smooth muscle contraction of
Guinea pig ileum
Human and guinea pig bronchi (small airways)
Human pulmonary artery and vein
Leukotriene D_4 causes
Dilation of small vessels from skin
Increase of mucus glycoprotein synthesis and decrease in rate of mucus clearance from asthmatic patients
Increased plasma leakage (enhanced by PGE_2)[a]
Leukotriene B_4 interaction with neutrophils induces
Chemotaxis
Aggregation
Degranulation
Expression of surface $C3_b$ receptors
Increases plasma leakage synergistically with PGE_2 and neutrophils

[a]PGE_2, prostaglandin E_2.

triene structure, the hydroxyl groups, and to modifications of the natural configuration resulting in shifts in activity of 50 to 100-fold or more (Naccache et al. 1982). The sensitivity of the biological activity of leukotriene B_4 to even subtle changes in its configuration coupled with competition experiments with analogs and derivatives of leukotriene B_4 have led to the suggestion that the effects of leukotriene B_4 were receptor mediated (Goetzl et al. 1982). This hypothesis has recently been confirmed upon the demonstration by direct binding assay of specific receptor sites on human neutrophils (Goetzl et al. 1982). The leukotriene B_4 receptors appear to be distinct from those of the formyl-methionyl peptides or of C5a (Goetzl et al. 1982). The location and detailed characterization of these putative receptors remains to be carried out (Goldman and Goetzl 1982).

Leukotriene B_4, at concentrations similar to those required for the expression of its biological activities, increases both ^{45}Ca uptake and efflux and causes a redistribution of intracellular calcium of a magnitude similar to that upon the addition of the chemotactic peptide, f-Met-Leu-Phe. The efficacy of leukotriene B_4 in the calcium uptake assay parallels closely its chemotactic, aggregatory, and secretory activities and displays the same stereospecificity (Fig. 11.9). Metabolic conversion to 20-OH-leukotriene B_4 or the dicarboxylic acid 20-COOH-leukotriene B_4 results in loss of both functional activity and ability to increase calcium uptake. The evidence that leukotriene B_4 increases the permeability of the cell membrane to calcium is indirect but is supported by several types of data, including the fact that the degranulation induced by leukotriene B_4 is enhanced by extracellular calcium (Showell et al. 1982). At the present time we believe that the stimulation of neutrophils by leukotriene B_4 is related to its ability to mobilize calcium.

Although the exact mechanism by which leukotriene B_4 increases the plasma membrane permeability to calcium is not known, suggestions have repeatedly been made that substances generated as a result of phospholipase activity or by the metabolism of arachidonic acid act as calcium ionophores. These putative ionophores include lysophospholipids, phosphatidic acid, prostaglandin endoperoxides, thromboxane A_2, and leukotrienes (for review see Feinstein and Sha'afi 1983). The evidence in favor of such action includes the ability to transport Ca^{2+} in model systems, release of calcium from membrane fractions that sequester the cation, and the promotion of Ca^{2+} uptake by intact cells. The concept of metabolites of lipid metabolism acting as ionophores for Ca^{2+} is appealing since it provides a way to link receptor-mediated cell stimulation to the mobilization of calcium necessary for subsequent responses. The ability of endoperoxides, thromboxane A_2, and leukotrienes to stimulate target cells is consistent with such an action. Other intermediates derived from phospholipase activity, such as lysophospholipids and lysophosphatidic acid, are also capable of directly stimulating smooth muscle, mast cells, and platelets or of potentiating the actions of other agonists (for review see Feinstein and Sha'afi 1983).

Despite the evidence cited above, the actions of the various lipid mediators differ significantly from those of classical ionophores such as A23187 or ionomycin. First, and perhaps most fundamentally, the effects of mediators, such as thromboxane A_2, and the various leukotrienes are cell specific, whereas A23187 or ionomycin will transport calcium across virtually any lipid bilayer, biological and synthetic, and activate Ca^{2+}-dependent processes in virtually all types of cells. Unlike ionophores, leukotrienes and thromboxane A_2 also lack the ability to affect transport in some isolated organelles; for example, phosphatidic acid

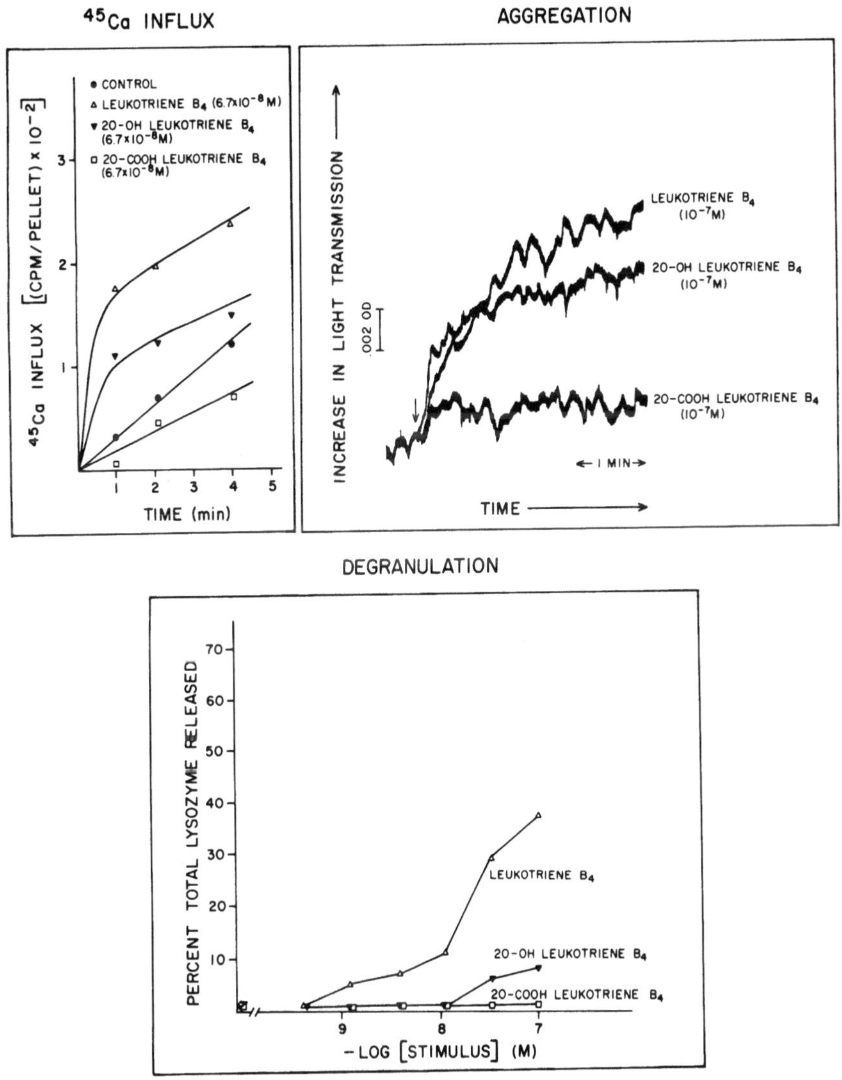

Figure 11.9. The effects of leukotriene B_4; 20-OH-leukotriene B_4 and 20-COOH-leukotriene B_4 on the initial rate of ^{45}Ca influx and on aggregation and degranulation in rabbit neutrophils.

does not release calcium from mitochondria (for review see Feinstein and Sha'afi 1983). Second, the great potency of mediators such as thromboxanes and leukotrienes, together with the extreme sensitivity of their various biological activities to relatively minor stereochemical modifications, argue strongly in favor of specific binding sites or receptors. Ionophores such as A23187 and ionomycin, on the other hand, are believed to diffuse and partition nonspecifically into hydrophobic environments of membranes to act as ion carriers. Third, neutrophils have been shown to become desensitized after exposure to high concentrations of either f-Met-Leu-Phe or leukotriene B_4, followed by washout. This is a commonly observed mechanism attending receptor-mediated stimulation (Sha'afi et al. 1981).

The reactions that ensue after stimulation generate, depending on cell type, several of the putative lipid "Ca^{2+} ionophores" almost simultaneously. It is not readily evident why multiple ionophores would have to be produced in the same cell in response to a stimulus; this would appear to be biochemical overkill. Even a single ionophorelike compound would seem to present cells with significant dangers. Two characteristics of ionophores are particularly unsuited for a physiological role in the regulation of cytoplasmic Ca^{2+} levels: (1) ionophores are not selective in their effects on Ca^{2+} permeability and would be expected to affect not only the plasma membrane, but also mitochondria, lysosomes, endoplasmic reticulum, and secretory granules indiscriminately; (2) ionophores can short circuit Ca^{2+} pumps, so that they have the potential to prevent restoration of the "resting state" of cells. Of course, a number of these objections can be countered. The production of putative ionophores could be highly localized and restricted and their existence transitory due to further metabolism. Furthermore, one of the putative lipid ionophores, phosphatidic acid, does exhibit some selectivity for certain membranes (Barritt et al. 1981), and since it can stimulate a Ca^{2+}-pump ATPase (Niggli et al. 1981) it might actually promote Ca^{2+} pumping.

We would suggest that numerous biological examples suggest another likely alternative for regulating Ca^{2+} permeability or "gating" by highly potent lipid mediators such as thromboxane A_2 and leukotrienes. A more specific mechanism is the linkage of a receptor protein with an existing ion-conductance channel protein, as in the case of the acetylcholine receptor. A receptor-linked ion channel provides very high specificity both with respect to the nature of the agonists capable of opening channel "gates" and with respect to sites of action, since only membranes containing receptor-linked channels can be affected. We would propose that highly potent arachidonic acid metabolites may control cytosolic activator-calcium levels in this way. This does not rule out the possibility that these lipid mediators may act as ionophores when used at high concentrations.

Two interrelated questions arise at this point. First, given that the activation of the lipoxygenase pathways is sufficient to initiate several neutrophil functions, is this pathway also a necessary condition for the cell's response? Second, what are the physiological roles of mediators such as leukotrienes? Are they "first messengers" and therefore sufficient, but not necessary, conditions designed to amplify the activation or recruitment of nearby cells. Or, on the other hand, are they intracellular mediators of the effects of receptor activators (f-Met-Leu-Phe, C5a, antigen, etc.), and therefore both necessary and sufficient for cellular activation? Although the evidence available to date does not allow an unambiguous answer to the above questions, the following lines of evidence argue in favor of an intracellular role for leukotriene B_4:

1. A number of clearcut cases exist in which the production of arachidonic acid metabolites is not essential for response to a receptor agonist (i.e., thrombin in platelets). Preincubation of neutrophils with arachidonic acid or leukotriene B_4 reduces the ability of the cells to respond to subsequent reexposure to the lipid factors, but the response to f-Met-Leu-Phe is unaffected. The direct implication of these results is that the chemotactic peptide is capable of stimulating neutrophils even when the arachidonic acid–leukotriene pathway is functionally ineffective (i.e., deactivated).
2. Leukotrienes and thromboxane are clearly able to escape readily from cells of origin and stimulate other tissues; that is, thromboxane from platelets can stimulate vascular smooth muscle, and leukotrienes can cause bronchoconstriction. Leukotriene B_4 when formed upon stimulation by chemotactic factors is recovered rapidly in extracellular medium. An in vivo correlate is the recovery of leukotriene B_4 from various inflammatory fluids (Kliestein et al. 1981). We would expect that a compound produced by cells for an essential intracellular role would be conserved by those cells and not be so readily lost to the external environment.

The ambiguities inherent in the interpretation of the results just summarized should, however, be kept in mind. It should prove difficult to distinguish clearly between the two hypotheses concerning the exact role of arachidonic acid-derived mediators, as most of the criteria used to define first and second messengers are at least experimentally, if not conceptually, related. A clear distinction awaits the identification and subcellular localization of putative receptors and the understanding of the biochemical mechanisms by which the levels of calcium are regulated.

Acknowledgments

This work was supported in part by NIH grants AI-13734-05 and AM-31000-01.

References

Aswanikumar, S., Corcoran, G., Schiffmann, E., Day, A.R., Freer, R.J., Showell, H.J., Becker, E.L., and Pert, C.B. (1977). Demonstration of a receptor on rabbit neutrophils for chemotactic peptides. *Biochem. Biophys. Res. Commun. 74,* 810–817.

Barritt, G.J., Dalton, K.A., and Whiting, J.A. (1981). Evidence that phosphatidic acid stimulates the uptake of calcium by liver cells but not calcium release from mitochondria. *FEBS Lett. 125,* 137–140.

Borgeat, P., Picard, S., Vallerand, P., and Sirois, P. (1981). Binding of the chemotactoc synthetic peptide [^{3}H]formyl-nor-len-len-phe to plasma membrane of rabbit neutrophils. *Prostaglandin Med. 6,* 557–566.

Chenoweth, D.E. and Hugli, T.D. (1979). Demonstration of specific C5a receptor on intact human polymorphonuclear leukocytes. *Proc. Natl. Sci. USA 75,* 3943-3947.

Feinstein, M.B. and Sha'afi, R.I. (1983). The structure-activity relations of synthetic peptides as chemotactic factors and producers of liposomal enzyme secretion for neutrophils. In *Calcium and Cell Function,* Cheung, W.Y., ed. Academic Press, New York, vol. 3, to be published.

Goetzl, E.J., Goldman, D.W., Naccache, P.H., Sha'afi, R.I., and Pickett, W.C. (1982). The formylpeptide chemotactic receptor on rabbit peritoned neutrophils. I. Evidence for two binding sites with different affinity. In *Leukotrienes and Other Lipoxygenase Products. Advances in Prostaglandins, Thromboxane and Leukotriene Research,* Samuelsson, B. and Peole Hi, R., eds. Raven Press, New York, vol. 9, p. 273.

Goldman, D.W. and Goetzl, E.J. (1982). Preparation of inside-out-membrane vesicles from neutrophils capable of actively transporting calcium. *J. Immunol. 129,* 1600–1609.

Gomperts, B.D., Bennett, J.P., and Allan, D. (1981). A synthetic ionophore for Ca^{2+}: Studies with model and biological systems. *Eur. J. Biochem. 117,* 559–562.

Hatch, G.E., Nochols, W.K., and Hill, H.R. (1977). A Ca^{2+} stimulated ATPase activity in rabbit neutrophil membranes. *J. Immunol. 119,* 450–456.

Hill, H.R., Estensen, R.D., Quie, P.G., Hogan, N.A., and Goldberg, N.D. (1975). *Metabolism 24,* 447–456.

Holub, B.J. and Piekarski, J. (1979). The formation of phosphatidylinositol by acylation of 2-acyl-*sn*-glycero-3-phosphorylinositol in rat liver microsomes. *Lipids 14,* 529–532.

Jackowski, S. and Sha'afi, R.I. (1979). Response of adenosine cyclic 3′,5′-monophosphate level in rabbit neutrophils to the chemotactic peptide formyl-methionyl-leucyl-phenylalanine. *Mol. Pharmacol. 16,* 473–481.

Jackowski, S., Petro, K., and Sha'afi, R.I. (1979). Stimulus-response coupling in the human neutrophil transmembrane potential and the role of extracellular Na^+. *Biochim. Biophys. Acta 558,* 348–352.

Jubiz, W., Radmark, O., Malmsten, C., Hansson, G., Lindgren, J.A., Palmblad, J., Uden, A.M., and Samuelsson, B. (1982). The effects of external K^+ and Na^+ on the chemostasis of rabbit peritoneal neutrophils. *J. Biol. Chem. 257,* 6106–6110.

Kliestein, L.B., Shapleigh, C., and Goetzl, E.J. (1980). The effects of extracellular K^+, Na^+ and Ca^+ on liposomal enzyme secretion from polymorphonuclearleukocytes. *J. Clin. Invest. 66,* 1166–1170.

Korchak, H.M. and Weissmann, G. (1980). Stimulus-response coupling in the human neutrophil, transmembrane potential and the role of extracellular Na^+. *Biochim. Biophys. Acta 601,* 180–194.

Mackin, W.M., Haung, C.K., and Becker, E.L. (1982). Chemotaxis by polymorphonuclear leukocytes. *J. Immunol. 129,* 1608–1611.

Molski, T.F.P., Naccache, P.H., and Sha'afi, R.I. (1983). Stimulus-dependent changes in calcium metabolism in rabbit neutrophils. *Cell. Calcium 4,* 57–68.

Murphy, P. (1976). Modulation of human neutrophil chemotactic responses, by cyclic 3:5 < guanonine monophosphate and cyclic 3:5-adenosine murophosphate. In *The Neutrophil.* Plenum Publishing, New York.

Naccache, P.H., Molski, T.F.P., Volpi, M., Mackin, W.M., Becker, E.L., and Sha'afi, R.I. To be published.

Naccache, P.H., Molski, T.F.P., Becker, E.L., Borgeat, P., Picard, S., Vallerand, P., and Sha'afi, R.I. (1982). Cyclic nucleotide changes in human neutrophils induced by chemoattractants and chemotactic modulators. *J. Biol. Chem. 257,* 8608–8611.

Naccache, P.H., Molski, T.F.P., Volpi, M., Borgeat, P., Showell, H.J., Becker, E.L., and Sha'afi, R.I. (1982). Induction of a transient elevation in intracellular levels of adenosin-3'5'-cyclic monophosphate by chemotactic factors: An early event in human neutrophil activation. *Fed. Proc. 41,* 577.

Niggli, V., Andrenyah, E.S., and Carafoli, E. (1981). Discovery of an arachidonyl coenzyme a synthetase in human platelets. *J. Biol. Chem. 256,* 8588–8592.

Samuelsson, B. (1982). Formation of phosphatidylinositol by activation of 2-acyl-SN-glycero-3-phosphorylinositol in rat liver microsomes. In *Leukotriens and Other Lipoxygenase Products. Advances in Prostaglandin, Thromboxane and Leukotriene Research,* Samuelsson, B. and Paoletti, R., eds. Raven Press, New York, vol. 9, pp. 1–17.

Samuelsson, B., Hammarstrom, S., and Borgeat, P. (1979). In *Advances in Inflammation Research,* Weissman, G., Samuelsson, B., and Paoletti, R., eds. Raven Press, New York, vol. 1, p. 405.

Sha'afi, R.I. and Naccache, P.H. (1981). A novel leukotriene produced by stimulation of leukocytes with formylmethionyleucylphenylamine. In *Advances in Inflammation Research,* Weissman, G., ed. Raven Press, New York, vol. 2, p. 115.

Sha'afi, R.I., Molski, T.F.P., and Naccache, P.H. Distribution of various enzyme activities in the rat brain. To be published.

Sha'afi, R.I., Molski, T.F.P., Borgeat, P., and Naccache, P.H. (1981). Deactivation of the effects of F-met-leu-phe and leukotriene B_4 on calcium mobilization in rabbit neutrophils. *Biochem. Biophys. Res. Comm. 103,* 766–773.

Sha'afi, R.I., Naccache, P.H., Molski, T.F.P., and Volpi, M. (1982). Biogenesis of chemotactic molecules by the arachidonic lipoxygenase system of platelets. In *Intracellular pH: Its Measurement, Regulation and Utilization in Cellular Function.,* Nuccitelli, R. and Deamer, D.W., eds. Alan R. Liss, New York, p. 513.

Sha'afi, R.I., Williams, K., Wacholatz, M.C., and Becker, E.L. (1978). Inhibition of neutrophil responsiveness by arachidonic-acid 15-HETE and 5,15-Di HETE. *FEBS Lett. 91,* 305–309.

Showell, H.J. and Becker, E.L. (1976). The effects of external K^+ and Na^+ on the chemotaxis of rabbit peritoneal neutrophils. *J. Immunol. 116,* 99–105.

Showell, H.J., Naccache, P.H., Sha'afi, R.I., and Becker, E.L. (1977). The leukotrienes: An introduction. *J. Immunol. 119,* 804–811.

Showell, H.J., Freer, R.J., Zigmond, S.H., Schiffman, F., Aswanikumar, S., Corcoran, B., and Becker, E.L. (1976). Specificity of the effect of lipoxygenase metabolites of arachidonic acid on calcium homeostasis in neutrophils. *J. Exp. Med. 143,* 1154–1169.

Showell, H.J., Naccache, P.H., Borgeat, P., Picard, S., Vallerand, P., Becker, E.L., and Sha'afi, R.I. (1982). Characterization of the secretory activity of leukotriene B_4 toward rabbit neutrophils. *J. Immunol. 128,* 811–816.

Simchowitz, L., Fischbein, L.C., Spilberg, I., and Atkinson, J.P. (1980). Specific binding of leukotriene B_4 receptors on human polymorphonuclear leukocytes. *J. Immunol. 124,* 1482–1491.

Stegel, M.C., McConnell, R.T., Bonser, R.W., and Cuatrecasas, P. (1982). Characterization of the secretory activity of leukotriene B_4 toward rabbit neutrophils. *Biophys. Res. Comm. 104,* 874.

Sugiyama, K. (1977). Histamine release from rat mast cells induced by Sendai virus. *Nature 270,* 614–615.

Turner, S.R., Tainer, J.A., and Lynn, W.S. (1975). Biogenesis of chemotactic molecules by the arachidonate lipoxygenase system of platelets. *Nature 257,* 680–681.

Volpi, M., Naccache, P.H., and Sha'afi, R.I. (1982). Deactivation of the effects of F-met-leu-phe and leukotriene B_4 on calcium mobilization in rabbit neutrophils. *Biochem. Biophys. Acta 558,* 348–352.

Wilson, D.B., Prescott, S.M., and Majerus, P.W. (1982). Evidence that phosphatidic acid stimulates the uptake of calcium by liver cells but not calcium release from mitochondria. *J. Biol. Chem. 257,* 3510–3515.

Zigmond, S.H. (1978). Acidic phospholipids, unsaturated fatty acids, and limited proteolysis mimic the effect of calmodulin on the purified erythrocyte Ca^{2+}-ATPase. *J. Cell Biol. 77,* 269–287.

Zurier, R.B., Weissmann, G., Hoffstein, S., Kammerman, S., and Tai, H.H. (1974). Lipoxygenation of arachidonic acid as a source of polymorphonuclear leukocyte chemotactic factors in synovial fluid and tissue in rheumatoid arthritus and spondyloarthritis. *J. Clin. Inv. 53,* 297–309.

12 Pathological Sequelae of Irritant-Induced Phagocytosis Exemplified by Pulmonary Emphysema

S. Smith and H. Baum

Many cell species contain large numbers of cytoplasmic membrane-bound granules. The number, size, and content of these granules are dependent upon the type and activity of the cell involved. Phagocytosis is not necessarily the major function of a granule-containing cell, but all cells can internalize material by this process.

The cells that are thought to be the most important in the pathogenesis of emphysema are the polymorphonuclear leukocytes (PMN) and alveolar macrophages (AM). The PMNs contain two types of granules, primary and azurophil, that contain potentially cytotoxic enzymes such as myeloperoxidase, cathepsin G' elastase, and acid hydrolases and smaller, more numerous secondary or specific granules that contain alkaline phosphatase, lysozyme, vitamin B_{12} binding protein, lactoferrin, and complement activating protease (Baggiolini et al. 1969; Spitznagel et al. 1974). It has been suggested that secondary granules of PMNs may have a more subtle function than the simple digestion of phagocytosed material, since they have been shown to discharge the majority of their contents extracellularly, rather than into phagosomes (Leffell and Spitznagel 1975). A possible function is in the mediation and regulation of inflammatory processes (Maallem et al. 1982), since lysozyme is a known bacteriostatic agent, and another granule constituent, lactoferrin, is known to increase the adhesive properties of neutrophils.

Alveolar macrophages contain abundant lysosomes, the contents of which include a number of acid hydrolases, proteases, nucleases, lipases, and acid phosphatase. All lysosomal enzymes are membrane bound to prevent the active centers lysing each other and the membrane (Reid and Leech 1980). The main function of these cells is to ingest and break down foreign material, including bacteria, as well as dead cells and other debris. Their activity is modulated by lymphokines, soluble factors produced by stimulated T-lymphocytes.

There are a number of mechanisms, both physiological and pathological, whereby a cell may be induced to discharge its granular contents. Some of these are discussed briefly below.

Cell Damage

It was once believed that lysosomal enzymes were released only after irreparable membrane damage when cell death was inevitable, for example, after osmotic shock or treatment with detergents (Metchnikoff 1892). It is now known that this is not the only process whereby granule contents may be lost. Certain compounds such as silica gain access to macrophage lysosomes by undergoing phagocytosis in the normal fashion, followed by fusion of the phagosome with a primary lysosome. Once within secondary lysosomes, the membranes are damaged from within. Silicic acid is a hydrogen-bonding agent and is thought to damage membranes by forming hydrogen-bonded complexes with phospholipid or protein (Nash et al. 1966), thereby releasing lysosomal enzymes into the cell cytoplasm. Support for this theory comes from Allison and co-workers (1966) who demonstrated that the hydrogen-bonding acceptor polyvinylpyridine-*N*-oxide protected against lysosomal damage by preferentially complexing with the silicic acid. However, although macrophage death follows intracellular release of lysosomal enzymes, it is by no means clear that the two events are directly linked. Recent evidence shows that death following silica ingestion is the result of a toxic influx of extracellular calcium ions (Kane et al. 1980) and that silica damages the plasma membrane directly by altering its permeability. Cells that have undergone postsilica-ingestion lysosomal enzyme release in the absence of extracellular calcium will survive and can be passaged and replated (Farber 1981).

This mechanism of enzyme release is not only provoked by xenobiotics but by endogenous substances such as monosodium urate (MSU) crystals, which are products of faulty nucleic acid metabolism (Shirahama and Cohen 1974). Weissmann and co-workers (1971a) demonstrated that MSU crystals are phagocytosed by PMNs but that they cause subsequent cell damage (as estimated by the release of lactate dehydrogenase). Tris-buffered saline afforded the cells greater protection from lysis than phosphate-buffered saline, possibly by interfering with hydrogen bonding between lysosomal membranes and urate (Weissmann et al. 1971). A "respiratory burst" of glucose oxidation was observed before enzyme release, indicating that the MSU was phagocytosed and was not simply damaging cells from without.

A wide variety of factors are known to alter the stability of lysosomal membranes. Nutritional status is one such factor (Wells et al. 1981). Elevation of cytosolic cyclic adenosine monophosphate (cAMP) has been shown to increase lysosomal stability and decrease enzyme release (Zurier et al. 1974) even though cAMP rises during phagocytosis (Smolen and Weissmann 1981). In contrast, raised cyclic guanosine monophosphate (cGMP) resulting from cigarette smoke exposure makes the lysosomal membrane more labile (Arnold et al. 1977). Many hormones, including oestrogens (Badenoch-Jones and Baum 1973) and glucocorticoids (Werb 1979), have been shown to affect lysosomal stability.

Secretion

Enzyme release by secretion is not the result of cell damage. There are several different modes of secretion discussed below.

Constitutive Secretion

Constitutive secretion involves a continuous synthesis and secretion of an enzyme irrespective of the presence or absence of any external stimuli. An example of this type of release is the continuous secretion of lysozyme by rabbit AM in culture (McCarthy and Henson 1979).

Nonconstitutive Secretion

Nonconstitutive secretion must be induced by a suitable stimulus such as the presence of immune complexes, opsonin-coated latex or zymosan particles, or complement fragments. Release of stored products follows the stimulus immediately, for example, the release of β-glucosaminidase from rabbit AM in response to opsonized zymosan. This release, which occurs in the presence of cycloheximide, was elegantly demonstrated by Henson (1980). Henson and others also illustrated a variant of this type of release (McCarthy et al. 1982), that is, protein-synthesis dependent, inductive release, which occurs in culture over a number of days in response to an initial stimulus, the release gradually increasing with time. This increase in enzyme secretion can be abolished by the use of cycloheximide. Unlike the nonsynthesis-dependent release, this leads to an increase in the total enzyme present in the culture (Henson 1980).

The nature of this stimulus–secretion coupling is not clearly understood. Smolen and Weissmann (1981) have identified two groups of secretagogues, which provoke slightly different responses. Complete secretagogues such as antigen–antibody complexes and formyl-methionyl-leucyl-phenylalanine (*N*-f-Met-Leu-Phe) chemotactic factors stimulate the release of azurophilic and specific granule contents as well as an elevation in total cellular cAMP. Incomplete secretagogues, such as concanavalin A, stimulate the release of specific granules only, without cAMP elevation. The elevation of cAMP is not the driving factor for lysosomal enzyme release because the removal of extracellular divalent cations and the substitution of Na^+ by K^+ abolishes release but not cAMP increases (Smolen et al. 1980).

The use of anion-channel blockers has shown anion fluxes to have an important role in stimulus-secretion coupling in PMNs. Blockers interfere with the secretion of lysosomal enzymes but do not affect superoxide generation (Korchak et al. 1980). The identity of the crucial anion and the direction of the flux are as yet unknown (Korchak et al. 1982), but it may be that the changes in flux cause either alterations in pH or inhibition of essential cation fluxes, perhaps of extracellular calcium (Edelson et al. 1982).

Before any discussion of the effects of immunological stimuli on secretion and phagocytosis it is important to note that the role of the macrophage in the immune response and its relationship to other immunological factors are complex and will not be discussed at length here. In brief, macrophages can phagocytose material prepared for them by other components of the immune system. They can also incorporate antigens into their plasma membranes and use them to activate lymphocytes. Activated T lymphocytes release lymphokines that modify the mobility and phagocytic and killing properties of macrophages. The PMN behavior is also modulated by humoral factors. Additionally, many interactions are specific to phagocytes of a particular type and to specific humoral factors. For a fuller discussion of the general relationship of the phagocyte to the immune system, see Roitt (1980).

To return to a more specific discussion of secretion by AM in response to immune stimuli, it has been found that release may occur in response to both antigen antibody-complexes and humoral factors such as C5 fragments (McCarthy and Henson 1979). These stimuli bind to specific receptors on the plasma membrane of the secreting cell. Several cell types possess receptors for the Fc fragment of some subclasses of IgG and for the C3b component of complement (Unkeless 1979), including macrophages, monocytes, PMNs, and lymphocytes (Grey and Anderson 1979). There has been a certain amount of controversy over the number and nature of Fc binding sites on the plasma membrane, but in the mouse at least two Fc receptors have been found, one specific to IgG2a and one to IgG2b (Grey and Anderson 1979). It has also been shown that binding is enhanced when the antibodies have interacted with antigen to form immune complexes. This enhancement increases as complex size increases (Leslie and Alexander 1979).

Although many of the stimuli that induce secretion from macrophages activate the alternative pathway of complement, weak bases appear to be able to bypass this pathway yet still stimulate release (Riches and Stanworth 1982).

A special form of stimulus-secretion coupling has been termed *reverse endocytosis* or *frustrated phagocytosis* (Henson 1971a). This can occur following contact between PMN surface receptors and a solid surface that is too large to ingest but that is coated with C3 or IgG (Henson 1971b). Such a mechanism could be relevant in vivo after damage to nonphagocytosable structures such as large stretches of collagen, elastin, or basement membrane. A probable example of this type of damage was shown by Shaw et al. (1980) who induced alveolar neutrophil recruitment with associated severe inflammation by intratracheal installation of complement-derived chemotactic fragments. Henson (1980) does not envisage the major enzyme release from phagocytes as being "accidental" but rather as a physiological response to a stimulus, for example, permitting cells to migrate through tissues to reach a site of inflammation (Keeling and Henson 1982), and this is envisaged in terms of an intracellular granule dissolution (Henson 1980).

Phagocytosis

All animal cells practice endocytosis as a mechanism to take up macromolecules from the extracellular fluid (Goldstein et al. 1979). A dozen or more systems have been described in which "nonphagocytic" cells can be shown to internalize proteins. These systems generally involve the attachment of a ligand to a membrane-bound receptor followed by internalization and subsequent degradation by lysosomes (Goldstein et al. 1979). The process of phagocytosis is a specialized form of endocytosis in which particulate matter, perhaps foreign to the host, is incorporated into intracellular vacuoles by invagination and pinching off of sacs of plasma membrane. It will be discussed here only in relation to AM and PMNs. The visible sequence of events occurring during this process can be summarized as follows:

1. Attraction of the phagocytic cell to the site of infection or inflammation
2. Contact of the phagocytic cell and particle(s), resulting in recognition and attachment
3. The appearance of cell pseudopodia in the region of the attached particle
4. Movement of the pseudopodia around the particle and fusion on its distal side
5. Fusion of lysosomes with the phagosomal membrane resulting in the emptying of granular contents on to the ingested particle

The first step in the process is the attraction of phagocytic cells to the site. The cells are attracted by chemotactic factors, the most important of which are the products of complement activation. Cleavage of C3 and C5 generates products C3a and C5a, which are strongly chemotactic. This production of C3a and C5a can be generated by a number of mechanisms, including direct proteolysis of C3 and C5, activation of complement by antigen–antibody interactions, or action of the alternative complement pathway (Smolen and Weissman 1981). Chemotactic factors can be released by bacteria as well as by phagocytosing cells.

The basis of recognition must be highly sensitive as phagocytes are very discriminating, rejecting healthy host cells but engulfing damaged and foreign ones and microorganisms. The process of recognition is known to be affected by a number of factors, including the surface charge and the hydrophobicity of the particle (Skutelsky and Danon 1969; Stossel et al. 1972). However, immunological factors play a more important role in recognition than do physical ones. Eighty years ago Wright and Douglas showed that the addition of human serum to incubation mixtures promoted microbial ingestion by human neutrophils and monocytes (Wright and Douglas 1903). This stimulation of phagocytosis was termed *opsonization,* and any serum component acting in this manner is known as an *opsonin.* The best characterized opsonins are the antibody IgG and part of a component of the complement system (C3). The sequential interaction of antibody and complement components C1, C4, C2, and C3 has been shown to opsonize microorganisms and red blood cells (Johnston et al. 1969). However, sera deficient in any component other than C3 maintain their opsonizing activity indicating that there is at least one other pathway capable of activating the system (Johnson et al. 1972; Jasin 1972; Winkelstein et al. 1972). Patients deficient in C3 lack any heat labile opsonic activity in their sera and tend to suffer from recurrent infection (Stossel 1975). IgG antibodies have the capacity to opsonize a variety of microorganisms and erythrocytes in the absence of C3. The antibody binds to particles by its Fab regions, but opsonic activity resides in the Fc region since papain- or pepsin-digested IgG binds to particles without opsonizing them (Quie et al. 1968).

Until recently it was thought that IgG activated the "classical" complement pathway, whereas C3 activated the alternative pathway. However, a recent publication has shown that IgG is not essential for activation of the classical pathway (Leist-Welsh and Bjornson 1982). Furthermore, AM can phagocytose particles in the absence of any serum opsonins (Czop et al. 1982). However, other data suggest that PMNs require a heat labile serum factor other than complement for killing pneumococci (Guckian et al. 1981a). Patients suffering from tuberculosis and sarcoidosis show a high incidence of a genetic defect of the alternative nonimmune complement pathway even though values of serum complement are normal (Johnson et al. 1980). It has also been shown that opsonin versus zymosan activity is different for different blood groups (Kobayashi and Usui 1982).

The pseudopodia that actually engulf the opsonized particles (Koerten et al. 1980) consist of hyaline ectoplasm, a material with a stiff, glassy appearance, which has been shown by electron microscopy to contain a network of filaments similar to that found in the peripheral cytoplasm of many cells (Reaven and Axline 1973; Wessells et al. 1971). The PMN have been shown to contain both actin and myosin (Crawford et al. 1980). Two forms of cytoplasmic actin have been found, neither of which is identical to muscle actins. The actin can polymerize to form microfilaments 5-7 nm in diameter (Dustin 1978), which are often attached to or forming a web immediately beneath the cell membrane (Gruenstein and Rich 1973; Allison et al. 1971). The actin:myosin ratio is very much higher than in muscle cells (Crawford et al. 1980), which is thought to result in the generation of a smaller force but a much greater shortening than in muscle.

The parts these contractile proteins play in the processes of cell locomotion, pseudopod formation, and enzyme secretion are incompletely understood.

Immunofluorescent techniques have shown that in resting peritoneal macrophages actin is distributed throughout the cell in a reticulate fashion, whereas myosin is localized in distinct dots (Painter et al. 1981). When a macrophage is engaged in particle ingestion, the contractile proteins have been shown to concentrate at the ingestion sites (Stendahl et al. 1980). Recent studies on platelets have shown that activation leads to an increase in the percentage of filamentous actin (Jennings et al. 1981). Phosphorylation of myosin is essential for secretion, which, it has been suggested, is a contractile process involving a force-generating interaction between actin and myosin (Fox and Phillips 1982). However, changes in cell shape can occur without myosin phosphorylation. Cytochalasin B is known to inhibit actin polymerization and block microfilament interactions. It has also been found to inhibit phagocytosis, so microfilaments must be essential for this process to occur in macrophages (Axline and Reaven 1974; Reaven and Axline 1973). Although it inhibits phagocytosis, it does not prevent secretion by PMNs (Smolen and Weissmann 1981). The fact that phagocytosing cells contain more actin-binding protein than resting ones also suggests that microfilaments play a role in phagocytosis (Hartwig and Stossel 1976).

Microtubules also appear to be involved in phagocytic processes, since colchicine will inhibit phagocytosis in some systems, but the evidence is not clear cut. It has been suggested that microtubules act as supports for actin filaments or levers to amplify their movements (Stossel 1975). Large doses of colchicine are necessary to inhibit phagocytosis by PMN or macrophages (Dustin 1978). It appears that although microtubules are important in the processes of exocytosis and secretion, they are less important in phagocytosis.

The final stage of pseudopod formation is the fusion of the two arms on the distal side of the particle. Contractile proteins appear to be required, but the nature of the interactions occurring between the two folds of membrane are not understood although they have been slightly better characterized for the fusion of mucocysts to the plasma membrane during mucus secretion by Tetrahymena pyriformis (Saicr and Stiles 1975).

In PMNs that have few mitochondria, the process of phagocytosis is energy requiring. Inhibitors of glycolysis such as sodium fluoride, iodoacetate, and deoxyglucose have been shown to impair ingestion by PMN leukocytes in a number of species, including humans (Cohn and Morse 1960; Sbarra and Karnovsky 1959; Giordano and Lichtman 1973). Ingestion in these cells is unaffected by inhibitors of mitochondrial activity, in contrast with AM, which derive much of their energy requirements from oxidative phosphorylation (Cohen and Cline 1971; Mason et al. 1973). In PMN, a respiratory burst occurs after the process of ingestion. This burst of oxygen consumption coincides with the generation of free radicals such as hydrogen peroxide, superoxide, and singlet oxygen (Kakinuma 1970; Johnston and Lehmeyer 1976). The chemistry of this free radical generation has been well reviewed by Klebanoff (1980) and Weiss and Lo Buglio (1982). The location of the oxidase responsible for this activity has been the subject of debate; but cytochrome *b* and myeloperoxidases are known to be involved, since failure of cytochrome *b* reduction leads to loss of the oxygen burst and of microbicidal activity (Segal and Jones 1980). However, it is possible that more than one cytochrome is involved.

The AM show a high level of resting oxygen metabolism that is elevated on stimulation and is necessary for the production of free radicals. It can be shown experimentally that patients suffering from certain respiratory disorders such as sarcoidosis do not show the same degree of free radical production as control subjects (estimated by chemiluminescence).

These highly reactive oxygen species, as well as possessing antimicrobial activity, are cytotoxic and may lead to cellular injury. This may be desirable, for example, as defence against neoplastic invasion, but it may also attack healthy tissue. A comprehensive review of the cytoxic effects of free radicals has been published recently (Weiss and Lo Buglio 1982).

Generally, a particle once engulfed is completely membrane bound and contained within a phagosome that undergoes fusion with the adjacent cytoplasmic granules. Normally, once fusion has occurred, the connecting membranes rupture, discharging the granular contents on to the particle that is digested. However, there is some evidence to suggest that on occasion fusion may occur before the phagosome has closed, or a phagosome containing granular enzymes may reopen to admit a further particle, resulting in leakage of lytic enzymes out of the cell. Weissmann and colleagues (1971) demonstrated that human PMNs will release lysosomal hydrolases without cytoplasmic leakage when they are treated with immune complexes or zymosan. He inferred that enzyme release occurred only when particles were ingested and argued that a lack of cytoplasmic enzymes could not be explained by any other mechanism (Weissman et al. 1971a, 1971b). Electron micrographs consistent with this hypothesis have been published by Henson (1971b, 1971c) and show phagosomes fused to granules and open to the extracellular fluid in order to admit a further particle. However, there is some doubt as to how major a contribution these mechanisms make to the release of cytotoxic enzymes in vivo (Henson 1980). Doubts arise because documented examples of "messy eating" occur infrequently and also because in PMNs and monocytes enzyme release continues after phagocytosis has ended, whereas one would expect the two events to finish together (Henson 1980). Finally, electron micrographs show that organelles are generally excluded from the area adjacent to the point of uptake, probably by the contractile apparatus discussed above. This is likely to pose a significant barrier to lysosome fusion with a developing phagosome (Berlin and Oliver 1978).

Enzyme Release, Antiproteases, and Emphysema

Pulmonary emphysema is defined as "an abnormal permanent increase in size of the respiratory portion of the lung distal to the terminal bronchioles, accompanied by destruction of their walls" (American Thoracic Society 1962).

The development of this disease generally follows one of two patterns. Centrilobular (centriacinar) emphysema generally affects the upper zones of the lung and is characterized by destruction of respiratory bronchioles with subsequent extension into the alveoli. The region of damage may show signs of particle deposition. In contrast, panlobular (panacinar) emphysema tends to affect the lower zones of the lung and involves destruction of large areas of acinar tissue. The normal acinar structure is destroyed and replaced by large air spaces.

The disease is confined almost exclusively to two overlapping groups of people, those who smoke tobacco and those who have a relatively rare genetic defect resulting in abnormally low plasma levels of the protease inhibitor α-$_1$-antitrypsin (α_1AT) (Eriksson 1964, 1965; Gadek et al. 1979; Auerbach et al. 1972).

The α_1AT is an acute-phase reactive glycoprotein, molecular weight of 54,000, with activity against a number of serine-dependent proteases, including elastase collagenase, trypsin, and chymotrypsin. It consists of a single polypeptide chain with three carbohydrate side chains incorporating a number of sialic acid residues. Variation of the side chains gives rise to three isoforms. The ratio of isoforms produced alters in favor of triantennary species during the acute reactive phase, for example, in response to inflammation (Vaughan et al. 1982). In 1967 Fagerhol and Laurell demonstrated that the α_1AT polypeptide chain also exists in a number of forms and postulated the existence of seven codominant alleles for the protein. There are now 26 known alleles, the gene products being identified by their different electrophoretic mobilities on starch gel or by isoelectric focusing (Breit and Penny 1980). The majority of the population (90%) possess the *MM* genotype, which gives rise to "normal" plasma levels of α_1AT (2.86 $\pm$ 0.75 mg/ml) (Lieberman et al. 1972), whereas a small minority possesses one or more variant genes. Mild deficiences of α_1AT have been associated with a number of inflammatory conditions and immune disorders, including rheumatoid arthritis, ankylosing spondylitis, asthma, and fibrosing alveolitis (Breit and Penny 1980). The frequency of the ZZ phenotype was estimated by Eriksson to be 0.024 (1965). This frequency is roughly the same in other populations studied (Cook 1974). Possession of the ZZ phenotype is associated with a severe deficiency in α_1AT. The circulating plasma levels of the inhibitor are reduced to 20% of normal, this being the result of poor secretion from the liver where it is synthesized. It has been suggested that liver transplant may be a way of combating severe α_1AT deficiency (Hood et al. 1980), although less drastic replacement therapy has also been tried (Gadek et al. 1980b). Hereditary α_1AT deficiency is strongly associated with the early development of pulmonary emphysema (Eriksson 1965). Other factors must be involved in the etiology of this disease, since the onset of development in ZZ subjects is earlier in smokers than nonsmokers and since virtually all nondeficient patients who contract emphysema have a history of smoking (Hutchison et al. 1971; Kueppers and Black 1974).

The strong predisposition of α_1AT-deficient subjects to emphysema led to the development of the protease–antiprotease theory, which was originally applied only to emphysema but has recently been applied to other lung diseases as well (Dunnill 1979; Bignon and de Crémoux 1980; Gadek et al. 1980; Dijkman et al. 1982). It is believed that damage to the respiratory bronchioles and alveoli is the result of uncontrolled proteolytic activity resulting from an excess of protease action or insufficient inhibitory capacity. Experimental emphysema can be induced in animals by intratracheal instillation of certain proteases including papain and porcine pancreatic and leukocytic elastase (Janoff et al. 1977; Snider et al. 1977; Karlinsky and Snider 1978). Proteases that lack any elastolytic activity, such as collagenase (Gadek et al. 1980), do not produce emphysema in experimental animals, so it seems that damage to the elastin network is a causal agent of emphysema. Circumstantial evidence for this is provided by "experiments of nature," such as cutis laxa patients who have an inborn error of elastin synthesis and often suffer from emphysema (Bignon and de Crémoux 1980). Likewise, the blotchy mouse, which also possesses an abnormal elastin structure, is predisposed to panlobular emphysema (Karlinsky and Snider 1978).

For an elastase to be involved in the etiology of emphysema it must have access to the elastin that is situated in the interstitium. Therefore, the enzyme may be derived from either the blood or the alveolar space. The cell types most likely to be involved are the PMN and the AM. In the normal lung, almost all the lung free cells are AM, with PMN constituting only a small percentage of the free cell profile (Warr 1979). However, PMN can be attracted into the lung by chemotactic agents when infection or inflammation is present (Roitt 1980). The PMN have been shown to synthesize a potent elastase (Janoff 1972). The activity of AM is less clear-cut. Human AM have been shown to secrete an elastolytic enzyme (Rodriguez et al. 1977), but it is not clear whether this enzyme is synthesized in the macrophage or whether the macrophage endocytoses and later secretes leukocyte elastase. In the mouse, there is evidence for the synthesis of a metal-dependent macrophage elastase (White et al. 1980; Banda and Werb 1981), whereas in humans the most recent evidence suggests that elastase is sequestered from leukocytes (Hinman et al. 1980; White et al. 1982). Certainly human AM carry leukocyte elastase receptors on their plasma membranes (Campbell et al. 1979).

Until recently α_1AT was the only antiprotease thought to be of significance in the etiology of emphysema. This is because α_1AT is the major antiprotease of the blood, accounting for 90% of the inhibitory capacity of the serum. However, α_1AT is only one of a group of antiproteases present in the lung. There are at least four antiproteases detectable in bronchoalveolar lavage fluid, α_1AT, α_2-macroglobulin (α_2M), α_1-antichymotrypsin (α_1A Ch), and antileukoprotease (ALP). The first three are serum derived, whereas ALP is locally produced in the upper airways (Ohlsson et al. 1977; Tegner and Ohlsson 1977; Hochstrasser et al. 1981). It appears that ALP is one of a family of acid-stable antiproteases with a molecular weight of 10,000–20,000 (Ohlsson 1979). It is produced in the serous glands of the upper airways (Ohlsson 1980) and possibly in the bronchioles by the Clara cells (Dijkman et al. 1982).

It remains for the contributions of these antiproteases to be fully evaluated. It has been suggested that ALP is of no significance in the alveolar region (Gadek et al. 1981), although recent studies have shown that this inhibitor is present at the periphery, albeit in small amounts (Tetley et al. 1983). ALP seems to be the predominant antiprotease of the upper airways (Ohlsson 1980) and may therefore be of more significance in diseases such as bronchitis than in emphysema.

The role of α_1A Ch is unknown. It has been suggested that it may be either locally produced or selectively concentrated from the serum (Stockley and Burnett 1980).

The α_2M is present only in very small amounts (Gadek et al. 1981) in the normal lung and not in all subjects because its great size precludes it from freely diffusing from the capillaries. However, where there is any sort of damage to the lung leading to leakage of serum proteins, α_2M can gain access to the interstitium and may be of greater significance. The molar ratio of α_2M:α_1AT in normal serum is about from 1:15 to 1:10. The relative avidities of α_1AT and α_2M for proteases are variable; for example, in the presence of an excess of normal serum, pancreatic elastase partitions predominantly into α_2M, whereas leukocytic elastase is preferentially bound to α_1AT. However, we have found that the latter binding is reversible so that any change in α_2M:α_1AT ratio or the presence of other competing proteases released as a result of inflammatory processes could lead to significant binding of leukocyte elastase in complexes with α_2M. A similar phenomenon has recently been shown by Beatty et al. (1982) in their studies on trypsin. This could be of pathological significance for two reasons. First, as we have shown (Baum and Sadeghi 1980) the α_2M complex of human elastase is fully active against substrates of low molecular weight and hence could influence the equilibrium between breakdown of native elastin and its resynthesis from precursors of low molecular weight (Galdston et al. 1979). Second, the α_2M complex with elastase in serum could be scavenged by circulating monocytes, eventually to reappear in the granules of AM as a latent form of elastase concentrated in the upper lobes and available for release on stimulation by smoke or could leak through a damaged epithelium thus establishing a positive feedback cycle. Similar conclusions have recently been drawn by Stone et al. (1982) as a result of their animal experiments. The physical characteristics and spectrum of activity of α_2M have been reviewed recently by van Leuven (1982).

Cigarette smoking and inflammatory disease cause a proliferation of macrophages in the lung (Harris et al. 1970; Warr 1979). Furthermore, cigarette smoke promotes elastase release from these cells (Rodriguez et al. 1977; Blue and Janoff 1978), perhaps as a result of particle phagocytosis (Weissman et al. 1971a) or secretion (Henson 1980). Baum has demonstrated significant release of lysosomal enzymes from human PMNs stimulated to phagocytose cigarette smoke particles (Hutchison et al. 1980).

A recent study has shown that protease levels are significantly greater at the lung surface in smokers than they are in nonsmokers (Smith et al. 1983). Under normal circumstances any proteases lost to the extracellular compartment should be inactivated by the antiproteases discussed earlier. However, Janoff and colleagues (1979) have demonstrated that exposure to cigarette smoke causes a marked reduction in the elastase inhibitory capacity (EIC) of α_1AT in rat lung and serum EIC in humans (Carp and Janoff 1978), findings that have been confirmed by Gadek and coworkers (1979) who found a dramatic decrease in the functional activity of α_1AT obtained from the bronchoalveolar lavage fluid of smokers. The available evidence suggests that this may be a result of oxidizing agents such as hydroxyl free radicals (Carp and Janoff 1979). These free radicals can be released from phagocytosing PMNs, and oxidizing agents are also present in cigarette smoke itself (Carp and Janoff 1978). Johnson and Travis (1979) showed that oxidation of the methionine active site of α_1AT led to complete and immediate loss of activity against a variety of proteases. The ALP is also prone to reduction of activity after cigarette smoke exposure (Janoff et al. 1980). However, antiproteases can also be damaged by factors other than oxidants, for example, proteases, which have been shown to cause fundamental changes to α_1AT. Release of large amounts of lysosomal enzymes could provide sufficient proteolytic challenge to overwhelm the antiproteases in the immediate area, thus causing localized damage and perhaps initiating a positive feedback loop.

The relationship between cigarette smoking and emphysema is a subtle one, since although cigarette smoke does reduce antiprotease potency (Janoff et al. 1979; Gadek et al. 1979) and elastase levels may be elevated (Lam et al. 1979; Harris et al. 1975), Baum has shown that cigarette smoke not only affects α_1AT but also inhibits elastase and other lysosomal hydrolases (Ejiofor et al. 1981). Furthermore, some studies have shown that tobacco smoking inhibits phagocytosis (Eichel and Shahrik 1969) and therefore presumably also reduces any phagocytosis-related enzyme release. Also, Baum has shown (Desai et al. 1978; Hutchison et al. 1980) that the soluble fraction of cigarette smoke inhibits particle-induced lysosomal enzyme release from PMNs of healthy, nonsmoking male subjects, whereas smoke apparently had no inhibitory effect on release from cells of emphysematous patients. It remains to be seen how general this effect is, what is its molecular basis, and whether it is an adaptive phenomenon of smokers or reflects a cellular feature predisposing certain individuals to emphysema. Chemotactic factors and chemotactic factor inhibitors are found in a variety of cell types and body fluid variations, and these factors could affect the numbers of PMNs drawn in response to a stimulus and therefore the amount of enzyme released (Abboud et al. 1979). Elastase and cathepsin G released from granulocytes have been shown to be chemotactic factor inactivators of variable potency, depending upon the factor inactivated (Brozna et al. 1977). Perhaps of more immediate relevance to the study of emphysema, patients deficient in serum α_1AT were also found to be deficient in a serum chemotactic factor inactivator active against both complement-dependent and -independent factors (Ward and Talamo 1973). Emphysema is relatively rare in women of the MM phenotype, and it is possible that this not only reflects the different smoking habits of women over the past 50 yr, but also a stabilizing effect of estrogens, resulting in a lowered release of elastase.

Any multifactorial description of how the balance of enzyme release and removal is disturbed in emphysema must explain why only some smokers of the MM phenotype develop the disease and with varying degrees of severity and in particular parts of the lung. It is tempting to ascribe lower zone emphysema to enzyme release from circulating leukocytes in the capillary bed and multilocular damage in the upper lobes to protease release from PAM. A recent hypothesis proposes that the localization of the disease is a function of the ventilation-perfusion (V–P) relationships in different zones. The V:P ratio varies from 3:1 in the lung apices to 0.6:1 at the bases. It is proposed that inhaled entities will be distributed parallel to ventilation and circulating α_1AT in parallel to perfusion. In centrilobular emphysema, which is the more common form in smokers who have normal α_1AT levels, the high V:P ratio of the upper zones would mean a high irritant:α_1AT ratio, thus overwhelming the protective capacity of the α_1AT. In the panlobular form, which is the more predominant among α_1AT-deficient patients, the authors propose that the area affected by the disease relates not to V:P but to ventilation alone, which is greater at the base of the lung than in the upper zone (Cockcroft and Horne 1982).

Summary

The research effort put into the study of emphysema, particularly in relation to cigarette smoking, is large, and the body of data accumulated is often conflicting, but certain factors seem clear. Emphysema patients are almost always smokers or exsmokers or they are α_1AT deficient, the latter factor suggesting an important role for antiproteases in pulmonary homeostasis. The involvement of PMN elastase has been demonstrated, and a role for AM elastase is also possible. Mechanisms for enzyme release have been postulated, but it is not clear how the balance of enzyme release and inhibition and of elastin breakdown and resynthesis are affected by the various components of tobacco smoke and by endogenous factors. Neither is it understood how this balance varies between individuals, so that some smokers are protected from this crippling disease. Once the complete mechanism has been fully revealed, the probability of evolving effective treatment will be improved and it may become possible to identify individuals at risk who can then be counseled against cigarette smoking or provided with appropriate prophylactic treatment.

References

Abboud, R.T., Rushton, J-M., and Grzybowski, S. (1979). Interrelationships between neutrophil elastase, serum alpha-antitrypsin, lung function and chest radiography in patients with chronic airflow obstruction. *Am. Rev. Respir. Dis. 120,* 31–40.

Allison, A.C., Davies, P., and De Petris, S. (1971). Role of contractile microfilaments in macrophage movement and endocytosis. *Nature New Biol. 232,* 153–155.

Allison, A.C., Harington, J.S., and Birbeck, M. (1966). An examination of cytotoxic effects of silica on macrophages. *J. Expl. Med. 124,* 141–153.

American Thoracic Society (1962). Chronic bronchitis, asthma, and pulmonary emphysema. *Am. Rev. Respir. Dis. 85,* 762–768.

Arnold, W.P., Aldred, R., and Murad, F. (1977). Cigarette smoke activates guanylate cyclase and increases guanosine 3′,5′-monophosphate in tissues. *Science 198,* 934-936.

Auerbach, O., Hammond, E.C., Garfinkel, L., and Benante, C. (1972). Relation of smoking and age to emphysema. Whole-lung section study. *N. Engl. J. Med. 286,* 853–857.

Axline, S.G. and Reaven, E.P. (1974). Exhibition of phagocytosis and plasma membrane mobility of the cultivated macrophage by cytochalasin B. *J. Cell Biol. 62,* 647–659.

Badenoch-Jones, P. and Baum, H. (1973). Progesterone-induced permeability changes in rat liver lysosomes. *Nature New Biol. 242,* 123–124.

Baggiolini, M., Hirsch, J.G., and de Duve, C. (1969). Resolution of granules from rabbit heterophil leukocytes into distinct populations by zonal sedimentation. *J. Cell Biol. 40,* 529–541.

Banda, M.J. and Werb, Z. (1981). Mouse macrophage elastase purification and characterization as a metalloproteinase. *Biochem. J. 193,* 589–603.

Baum, H. and Sadeghi, P. (1980). In Proc. 13th FEBS Meeting, Jerusalem, Israel. C2-P16, 152 Interactions of elastase with protease inhibitors of human sera.

Beatty, K., Travis, J., and Bieth, J. (1982). The effect of α_2 macroglobulin on the interaction of α-proteinase inhibitor with porcine trypsin. *Biochim. Biophys. Acta 704,* 221–226.

Berlin, R.D. and Oliver, J.M. (1978). Analogous ultrastructure and surface properties during capping and phagocytosis in leukocytes. *J. Cell Biol. 77,* 789–804.

Bignon, J. and de Crémoux, H. (1980). Pathological and pathogenetic aspects of chronic obstructive lung disease. *Bull. Europ. Physiopath. Resp. 16* (Suppl.), 13–25.

Blue, M-L. and Janoff, A. (1978). Possible mechanisms of emphysema in cigarette smokers: Release of elastase from human polymorphonuclear leukocytes by cigarette smoke condenate in vitro. *Am. Rev. Respir. Dis. 117,* 317-325.

Breit, S.N. and Penny, R. (1980). The role of alpha$_1$ protease inhibitor (alpha$_1$ antitrypsin) in the regulation of immunologic and inflammatory reactions. *Aust. N.A. J. Med. 10,* 449-453.

Brozna, J.P., Senior, R.M., Reutzer, D.L.K., and Ward, P.A. (1977). Chemotactic factor inactivators of human granulocytes. *J. Clin. Invest. 60,* 1280-1288.

Burnett, D. and Stockley, R.A. (1981). Serum and sputum alpha-2-macroglobulin in patients with chronic obstructive airways disease. *Thorax 36,* 512-516.

Campbell, E.J., White, R.R., Senior, R.M., Rodriguez, R.J., and Kuhn, C. (1979). Receptor-mediated binding and internalization of leukocyte elastase by alveolar macrophages in vitro. *J. Clin. Invest. 64,* 824-833.

Carp, H. and Janoff, A. (1978). Possible mechanisms of emphysema in smokers: In vitro suppression of serum elastase inhibitory capacity by fresh cigarette smoke and its prevention by antioxidants. *Am. Rev. Respir. Dis. 118,* 617-621.

Carp, H. and Janoff, A. (1979). In vitro suppression of serum elastase inhibitory capacity by reactive oxygen species generated by phagocytosing polymorphonuclear leukocytes. *J. Clin. Invest. 63,* 793-797.

Cockcroft, D.W. and Horne, S.L. (1982). Localization of emphysema within the lung: An hypothesis based upon ventilation/perfusion relationships. *Chest 82,* 483-487.

Cohen, A.B. and Cline, M.J. (1971). The human alveolar macrophage: Isolation, culturation in vitro, and studies of morphologic and functional characteristics. *J. Clin. Invest. 50,* 1390-1398.

Cohn, Z.A. and Morse, S.I. (1960). Functional and metabolic properties of polymorphonuclear leukocytes. I. Observations on the requirements and consequences of particle ingestion. *J. Exptl. Med. 111,* 667-687.

Cook, P.J.L. (1974). Genetic aspects of Pi system. *Postgrad. Med. J. 50,* 362-364.

Crawford, N., Chahal, H., and Jackson, P. (1980). The isolation and characterization of guinea-pig polymorphonuclear leucocyte actin and myosin. *Biochim. Biophys. Acta 626,* 218-233.

Czop, J.K., McGowan, S.E., and Center, D.M. (1982). Opsonin-independent phagocytosis by human alveolar macrophages: Augmentation by human plasma fibronectin. *Am. Rev. Respir. Dis. 125,* 607-609.

Desai, R., Baum, H., Bellamy, D., and Hutchison, D.C.S. (1978). Induction of lysosomal enzyme-release from leukocytes of normal and emphysematous subjects and effects of cigarette smoke. *Clin. Sci. Mol. Med. 54,* 15P (Abstract).

Dijkman, J.H., Franken, C., Kramps, J.A., and Meijer, C.J.L.M. (1982). Enzymes and enzyme-inhibitors in the small airways. *Eur. J. Respir. Dis. 63* (Suppl.) *121,* 53-59.

Dunnill, M.S. (1979). Etiology of emphysema. *Bull. Eur. Physiopath. Resp. 15,* 1015-1029.

Dustin, P., Tondeur, M., and Libert, J. (1978). Metabolic and storage diseases. In *Microtubules.* Springer-Verlag, New York, vol. 2, pp. 151–245.

Edelson, H.S., Kaplan, H.B., Korchak, H.M., Smolen, J.E., and Weissmann, G. (1982). Dissociation by piroxicam of degranulation and superoxide anion generation from decrements in chlortetracycline fluorescence of activated human neutrophils. *Biochem. Biophys. Res. Comm. 104,* 277–283.

Eichel, B., and Shahrik, H.A. (1969). Tobacco smoke toxicity: Loss of human oral leukocyte function and fluid cell metabolism. *Science 166,* 1424–1428.

Ejiofor, A., Hutchison, D.C.S., and Baum, H. (1981). Inhibition of human leukocyte elastase activity by cigarette smoke. *Biosci. Rep. 1,* 715–720.

Eriksson, S. (1964). Pulmonary emphysema and alpha$_1$-antitrypsin deficiency. *Acta Med. Scand. 175,* 197–205.

Eriksson, S. (1965). Studies in alpha$_1$-antitrypsin deficiency. *Acta Med. Scand.* Suppl. *177,* 432.

Farber, J.L. (1981). Mini review: The role of calcium in cell death. *Life Sci. 29,* 1289–1295.

Fagerhol, M.K. and Laurell, C.-B. (1967). The polymorphism of "prealbumins" and alpha$_1$-antitrypsin in human sera. *Clin. Chim. Acta 16,* 199–203.

Fox, J.E.B. and Phillips, D.R. (1982). Role of phosphorylation in mediating the association of myosin with the cytoskeletal structures of human platelets. *J. Biol. Chem. 257,* 4120–4126.

Gadek, J.E., Fells, G.A., and Crystal, R.G. (1979). Cigarette smoking induces functional antiprotease deficiency in the lower respiratory tract of humans. *Science 206,* 1315–1316.

Gadek, J.E., Fells, G.A., Zimmerman, R.L., Rennard, S.I., and Crystal, R.G. (1981). Antielastases of the human alveolar. Structures implications for the protease-antiprotease theory of emphysema. *J. Clin. Invest. 68,* 889–898.

Gadek, J.E., Fulmer, J.D., Gelfand, J.A., Frank, M.M., Petty, T.L., and Crystal, R.G. (1980b). Danazol-induced augmentation of serum α_1-antitrypsin levels in individuals with marked deficiency of this antiprotease. *J. Clin. Invest. 66,* 82–87.

Gadek, J.E., Hunninghake, G.W., Fells, G.A., Zimmerman, R.L., Keogh, B.A., and Crystal, R.G. (1980). Evaluation of the protease-antiprotease theory of human destructive lung disease. *Bull. Eur. Physiopath. Resp.* (Suppl.) *16,* 27–40.

Galdston, M., Levytska, V., Leiner, I.E., and Twumasi, D. (1979). Degradation of tropoelastin and elastin substrates by human neutrophil elastase, free and bound to α_2-macroglobulin in serum of the M and Z (Pi) phenocytes for alpha-antitrypsin. *Am. Rev. Respir. Dis. 119,* 435–441.

Giordano, G.F., and Lichtman, M.A. (1973). The role of sulfhydryl groups in human neutrophil adhesion, movement and particle ingestion. *J. Cell Physiol. 82,* 387–396.

Goldstein, J.L., Anderson, R.G.W., and Brown, M.S. (1979). Coated pits, coated vesicles, and receptor-mediated endocytosis. *Nature 279,* 679–685.

Grey, H.M. and Anderson, C.L. (1979). Structural characteristics of F_c receptors on macrophages. In *Mononuclear Phagocytes. Part I. Functional Aspects.* van Furth, R., ed. Martinus Nijhoff Pub., pp. 757-78.

Gruenstein, E. and Rich, A. (1973). Actin associated with plasma membranes from 3T3 and HeLa-cells. *J. Cell Biol. 59,* 127a.

Guckian, J.C., Cavallo, T., and Fine, D.P. (1981a). Opsonization of pneumococci. II. Metabolic effects of a "third" human activity that mediates intracellular killing. *J. Immunol. 127,* 1659-1665.

Guckian, J.C., Christensen, G.D., Schweinle, J.E., and Fine, D.P. (1981b). I. Heat labile serum activity other than complement is required for killing by human polymorphonuclear leukocytes. *J. Immunol. 127,* 1666-1670.

Harris, J.O., Olsen, G.N., Castle, J.R., and Maloney, A.S. (1975). Comparison of proteolytic enzyme activity in pulmonary alveolar macrophages and blood leukocytes in smokers and nonsmokers. *Am. Rev. Respir. Dis. 111,* 579-586.

Harris, J.O., Swenson, E.W., and Johnson, J.E. (1970). Human alveolar macrophages: Comparison of phagocytic ability, glucose utilization and ultrastructure in smokers and non-smokers. *J. Clin. Invest. 49,* 2086-2096.

Hartwig, J.H. and Stossel, T.P. (1976). Interactions of actin, myosin, and an actin-binding protein of rabbit pulmonary macrophages. III. Effects of cytochalasin B. *J. Cell Biol. 71,* 295-303.

Henson, P.M. (1971a). The immunologic release of constituents from neutrophil leukocytes. I. The role of antibody and complement on non-phagocytosable surfaces or phagocytosable particles. *J. Immunol. 107,* 1535-1546.

Henson, P.M. (1971b). The immunologic release of constituents from neutrophil leukocytes. II. Mechanisms of release during phagocytosis and adherence to non-phagocytosable surfaces. *J. Immunol. 107,* 1547-1557.

Henson, P.M. (1971c). Interaction of cells with immune complexes: Adherence, release of constituents, and tissue injury. *J. Exp. Med. 134,* 114s-135s.

Henson, P.M. (1980). Mechanisms of exocytosis in phagocytic inflammatory cells. *Am. J. Pathol. 101,* 494-511.

Hinman, L.M., Stevens, C.A., Matthay, R.A., and Gee, J.B.L. (1980). Elastase and lysozyme activities in human alveolar macrophages. Effects of cigarette smoking. *Am. Rev. Respir. Dis. 121,* 263-271.

Hochstrasser, K., Albrecht, G.J., Schönberger, Ö.L., Rasche, B., and Lempart, K. (1981). An elastase-specific inhibitor from human bronchial mucus. I. Isolation and characterization. *Hoppe-Seylers Z. Physiol. Chem. 362,* 1369-1375.

Hood, J.M., Koep, L.J., and Peters, R.L. (1980). Liver transplantation for advanced liver disease with alpha-1-antitrypsin deficiency. *N. Eng. J. Med. 302,* 272-275.

Hutchison, D.C.S., Cook, P.J.L., Barter, C.E., Harris, H., and Hugh-Jones, P. (1971). Pulmonary emphysema and alpha-1-antitrypsin deficiency. *Brit. Med. J. 1,* 689-694.

Hutchison, D.C.S., Desai, R., Bellamy, D., and Baum, H. (1980). The induction of lysosomal enzyme release from leukocytes of normal and emphysematous subjects and the effects of tobacco smoke upon phagocytosis. *Clin. Sci. 58,* 403–409.

Janoff, A. (1972). Human granulocyte elastase: Further delineation of its role in connective tissue damage. *Am. J. Pathol. 68,* 579–592.

Janoff, A., Carp, H., Lee, D.K., and Drew, R.T. (1979). Cigarette smoke inhalation decreases antitrypsin activity in rat lung. *Science 206,* 1313–1314.

Janoff, A., Carp, H., and Lee, D.K. (1980). Inactivation of alpha$_1$-proteinase inhibitor and bronchial mucous proteinase inhibitor by cigarette smoke in vitro and in vivo. *Bull. Eur. Physiopathol. Resp.* (Suppl.) *16,* 321–338.

Janoff, A., Sloan, B., Weinbaum, G., Damiano, V., Sandhaus, R.A., Elias, J., and Kimbel, P. (1977). Experimental emphysema induced with purified human neutrophil elastase: Tissue localization of the instilled protease. *Am. Rev. Respir. Dis. 115,* 461–478.

Jasin, H.E. (1972). Human heat labile opsonins: Evidence for their mediation via the alternative pathway of complement activation. *J. Immunol. 109,* 26–31.

Jennings, L.K., Fox, J.E.B., Edwards, H.H., and Phillips, D.R. (1981). Changes in the cytoskeletal structure of human platelets following thrombin activation. *J. Biol. Chem. 256,* 6927–6932.

Johnson, D. and Travis, J. (1979). The oxidative inactivation of human α-1-proteinase inhibitor. Further evidence for methionine at the reactive center. *J. Biol. Chem. 254,* 4022–4026.

Johnson, F.R., Agnello, V., and Williams, R.C. (1972). Opsonic activity in human serum deficient in C2. *J. Immunol. 109,* 141–145.

Johnson, N. McI., McNichol, M.W., Kapaer, A., Burtonkee, J.E., and Mowbray, J.F. (1980). Defective yeast opsonisation of serum in tuberculosis. *Thorax 35,* 523–525.

Johnston, R.B. and Lehmeyer, J.E. (1976). Elaboration of toxic oxygen byproducts by neutrophils in a model of immune complex disease. *J. Clin. Invest. 57,* 836–841.

Johnston, R.B., Klemperer, M.R., Alper, C.A., and Rosen, F.S. (1969). The enhancement of bacterial phagocystosis by serum: The role of complement components and two cofactors. *J. Expl. Med. 129,* 1275–1290.

Kakinuma, K. (1970). Metabolic control and intracellular pH during phagocytosis by polymorphonuclear leukocytes. *J. Biochem. (Tokyo) 68,* 177–185.

Kane, A.B., Stantron, R.P., Raymond, E.G., Dobson, M.E., Knafelc, M.E., and Farber, J.L. (1980). Dissociation of intracellular lysosomal rupture from the cell death caused by silica. *J. Cell Biol. 87,* 643–657.

Karlinsky, J.B. and Snider, G.L. (1978). Animal models of emphysema. *Am. Rev. Respir. Dis. 117,* 1109–1133.

Keeling, P.J. and Henson, P.M. (1982). Lysosomal enzyme release from human monocytes in response to particulate stimuli. *J. Immunol. 128,* 563–567.

Klebanoff, S.J. (1980). Oxygen metabolism and the toxic properties of phagocytes. *Ann. Intern. Med. 93,* 480-489.

Kobayashi, Y. and Usui, T. (1982). Opsonic activity against zymosan of normal human serum of different blood types. *Exper. 38,* 131-132.

Koerten, H.K., Ploem, J.S., and Daems, W.T. (1980). Ingestion of latex beads by filopodia of adherent mouse peritoneal macrophages. *Exp. Cell Res. 128,* 470-475.

Korchak, H.M., Elsenstat, B.A., Hoffstein, S., Dunham, P.B., and Weissmann, G. (1980). Anion channel blockers inhibit lysosomal enzyme secretion from human neutrophils with affecting generation of superoxide anions. *Proc. Natl. Acad. Sci. USA 77,* 2721-2725.

Korchak, H.M., Elsenstat, B.A., Smolen, J.E., Rutherford, L.E., Dunham, P.B., and Weissmann, G. (1982). Stimulus-response coupling in the human neutrophil: The role of anion fluxes in degranulation. *J. Biol. Chem. 257,* 6916-6922.

Kueppers, F. and Black, L.F. (1974). α_1-Antitrypsin and its deficiency. *Am. Rev. Respir. Dis. 110,* 176-194.

Lam, S., Abboud, R.T., Chan-Yeung, M., and Rushton, J-M. (1979). Neutrophil elastase and pulmonary function in subjects with intermediate alpha-$_1$-antitrypsin deficiency (MZ phenotype). *Am. Rev. Respir. Dis. 119,* 941-951.

Leffell, M.S. and Spitznagel, N.K. (1975). Fate of human lactoferrin and myeloperoxidase in phagocytizing human neutrophils: Effects of immunoglobulin G subclasses and immune complexes control on latex beads. *Infect. Immunol. 12,* 813-820.

Leist-Welsh, P. and Bjornson, A.B. (1982). Immunoglobulin-independent utilisation of the classical complement pathway in opsonophagocytosis of Escherichia coli by human peripheral leukocytes. *J. Immunol. 128,* 2643-2651.

Leslie, R.G.Q. and Alexander, M.D. (1979). Quantitative studies on the interaction of macrophages with antibodies and immune complexes. In *Mononuclear Phagocytes. Part I. Functional Aspects.* van Furth, R., ed. Martinus Nijhoff Pub., pp. 783-802.

Lieberman, J., Gaidulis, L., Garoutte, B., and Mittman, C. (1972). Identification and characteristics of the common alpha$_1$-antitrypsin phenotypes. *Chest 62,* 557-554.

McCarthy, K. and Henson, P.M. (1979). Induction of lysosomal enzyme secretion by alveolar macrophages in response to the purified complement fragments C5a and C5a des-arg. *J. Immunol. 123,* 2511-2517.

McCarthy, K., Musson, R.A., and Henson, P.M. (1982). Protein synthesis-dependent and protein synthesis-independent secretion of lysosomal hydrolases from rabbit and human macrophages. *J. Reticuloendothel. Soc. 31,* 131-144.

Maallem, H., Sheppard, K., and Fletcher, J. (1982). The discharge of primary and secondary granules during immune phagocytosis by normal and chronic granulocytic leukemia polymorphonuclear neutrophils. *Br. J. Haematol. 51,* 201-208.

Mason, R.J., Stossel, T.P., and Vaughan, M. (1973). Quantitative studies of phagocytosis by alveolar macrophages. *Biochim. Biophys. Acta 304,* 864-870.

Metchnikoff, E. (1892). Lecons sur la pathologic comparée de l'imflammation. Paris: G. Masson.

Nash, T., Allison, A.C., and Harington, J.S. (1966). Physio-chemical properties of silica in relation to its toxicity. *Nature 210,* 259-261.

Ohlsson, K. (1979). The low molecular weight proteinase inhibitor in bronchial mucus. *Les Coll. d'Inserm. 84,* 105-118.

Ohlsson, K. (1980). Interactions between granulocyte proteases and protease inhibitors in the lung. *Bull. Eur. Physiopath. Resp. 16* (Suppl.), 209-222.

Ohlsson, K., Tegner, H., and Åkesson, U. (1977). Isolation and partial characterization of a low molecular weight acid stable protease inhibitor from human bronchial secretion. *Hoppe Seylers Z. Physiol. Chem. 358,* 583-589.

Painter, R.G., Whisenand, J., and McIntosh, A.T. (1981). Effects of cytochalasin B on actin and myosin in association with particle binding sites in mouse macrophages: Implications with regard to the mechanism of action of the cytochalasins. *J. Cell Biol. 91,* 373-384.

Quie, P.G., Messner, R.P., and Williams, R.C. (1968). Phagocytosis in subacute bacterial endocarditis: Localization of the primary opsonic site to F_c fragment. *J. Expl. Med. 128,* 553-570.

Reaven, E.P. and Axline, S.G. (1973). Subplasmalemmal microfilaments and microtubules in resting and phagocytozing cultivated macrophages. *J. Cell Biol. 59,* 12-27.

Reid, R.A. and Leech, R.M. (1980). Biochemistry and structure of cell organelles. Publisher Blackie.

Riches, D.W.H. and Stanworth, D.R. (1982). Evidence for a mechanism for the initiation of acid hydrolase secretion by macrophages that is functionally independent of alternative pathway complement activation. *Biochem. J. 202,* 639-645.

Rodriguez, R.J., White, R.R., Senior, R.M., and Levine, E.A. (1977). Elastase release from human alveolar macrophages: Comparison between smokers and nonsmokers. *Science 198,* 313-314.

Roitt, I. (1980). Essential immunology. Blackwell Scientific Publications.

Saier, M.H. and Stiles, C.D. (1975). Molecular dynamics in biological membranes. *Heidelberg Science Library 22.* New York: Springer-Verlag.

Sbarra, A.J. and Karnovsky, M.L. (1959). The biochemical basis of phagocytosis. I. Metabolic changes during the ingestion of particles by polymorphonuclear leukocytes. *J. Biol. Chem. 234,* 1355-1362.

Segal, A.W. and Jones, O.T.G. (1980). Absence of cytochrome-b reduction in stimulated neutrophils from both female and male patients with chronic granulomatous disease. *FEBS Lett. 110,* 111-114.

Shaw, J.O., Henson, P.M., Henson, J., and Webster, R.O. (1980). Lung inflammation induced by complement-derived chemotactic fragments in the alveolus. *Lab. Invest. 42,* 547-558.

Shirahama, T. and Cohen, A. (1974). Ultrasound evidence for leakage of lysosomal contents after phagocytosis of monosodium urate crystals. *Am. J. Pathol. 76,* 501-520.

Skutelsky, E. and Danon, D. (1969). Reduction in surface charge as an explanation of the recognition by macrophages of nuclei expelled from normoblasts. *J. Cell Biol. 43,* 8-15.

Smith, S., Burton, G., Cooke, N., Guz, A., and Tetley, T. (1983). Enhanced free protease activity in the lungs of smokers. *Clin. Sci. 64* 32P (Abstract).

Smolen, J.E., and Weissmann, G. (1981). Stimuli which provoke secretion of azurophilic enzymes from human neutrophils induce increments in adenosine cyclic 3′,5′-monophosphate. *Biochim. Biophys. Acta 672,* 197-206.

Smolen, J.E., Korchak, H.M., and Weissmann, G. (1980). Increased levels of cyclic adenosine 3′,5′-monophosphate in human polymorphonuclear leukocytes after surface-stimulation. *J. Clin. Invest. 65,* 1077-1085.

Snider, G.L., Sherter, C.B., Koo, K.W., Karlinsky, J., Hayes, J.A., and Franzblau, C. (1977). Respiratory mechanics in hamsters following treatment with endotracheal elastase or collagenase. *J. Appl. Physiol.: Resp. Environ. Exercise Physiol. 42,* 206-215.

Spitznagel, J.K., Dalldorf, F.G., Leffell, M.S., Folds, J.D., Welsh, I.R.H., Cooney, M.H., and Martin, L.E. (1974). Character of azurophil and specific granules purified from human polymorphonuclear leukocytes. *Lab. Invest. 30,* 774-785.

Stendahl, O.I., Hartwig, J.H., Brotschi, E.A., and Stossel, T.P. (1980). Distribution of actin-binding protein and myosin in macrophages during spreading and phagocytosis. *J. Cell Biol. 84,* 215-224.

Stockley, R.A. and Burnett, D. (1980). Alpha$_1$-antichymotrypsin in infected and noninfected sputum. *Am. Rev. Respir. Dis. 122,* 81-88.

Stone, P.J., Calore, J.D., Snider, G.L., and Franzblau, C. (1982). Role of α_2-macroglobulin-elastase complexes in the pathogenesis of elastase-induced emphysema in hamsters. *J. Clin. Invest. 69,* 920-931.

Stossel, T.P. (1975). Phagocytosis: Recognition and ingestion. *Seminar in Haematol. 12,* 83-116.

Stossel, T.P., Mason, R.J., Hartwig, J., and Vaughan, M. (1972). Quantitative studies of phagocytosis by polymorphonuclear leukocytes: Use of emulsions to measure the initial rate of phagocytosis. *J. Clin. Invest. 51,* 615-624.

Tegner, H. and Ohlsson, K. (1977). Localization of a low molecular weight protease inhibitor to tracheal and maxillary sinus mucosa. *Hoppe-Seylers Z. Physiol. Chem. 358,* 425-429.

Tetley, T., Cooke, N., Burton, G., Guz, A., and Smith, S. (1983). Detection of antileucoprotease and elastase activity in peripheral lung lavage. *Clin. Sci. 64* 11P (Abstract).

Unkeless, J.C. (1979). F_c receptors on mouse macrophages. In *Mononuclear Phagocytes. Part I. Functional Aspects.* van Furth, R., ed. Martinus Nijhoff Pub., pp. 735–751.

Van Leuven, F. (1982). Human alpha$_2$ macroglobulin structure and function. *Trends Biochem. Sci. 7,* 185–187.

Vaughan, L., Lorier, M.A., and Carrell, R.W. (1982). α_1-antitrypsin microheterogeneity. Isolation and physiological significance of isoforms. *Biochim. Biophys. Acta 701,* 339–345.

Ward, P.A. and Talamo, R.C. (1973). Deficiency of the chemotactic factor inactivator in human sera with α_1-antitrypsin deficiency. *J. Clin. Invest. 52,* 516–519.

Warr, G.A. (1979). The biology of normal human bronchoalveolar cells. *Les Coll. d'Inserm. 84,* 137–158.

Weiss, S.J. and LoBuglio, A.F. (1982). Phagocyte-generated oxygen metabolites and cellular injury. *Lab. Invest. 47,* 5–18.

Weissmann, G., Zurier, R.B., Spieler, P.J., and Goldstein, I.M. (1971). Mechanisms of lysosomal enzyme release from leukocytes exposed to immune complexes and other particles. *J. Expl. Med. 134,* 149s–165s.

Weissmann, G., Dukor, P., and Zurier, R.B. (1971b). Effect of cyclic AMP on release of lysosomal enzymes from phagocytes. *Nature New Biol. 231,* 131–135.

Wells, W.W., Collins, C.A., and Kurtz, J.W. (1981). Metabolic regulation of lysosome activity. In *Lysosomes and Lysosomal Storage Diseases.* Callahan, J.W. and Lowden, J.A., eds. Raven Press.

Werb, Z. (1979). Hormone receptors and hormonal regulation of macrophage physiological functions. In *Mononuclear Phagocytes. Part I. Functional Aspects.* van Furth, R., ed. Martinus Nijhoff Pub., pp. 809–829.

Wessells, N.K., Spooner, B.S., Ash, J.F., Bradley, M.O., Luduena, M.A., Taylor, E.L., Wrenn, J.T., and Yamada, K.M. (1971). Microfilaments in cellular and developmental processes. *Science 171,* 135–143.

White, R., Janoff, A., Gordon, R., and Campbell, E. (1982). Evidence for in vitro internalization of human leukocyte elastase by alveolar macrophages. *Am. Rev. Respir. Dis. 125,* 779–781.

White, R., Norby, D., Janoff, A., Dearing, R. (1980). Partial purification and characterization of mouse peritoneal exudative macrophage elastase. *Biochim. Biophys. Acta 612,* 233–244.

Winkelstein, J.A., Shin, H.S., and Wood, W.B. (1972). Heat labile opsonins to pneumococcus. III. The participation of immunoglobulin and of the alternate pathway of C_3 activation. *J. Immunol. 108,* 1681–1689.

Wright, A.E., and Douglas, S.R. (1903). An experimental investigation of the role of the blood fluids in connection with phagocytosis. *Proc. Roy. Soc. (London) B. Biol. Sci. 72,* 357–362.

Zurier, R.B., Weissmann, G., Hoffstein, S., Kammersman, S., and Tai, H-H. (1974). Mechanisms of lysosomal enzyme release from human leukocytes. II. Effects of cAMP and cGMP. Autonomic agonists, and agents which affect microtubulic function. *J. Clin. Invest. 53,* 297–309.

13

Regulation of Erythropoiesis and Possible Implications of the Cell Membrane

I. Baciu

The regulation of erythropoiesis and, finally, of the erythrocyte count and of the amount of circulating hemoglobin on which the volume of oxygen transported by the blood from the lungs toward the tissues is directly dependent, represents a fundamental process of the system of oxygen homeostasis. Since 1954 our research group has investigated the mechanism by which erythropoiesis is activated by hypoxia. Originally two contradicting explanations existed: one, implicating the direct effect of hypoxia, without any mediation of the nervous system, in the release of certain substance exerting a stimulating effect upon the hematogenic bone marrow (Carnot and Deflandre 1906; Grant and Root 1952; Stohlman et al. 1954) and the other, suggesting hypoxia activates erythropoiesis by means of the nervous system (Schulhof and Mathies 1927; Hoff 1933; Seip 1953).

Results and Discussion

The Nervous Regulation of the Erythropoiesis

In the first investigations (Baciu 1957) it was reported that following isolation of the encephalic centers of rats by sectioning the spinal cord in the cervical area, hypoxia simulating an altitude of 5 km has no effect on reticulocytosis and polyglobulia after 3 and 4 days, respectively (Fig. 13.1). Marked effects on reticulocytosis and polyglobulia, in contrast, are observed in control animals subject to hypoxia.

The nervous regulation was analyzed by our group (Baciu et al. 1960) in experiments performed on dogs with surgical separation of the somatocephalic blood flow (chronic isolated head). This preparation is illustrated in Fig. 3.2. Cephalic ischemic hypoxia, obtained in the dogs prepared in this way, 8 hr daily for 3 days, by partial compression of the tunnelized common carotid arteries, produced an erythrocytic response with a progressive increase of the reticulocyte count beginning on the second day (Fig. 13.3). Arterial cephalic

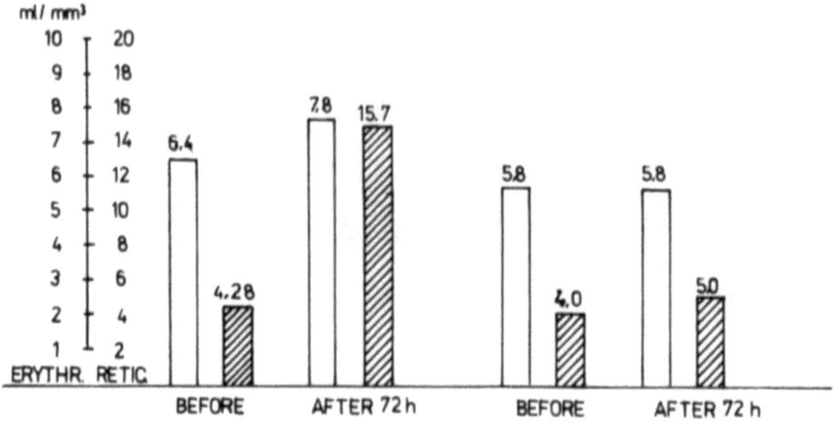

Figure 13.1. Erythrocytic reaction to hypoxia (5-km simulated altitude in a hypobaric chamber) in controls ☐ and in rats with sectioned spinal cord ▨ in the cervical area.

hypoxia required for this response involves a reduction of between 40 and 60% in the blood flow in the isolated head such that the tissue PO_2 in the region of the hypothalamus reaches a critical value of about 20 mm Hg (Baciu 1977). If the sinocarotid chemoceptive areas are denervated by sectioning Hering's nerves, hypoxia induced in these animals as before still results in an erythrocytic response, but this is only manifest after a latent period of 3 days. In these animals the response is invariably of greater intensity. Erythropoietin increases more and for a longer period in the peripheral blood (Fig. 13.4).

These results suggest that hypoxia exerts both a chemoceptive action, in a reflex manner, as well as a direct action on the central nervous system, possibly

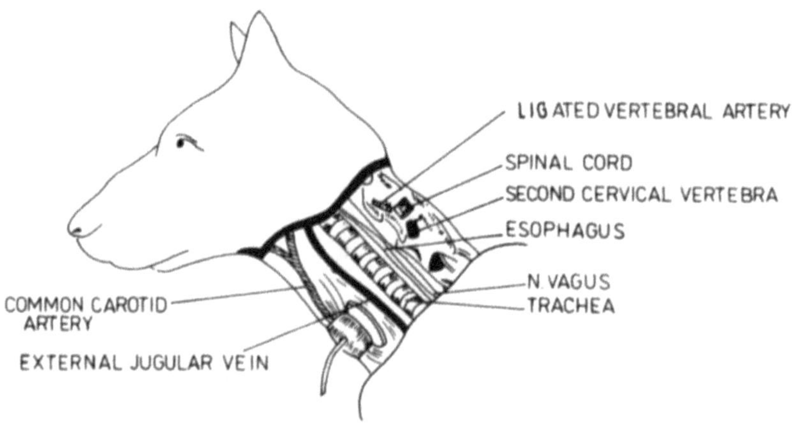

Figure 13.2. Schematic representation of chronic separation of the somatocephalic blood flow in dogs.

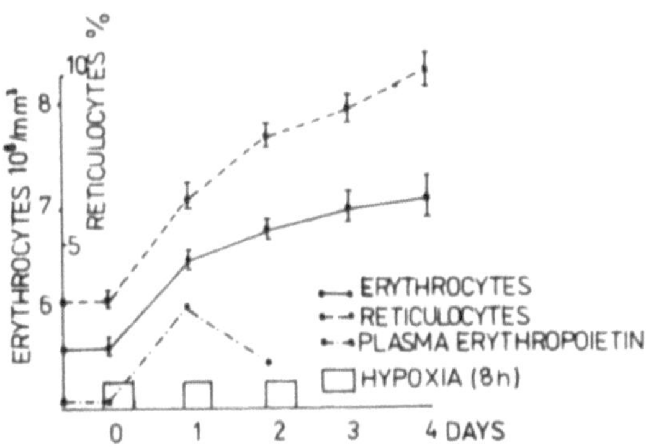

Figure 13.3. Erythrocytic reaction induced by cephalic hypoxia achieved for 3 days (8 h daily) by compression of the carotid arteries in dogs with intact sinocarotid reflexogenic areas (mean levels). •–•, Erythrocytes; o– –o, reticulocytes; •– • –•, plasma erythropoietin; □, hypoxia.

on some selected neuronal groups. By direct stimulation of the chemoceptive areas, a number of immediate reflexes are released such as the adjustment to hypoxia, including hypertension, changes in the pulse rate, polypnea, yawning reflexes, and intensifications of the cytodiabasis. Obviously, erythropoietin secretion is achieved through stimulation of the chemoreceptors.

In the absence of the reflex reponses following sinocarotid denervation, however, the same hypobaric hypoxia results in a greater and more prolonged increase of erythropoietin secretion followed by a more marked polyglobulia. In subsequent experiments Něcas and Thorling (1972) infused cyanide intravenously into rats and rabbits exposed to hypoxia and found out that, although the sensitivity of the sinusal and aortic chemoreceptors was blocked, erythropoietin secretion was unaffected. Paulo et al (1973) on the other hand, described a persistent erythrocytic response to hypoxia following sinocarotid denerva-

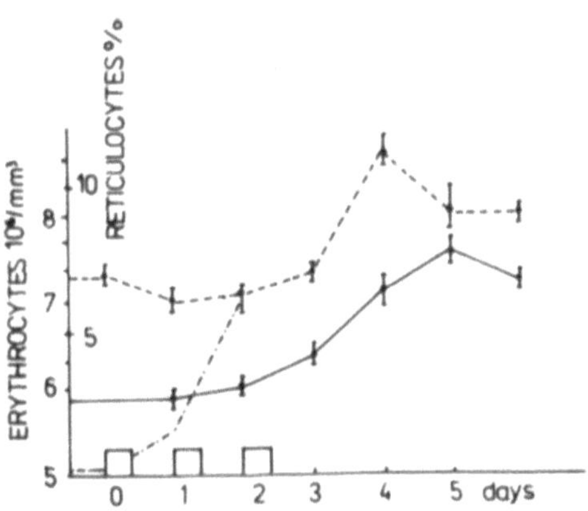

Figure 13.4. Erythrocytic reaction induced by cephalic hypoxia achieved for 3 days (8 h/day) by compression of the carotid arteries in dogs with sinocarotid reflexogenic areas denervated by sectioning the sinusal nerves (mean levels).

tion. This suggests that erythropoietin secretion is not initiated in a reflex way through excitation under the effect of hypoxia of the sinocarotid chemoceptive zone.

That vegetative areas of the central nervous system played some role in the regulation of erythropoiesis was established by experiments with rats. Symmetrical areas were obtained by electrocoagulation in the anterior and posterior hypothalamus in the mesencephalic and bulbar reticular formation by using bipolar electrodes and stereotaxic techniques (Baciu et al. 1965, 1968) (Fig. 13.5). After injury, the animals were subjected to hypoxia (simulating an altitude of 5 km) for at least 12 h. After the fourth to fifth day the reticulocytic and erythrocytic responses were estimated. The bilateral symmetrical lesions in the mesencephalic reticular formation do not significantly reduce the amplitude of the erythrocytic response. Destruction of the anterior hypothalamus in the same way diminishes the reticulocytic response. The bilaterial injuries of the posterior hypothalamus in the substantia grisea periventricularis and in the area mamilaris and supramamilaris, besides other important functional disturbances, clearly blocks the erythrocytic response to hypoxia. An inhibiting effect also occurs following injury to the bulbar reticular formation. Other authors have shown that either direct electrical stimulation or lesioning the nuclei at the basis of the brain supports a hypothalamic central nervous control of erythropoiesis (Halvorsen 1961; Natscheff 1961; Mirand et al. 1964).

There appear to be at least two components of the central nervous system involved in the response to hypoxia. First, the immediate response to hypoxia is related to the sinocarotid afferences through which the great respiratory, circulatory, and cytodiabasis reflexes are achieved. The second component is primarily involved with erythropoietin secretion and is mediated through other

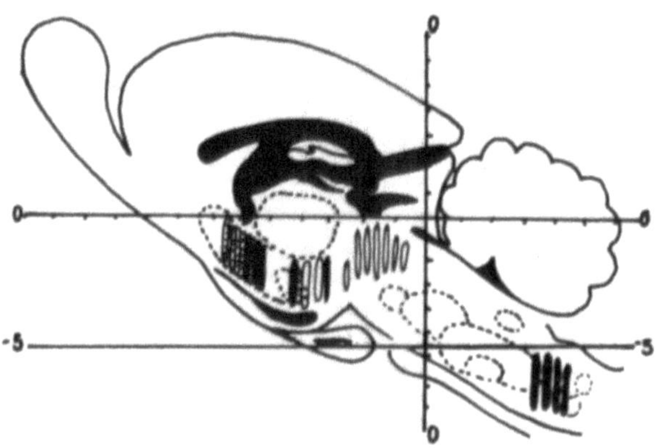

Figure 13.5. Schematic representation of the areas with electrolytic lesions and their influence on erythropoiesis. □, inactive lesions (control); ▨, with moderate effects; ■, with inhibitory effects.

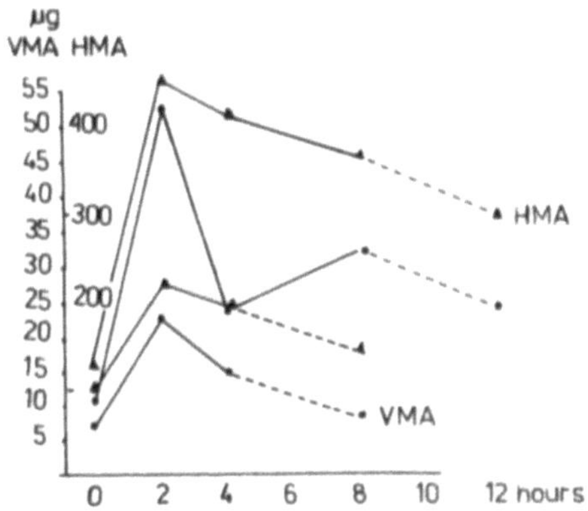

Figure 13.6. Typical changes of VMA and HMA secretion following 4–8 h exposures to 7 km. •–•–•, vanylmandelic acid (VMA); ▲–▲–▲, *p*-hydroxymandelic acid (HMA), – – –, posthypoxic acid.

central receptors. Thus erythropoietin secretion persists following sinocarotid denervation.

The efferent nerve pathway is represented by the sympathoadrenal system as exemplified by the intense sympathoadrenergic response of dogs to 1-h extrarenal hypoxia (simulated altitude in a hypobaric chamber of 5 km). This produces a three- to fourfold increase in vanyl mandelic acid (VMA) urinary excretion (Baciu et al. 1969) (see Fig. 13.6). Furthermore, there is a reduction of the erythrocytic response to hypoxia in adrenalectomized rats treated with glycocorticoid hormones (Fig. 13.7); this is especially marked following administration of β blockers such as propranolol (Baciu 1974, 1977) (Fig. 13.8). Concomitantly, Fink and Fisher (1974) showed that the animals treated with β blockers produced less erythropoietin than untreated controls.

Since the nervous system is primarily responsible for control of erythro-

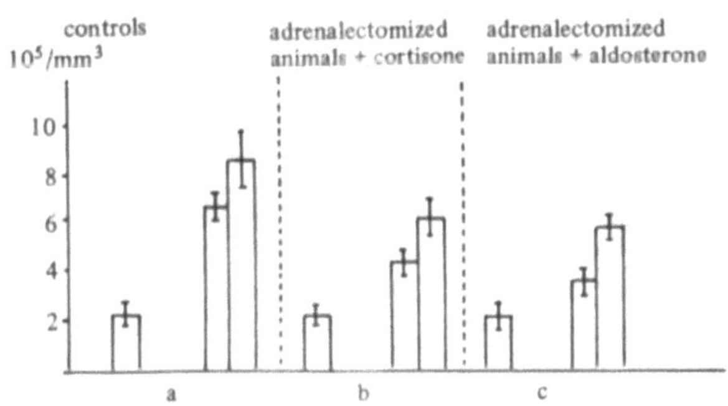

Figure 13.7. Reticulocytic reaction of rats to hypobaric hypoxia (5.5 km) in controls (*a*) and in adrenalectomized animals, treated with cortisone (*b*) or aldosterone (*c*) on day 4 and 5 from exposure. Column 1 represents the initial values.

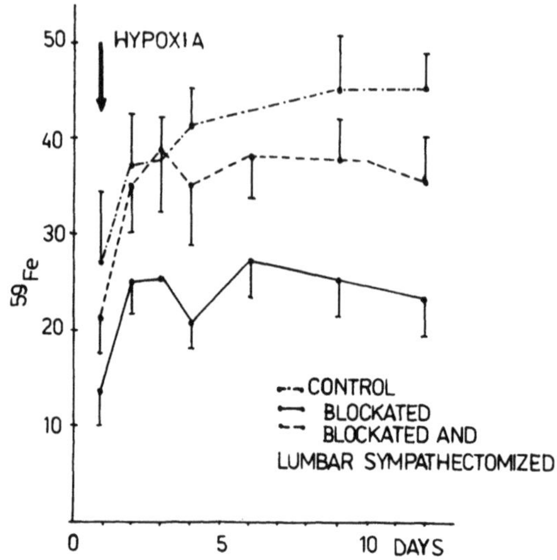

Figure 13.8. Effect of the rat serum kept under hypobaric hypoxia (simulated altitude barochamber 5 km) at various intervals of hypoxia on the incorporation of ^{59}Fe into the erythrocytes of hypertransfused polycythemic rats. •– • –•, Control; •–•, blockated; •– –•, blockated and lumbar sympathectomized.

poiesis, the question arises as to how newborn animals with still undeveloped nervous systems respond to hypoxia. This was examined by Baciu et al. (1971), who exposed a bitch and her five newborn puppies to hypoxia immediately after birth. The erythropoietic response of the mother (measured by ^{59}Fe uptake, reticulocyte, and erythrocyte counts) conformed to established patterns within 3 to 5 days, but the response of the puppies appeared delayed after a longer period of hypoxia (7-9 days). The amplitude of the response is also reduced. This is consistent with the response of spinal animals to hypoxia, which is similar to that of puppies with immature nervous systems (Baciu 1977). It may be concluded, therefore, that hypoxia is able to induce the release of erythropoietic substances and mediate an erythrocytic response independently of the nervous system, presumably by a direct action upon certain cells. In this case, the response is reduced and requires a longer latent period. Erslev et al. (1965) obtained similar results by perfusing dog kidneys with hypoxic blood (5% oxygen). In these animals renal cell necrosis is observed in the perfused fluid, and active erythropoietin, presumably released by the direct cellular action of hypoxia, can be detected.

Cellular Mechanisms of Erythropoietin Formation

The site and mode of erythropoietin formation is still controversial. Jacobson et al. (1957), for example, demonstrated absence of erythropoietin secretion in binephrectomized animals exposed to hypoxia or treated with cobalt cloride

($CoCl_2$). Furthermore, Kuratowska et al. (1961) showed an increase of the erythropoietic activity of the blood perfusing the isolated kidney in conditions of advanced hypoxia. Both studies suggested that the kidneys were the major or even the only erythropoietin-producing organ. More recent evidence (Gordon et al. 1968), however, suggests that erythropoietin originates from extrarenal sources. It is now believed that the hormone is synthesized by combination of a renal factor (REF) produced in the kidney, an enzyme, erythrogenia, and a serum factor referred to as erythropoietinogen.

To establish the role of the kidney in erythropoietin formation, we carried out investigations on rats and dogs (Baciu et al. 1963). We found that the serum of binephrectomized rats that had been exposed, after at most 2 h, to hypoxia (simulated altitude of 5 km) for 24 h, contained pseudoglobulins, including erythropoietin, but no apparently toxic products of renal retention. The pseudoglobulins isolated by precipitation when tested on rats produced a significant erythrocytic response.

Another series of experiments was made on dogs with separation of the somatocephalic blood flow on which renal autotransplant was made on one of the carotid arteries and the corresponding jugular vein. The other kidney was removed (Fig. 13.9). Renal ischemic hypoxia through compression of the carotid-renal anastomosis, 8 h/day for 3 days, was not associated with an increase in serum erythropoietin or by the characteristic erythrocytic response (reticulocytosis and polyglobulia) (Fig. 13.10b). On the contrary, this occurs regularly at cephalic ischemic hypoxia achieved by compression of the other carotid artery (Fig. 13.10a). This suggests that renal hypoxia does not directly stimulate the production of erythropoietin but that this can nevertheless be induced by cephalic ischemia. This also established that erythropoietin secretion in an organ other than the kidney was under nervous control.

The demonstration of an extrarenal area releasing erythropoietin generated numerous contradicting investigations. Baciu et al. (1970) reported, for example, that the intravenous administration of charcoal to rats in an amount of 32 mg/ml arrests the erythrocytic response to hypoxia, implicating participation of the reticuloendothelia system, more particularly the macrophages, in erythropoietin biogenesis.

The cellular mechanism of erythropoietin formation, its transport by the blood, and its mode of action in the stimulation of erythropoiesis includes a number of membranal processes, some of which have been studied by our group. Baciu et al. (1973) reported the effect of proteases and esterases and their inhibitors upon erythropoiesis (Figs. 13.11, 13.12, 13.13). Smith and Fisher (1973) showed that cobalt produces a significant increase of plasma erythropoietin simultaneously with an increase in the activity of renal hydrolases, cathepsine A and B, involving a possible relationship between the two processes. Furthermore, Libbin (1974) by in vitro incubation of normal rat serum with lysosomal fraction of hypoxic rat kidneys obtained an active erythropoietic factor.

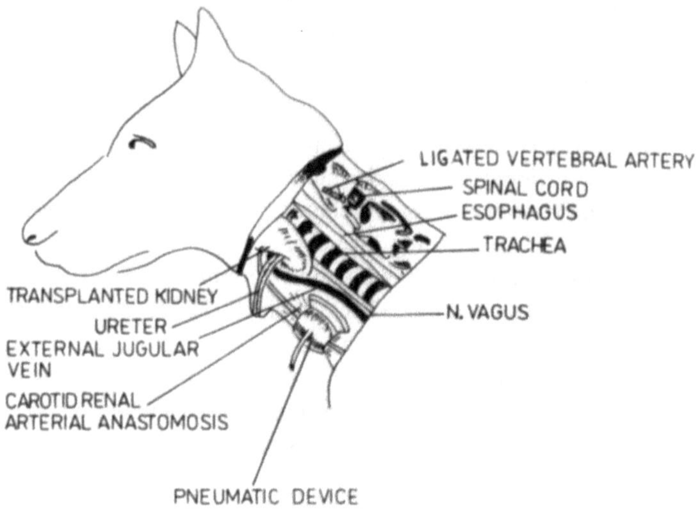

Figure 13.9. Schematic representation of the animal with chronically "isolated head" and kidney transplant in the neck area, the other kidney having been removed.

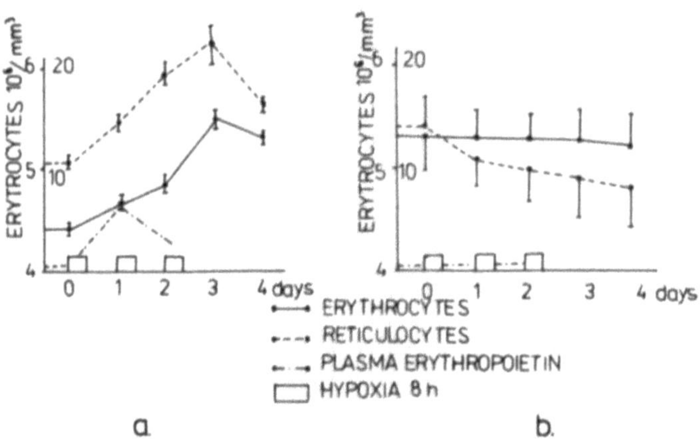

Figure 13.10(*a*). Effect of cephalic ischemic arterial hypoxia by compression of the common artery supplying the brain, with a pressure 10 mm Hg below the value of the systolic pressure, 8 h/day for 3 days, on erythropoietin secretion and erythrocytic response on dogs with chronically "isolated head." (*b*) Renal ischemic hypoxia obtained by pneumatic compression of the carotid renal anastomosis, 8 h/day for 3 days. It is not followed by increase of serum erythropoietin reticulocytosis and polyglobulia. •—•, Erythrocytes; •– – –•, reticulocytes; •– • –•, plasma erythropoietin; □, hypoxia 8 h.

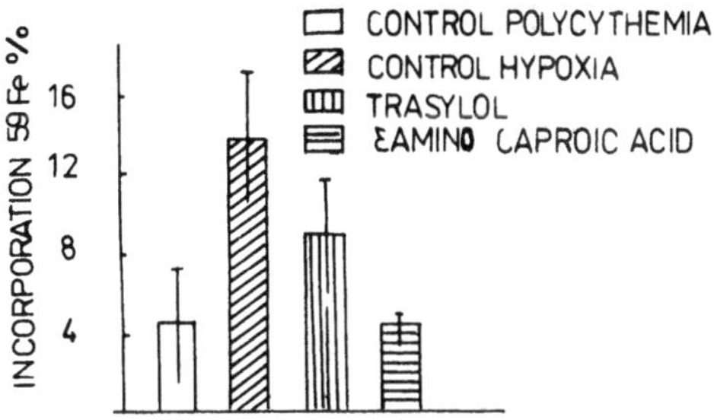

Figure 13.11. The ^{59}Fe incorporation in the circulating erythrocytes of posthypoxic polycitemic mice injected with pseudoglobulins f II obtained from rats treated before exposure to hypoxia (5 h at 0.4 atm) with Trasylol (aprotinin) 25,000 units/kg body weight or E aminocaproic acid (EACA) 1 g/kg body weight. For comparison, the results obtained in a control group of mice treated with pseudoglobulins f II extracted from the serum of control rats kept under the same conditions of hypoxia are given. □, Control polycythemia; ▨, control hypoxia; ▥, Trasylol; ▤, EACA.

A first involvement of the membranes in the formation of erythropoietin derives from the possible lysosomal origin of the enzymatic factor, protease or esterase, referred to as erythrogenin. In order to check the lysosomal origin of erythrogenin and to demonstrate the participation of the lysosomal membranes, we carried out investigations on rats (Baciu et al. 1978). Lysosomes were isolated according to Shibko and Tappel (1965) from the kidneys of rats exposed

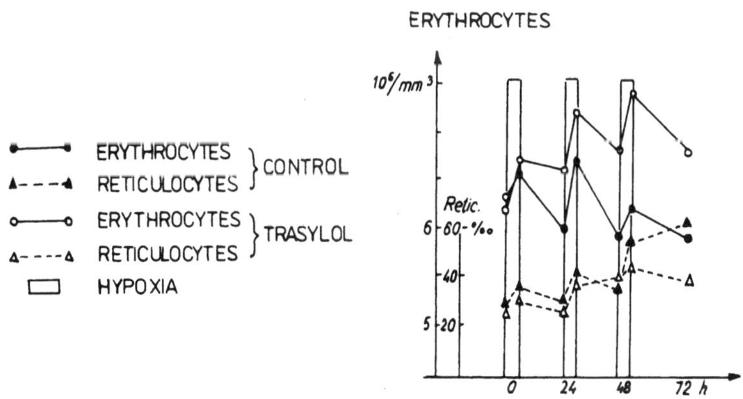

Figure 13.12. Reticulocyte and erythrocyte changes under the influence of hypoxia (5 h/day at 0.4 atm for 30 days) in rats injected daily, before exposure to hypoxia, with Trasylol 25 units/kg body weight. For comparison, the results obtained in a control group of rats kept under the same conditions of hypoxia, but not injected, are given. control: •–•, erythrocytes; ▲– –▲, reticulocytes. Trasylol: o–o, erythrocytes; △– –△, reticulocytes. □, Hypoxia.

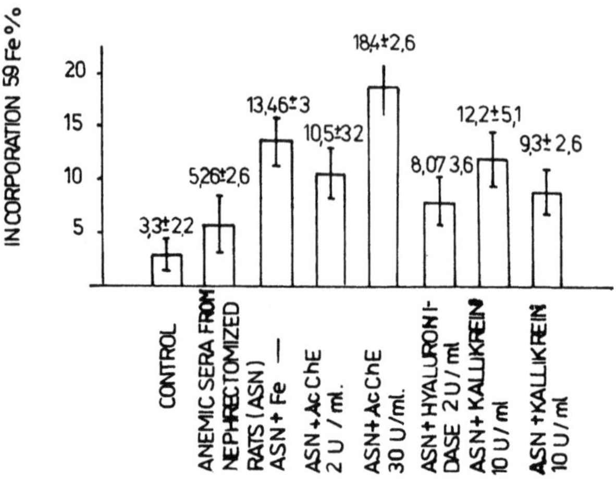

Figure 13.13. Incorporation of ^{59}Fe in the circulating erthrocytes of post-hypoxic polycithemic mice under the influence of the serum of rats with phenyl-hydrasine induced anemia and nephrectomized (ANF) incubated with renal erythropoietic factor, acethylcholinesterase or hialuronidase in a dose of 2 units and 30 units/ml and kallicrein 10 units/ml and of sodium chloride isotonic solution (control group).

to hypobaric hypoxia (5-km simulated altitude) for 1.5, 4, 8, and 12 h, following up the effect of permeability changes through hypoxia on the release of this factor. The lysosome suspensions were incubated at 37° for 30 min with normal serum and then injected intravenously into polycythemic mice in amounts equivalent to 1.4 mg lysosomal protein. The uptake of ^{59}Fe was measured to assess the possible erythro-stimulating activity. At the same time the erythropoietic activity of the serum obtained from donors of hypoxic kidneys was tested in the same way. The erythropoietic activity of the serum and lysosomes-normal serum mixture is presented in Fig. 13.14, which shows the variation of ^{59}Fe uptake in the red blood cells of polycythemic mice injected with these products. Both the serum injected and the lysosomes–normal serum mixture caused significant changes in the incorporation of ^{59}Fe beginning with 1.5 h of hypoxia. The maximal level is reached in both cases after 9 h of hypoxia. The uptake shows a marked decrease of 12 h. These results clearly show that renal lysosomes obtained from hypoxic animals contain an erythropoietin-forming factor.

In the serum a progressive and significant increase was observed in acid phosphatase activity, as seen in Fig. 13.15. In the renal lysosomes a decrease of the latent activity and an increase of the free activity was found, this being insignificant and reaching maximal values at 4 h (Fig. 13.15). The fact that the values of the latency changes of the renal lysosomes are not significant and reach maxi-

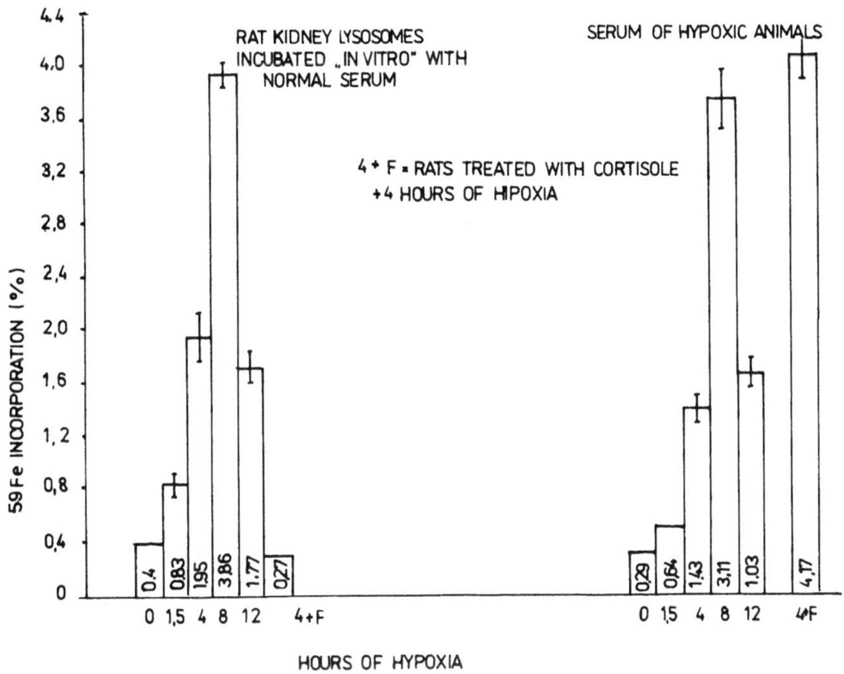

Figure 13.14. The variation of ^{59}Fe uptake in the red blood cells of polycythemic mice injected with lysosomes–normal serum mixture and hypoxic serum.

mal values after 4 h rather than 8 h suggests that lysosomal enzymes of extrarenal origin are involved. In the same series of investigations the rats were treated with cortisol hemisuccinate as a stabilizer of the lysosomal membrane (5 mg/100 kg body weight i.p.) and then subjected to hypoxia of 4 h. The lysosomes of the animals treated with cortisol, then exposed to hypoxia, and incubated with serum, according to above techniques, do not confer erythropoietic activity

Another concern is whether the permeabilization of the lysosomal membranes that is altered by hypervitaminosis A (2000 units/kg i.m. for 7 days) causes an increased erythropoietin release following exposure to hypobaric hypoxia. Control animals, not subjected to hypoxia, showed increases of acid phosphatase in the serum but no change in erythropoietin release. Hypoxia in these animals, however, results in determining significantly greater increases of erythropoietin as compared with the controls. Thus permeabilization of the lysosomal membranes through hypervitaminosis A, although alone has no effect on the release of erythropoietin, however, cumulates its effect with that of hypoxia, determining a significantly greater increase of erythropoietin as compared to controls. Cortisol hemisuccinate (administered for 8 days by i.p. injections, 5 mg/100 g body weight), in contrast, caused a marked inhibition of the

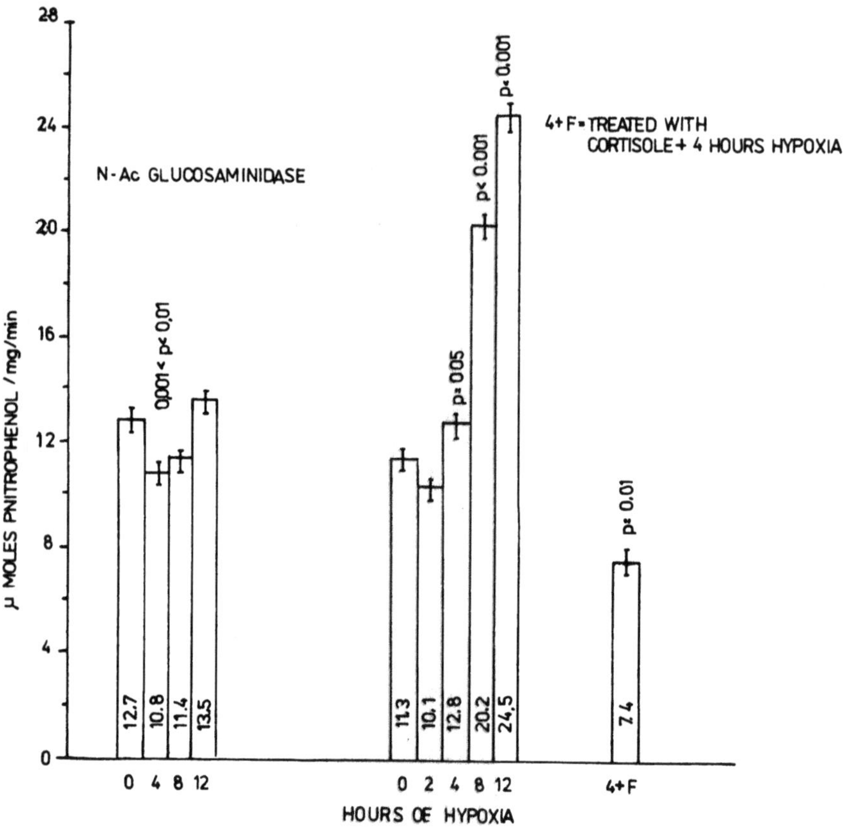

Figure 13.15. The release of lysosomal enzymes into rat serum during hypoxia.

release of lysosomal enzymes into the serum and also resulted in a marked inhibition of the stimulating effects of hypobaric hypoxia on the occurrence of erythropoietin consistent with its action as a lysosomal stabilizer.

Erythropoietinogen is not exclusively produced in the kidney. Unpublished observations by our group have indicated that following bilateral nephrectomy and exposure to hypoxia, the erythropoietinogenetic activity of the splenic lysosomes is stimulated. This is consistent with the notion that reticuloendothelial system is also involved in biogenesis of the hormone.

Interaction of Erythropoietin with the Red Blood Cell Membranes

The plasma concentration of erythropoietin in subjects exhibiting polyglobulia with activated erythropoiesis is inversely related to the hematocrit value, suggesting some connection between erythropoietin formation and the number of red blood cells (membranes) upon which the hormone acts. This has been clearly shown by Prentice and Mirand (1962), who showed that a decrease in hematocrit was associated with an increase in the titer of plasma erythropoietin. This was explained as an inactivation of erythropoietin in the presence of increased numbers of erythrocytes.

In initial studies of the effects of erythropoietin on the metabolism of medullary cells, Baciu et al. (1977) demonstrated an increase in the uptake of ribose from the incubation medium in the presence of the hormone. The effect does not depend exclusively on ribonucleic acid (RNA) synthesis regulated by erythropoietin since the effect is observed in erythrocyte suspension both in vitro and in vivo, suggesting that erythropoietin exerts its effect through an action at the level of erythrocyte membrane (Baciu et al. 1980).

Studies have been undertaken to investigate the binding of erythropoietin to red blood cell membranes and the effect of binding on ribose uptake. It is assumed that the erythrocyte membrane possesses a fixed number of erythropoietin binding sites and that relatively few of these receptors are occupied in erythrocytes from normal animals compared with cells from animals subjected to hypoxia. Furthermore, the amount of hormone binding sites occupied will influence the rate of ribose uptake by the cells. Erythropoietin tests were made on a suspension of normal erythrocytes to which various sera were added as follows: (a) normal, (b) hypoxic, (c) hypoxic after preincubation with hypoxic erythrocytes, (d) hypoxic after preincubation with normal erythrocytes. The ribose consumption of normal erythrocytes was also tested after contact with hypoxic serum (e). The results of the test (presented in Fig. 13.16) confirm the hypothesis that the erythrocytes modify their ribose metabolism as a consequence of the erythropoietin adsorbtion on the membranes.

In further studies of the optimal conditions of erythropoietin binding on the membrane it was found that the degree of adsorbtion is dependent upon the medium:cells ratio. The higher the ratio, the more intense will be the adsorbtion, so that a suspension with a hematocrit of 67–70% binds most of the circulating erythropoietin. Adsorbtion takes place at an optimal pH between 7.4 and 7.8 and a temperature of 37°C; at a pH of 6–7 or a temperature of +4°C, binding is weak.

In order to establish the binding affinity of erythropoietin to the membranes, we have recently initiated experiments in which erythropoietin adsorption to normal erythrocytes obtained from rats or mice was measured on isolated membrane preparations. Membranes were isolated by using various lysis techniques at controlled pH, temperature, and so on, to determine the optimal conditions at which the hormone remains attached to the membrane. Baciu and Ivanoff (1983) concluded from preliminary data that erythropoietin remains in a great part attached to the membranes when hemolysis of the cells takes place by treatment at alkaline pH and at room temperature (these injected membranes of polycythemic mice determining a stimulation of ^{59}Fe uptake).

If we take into account the capacity of the erythrocyte membranes of binding other hormones such as progesterone, thyroxine (DeVenuto 1967; DeVenuto

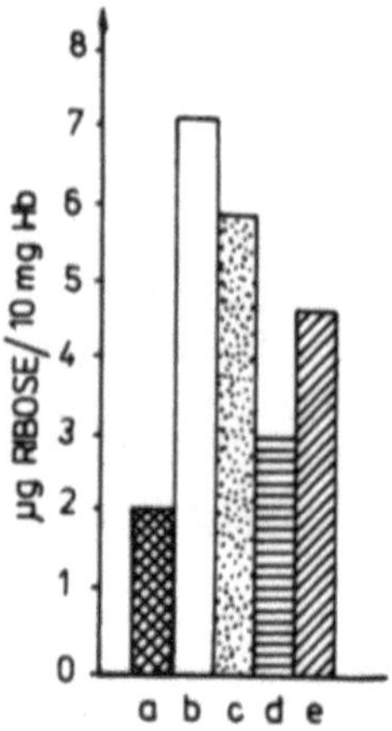

Figure 13.16. Binding of erythropoietin on the erythrocytes membranes. Ribose consumption was tested on a suspension of control erythrocytes to which were added (a) control sera, (b) hypoxic sera, (c) hypoxic sera after incubation with hypoxic erythrocytes, (d) hypoxic sera after incubation with normal erythrocytes, (e) normal erythrocytes after incubation with hypoxic sera.

et al. 1969), and insulin (Robinson et al. 1979), erythropoietin binding to the membranes is not suprising even if its physiological significance is still unclear. We can, however, suppose that ribose, as an important source in the synthesis of triosophosphates and hence of 2,3-diphosphoglyceric acid, the major phosphorylated metabolic intermediate of the red blood cell with a role in the establishment of the affinity of hemoglobin for oxygen through the control exerted by erythropoietin, could be involved in a complex regulation process at the cellular level of oxygen homeostasis. The increase of 2,3-diphosphoglyceric acid in hypoxic states, which until now could not be plausibly explained, as well as the possibility of the stimulation of certain enzymes of the type of phosphoribosyl pyrophosphatase involved both in RNA synthesis at the level of the bone marrow and in the synthesis of the nucleotides required by the erythrocyte metabolism could constitute explanations for the intensification of ribose metabolism in the presence of erythropoietin.

From these studies of erythropoietin binding to the red blood cell membrane we suggest that, in the blood, the hormone is distributed between two compartments, plasma and erythrocytes, and that this distribution is influenced by the hematocrit level. The existence of an inverse correlation between the hematocrit level and the concentration of plasma or urinary erythropoietin has been reported by other authors (Adamson 1968; Adamson and Finch 1968; Hammond et al. 1962, 1968; Alexanian 1977). The main evidence has come from studies of urinary excretion of erythropoietin in various forms or in anemia, where elevated hormone levels do not always accord with the degree of anemia and erythropoietin can be detected only when the hematocrit decreases below 30%. Even more interesting from this point of view are observations made on polycythemic subjects. Here the high erythrocyte count is associated with normal or even subnormal erythropoietin concentrations and there is a sudden increase in erythropoietin titer when the hematocrit is reduced through repeated phlebotomies. The current interpretations of these observations attribute these discrepancies to either a lack of erythropoietin synthesis or to a reduced sensitivity of the erythroid cells to the action of this hormone. In a still unpublished investigation Baciu and Ivanof (1983), checked the possibility that erythropoietin was distributed differently between the plasma and on the erythrocytes; for this purpose two groups of seven rats each were used. Of these a first group of polycythemic hypertransfused animals received at the beginning of the experiment a single 9 unit erythropoietin (Ep) dose per animal and then was rendered anemic gradually by bleeding down to a hematocrit of 54%; the second group of control animal (hematocrit 44-45%) was rendered anemic successively for 4 days down to a hematocrit of 20%, replacing the volume of the plasma lost with normal plasma. Erythropoietin was determined in the plasma by the test of ^{59}Fe uptake and that adsorbed to erythrocytes by the estimation of ribose consumption. The results showed that most hormone was bound to the cells when the hematocrit levels were between 62 and 54%; the plasma, conversely, contains only trace amounts of hormone. Anemiation induces an increase of erythropoietin both in the plasma and on the erythrocytes in normal rats down to a hematocrit of 53%. Below this level it increases in the plasma only (Fig. 13.17). This suggests that below a hematocrit of 35% the cell count is already too low to be able to bind the large amounts of hormone released by anemiation.

The in virto investigations with erythrocyte suspension of various densities revealed similar results, demonstrating that the erythrocytic response for a certain dose of erythropoietin is greater the lower the density of the suspension.

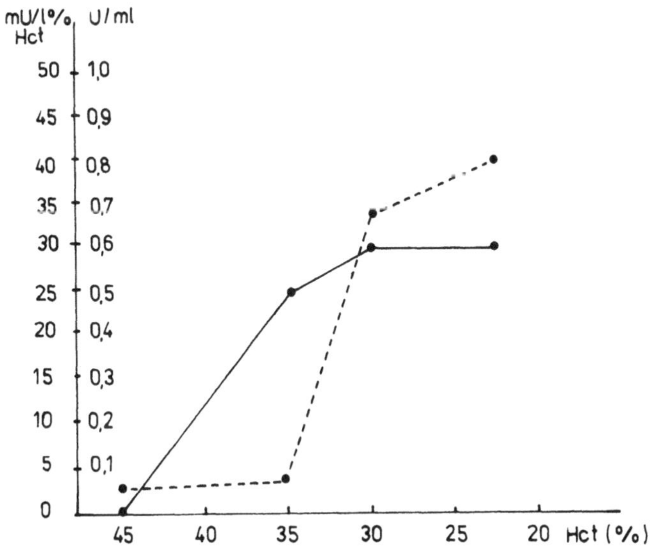

Figure 13.17. Distribution between plasma and erythrocytes of erythropoietin formed in the course of gradual anemia of normal rats for 5 days.

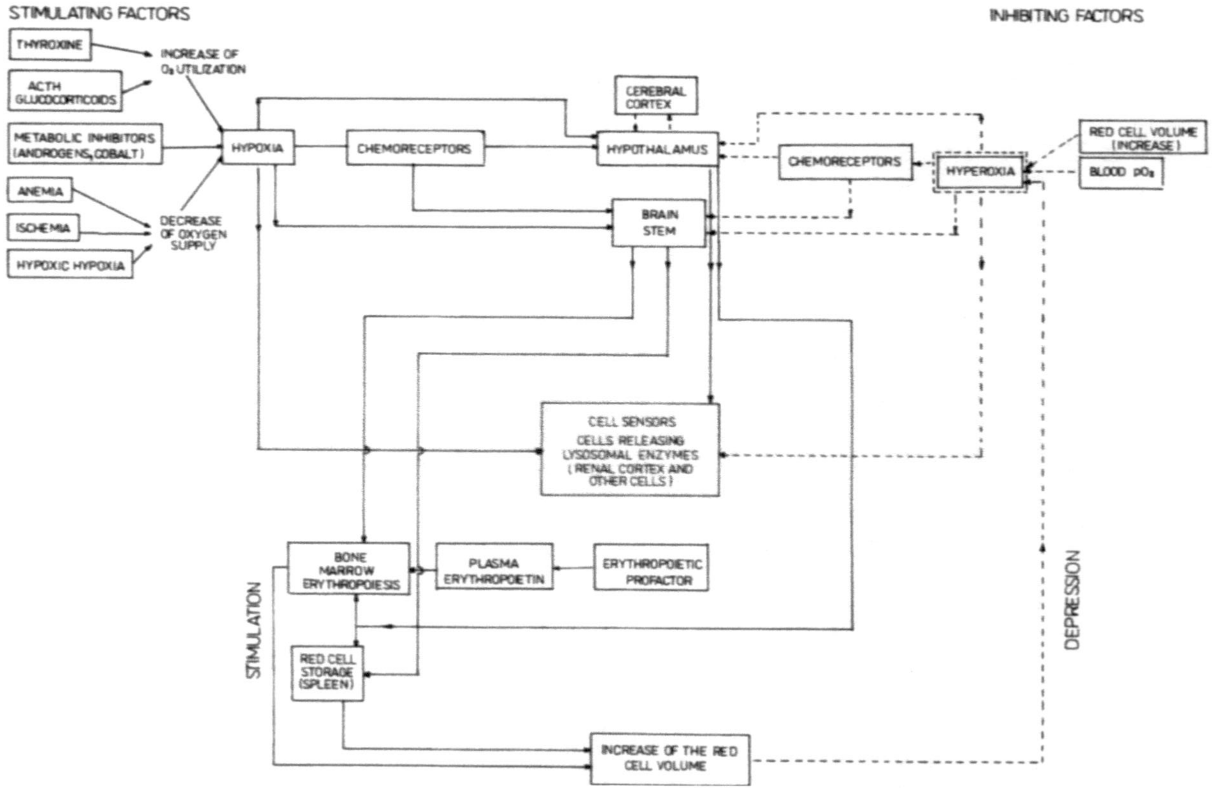

Figure 13.18. Proposed model for the regulation of erythropoiesis.

The results obtained by our group concerning the effects of erythropoietin on erythrocytes metabolism, besides the theoretical aspect described above, also have some practical implications. First, we have elaborated a biochemical method for the assay of erythropoietin both in the serum and directly on erythrocytes. The method is sensitive, relatively simple, and requires shorter time than the methods used at present. Second, erythropoietin binding to erythrocytes allows the assay of the hormone even in the cases of polycythemia in which the presence of erythropoietin cannot be revealed in the blood plasma. There remain two other membrane involvements in the regulation of erythropoiesis that are of importance in the understanding of the process as such but that have not yet been studied. One is the way hypoxia acts upon certain possible cellular receptors, initiating the release of erythropoietin by means of the nervous system. The other is that it is unclear whether the erythropoietin adsorbed on the stem and primordial cells of the erythrocytic series could induce synthesis of nucleic acids through the membrane system of synthesizing cyclic nucleotides or whether penetrating into the cell influences this synthesis in a direct manner.

Figure 13.18 is a schematically reproduced stage model, representing the main mechanisms regulating erythropoiesis in the light of the experimental observations reported above.

Conclusions

Experimental investigations carried out at our department of physiology since 1954 on the nervous and humoral regulation of erythropoiesis have demonstrated the following:

1. Hypoxia (5-km simulated altitude for 3 days) determines reticulocytosis and polyglobulia that are attenuated following spinal cord sectioning.

   By somatocephalic isolation of the blood flow (chronically isolated head in experimental dogs), ischemic arterial hypoxia of the head produced an increase in erythropoietin and reticulocytes with polyglobulia. These results were also observed following carotid zone denervation, however, with a greater increase in erythropoietin and a delayed erythrocytic response.

   These results suggest that hypoxia exerts a direct action on the central nervous system, possibly on a certain determined neuronal group generating an impulse transmitted to the periphery by the way of the spinal cord.

   Erythropoietin is not released by the action of hypoxia on the chemoceptive sinocarotid zones. Their denervation, however, produces the disappearance of the circulatory and respiratory reflexes to hypoxia as well as the release in the blood flow of the erythrocytes from the hematopoietic organs.
2. The nervous centers, through which erythropoietin release occurs under the action of hypoxia, are situated in the posterior hypothalamus and the bulbar reticular formation. Following their bilateral lesioning in rats, the erythropoietic response to hypoxia is attenuated to a great extent.

3. The efferent nervous pathway is represented by the sympathoadrenal system. As the cathecolamine metabolites increase following exposure to hypoxia, the erythropoietic response is attenuated in adrenalectomized rats kept alive with glycocorticoid hormones as well as after β-blocker administration.
4. In the release of erythropoietin in response to hypoxia, a cellular mechanism also intervenes. This was demonstrated in newborn rats and puppies, with immature nervous system, as well as in rats with sectioned spinal cords. The response is reduced and requires a much longer latency period.

The cellular mechanism of erythropoietin formation was studied under the aspects of organ formation and the release of erythrogenin, a protease enzyme that plays a certain role in erythropoiesis formation.

1. Erythrogenin may be produced extrarenally. Following binephrectomy of rats, the exposure to hypoxia generates in the blood of these animals amounts of erythropoietin that are statistically identical to those found in controls.

   Following separation of the somatocephalic blood flow and renal autotransplant in the neck area, with the removal of the other kidney, renal arterial hypoxia in this animal does not determine an erythropoietic response; cephalic hypoxia, on the contrary, produces such a response. Therefore, erythropoietin production can be induced by cephalic ischemia, its release taking place in another organ than the kidney, which was denervated by transplantation.

   In erythropoietin formation the participation of the macrophages of the reticuloendothelial system is possible because colloidal charcoal injected intravenously in high amounts blocks the erythropoietic response.
2. The subcellular mechanism of erythropoietin formation consists in the release in the blood flow of erythrogenin, a proteolytic enzyme whose effect, according to our researches, can be inhibited by proteases, especially Trasylol.

   The participation of the lysosomes suspended in vitro in normal serum at 37°C for 30 min in the release of erythropoietin was also demonstrated.

   The effect is positive in vivo, too, following intravenous injection of renal lysosomes. The lysosomes extracted from the spleen exert the same effects, elucidating the extrarenal mechanism of erythrogenin release.

Investigations on the interaction of erythropoietin with red blood cells membranes revealed the following:

1. Erythropoietin acts on these membranes producing an increased ribose uptake in vitro from the incubation medium proportional to the amount of added hormone. This finding constitutes the basis of the elaboration of a biochemical method of in vitro erythropoietin assay on red blood cell suspensions incubated with erythropoietin and ribose.
2. It was demonstrated that erythropoietin binding to the membranes takes place by means of certain receptors. The hormone distribution between

plasma and the erythrocyte surface is dependent on the erythrocyte:plasma ratio, reaching maximal levels in the plasma at very low hematocrits and decreasing to low levels in polyglobulia.

The erythrocyte-plasma equilibrium of erythropoietin can be a peripheral cellular mechanism of erythropoiesis regulation.

Acknowledgments

The author thanks Dr. P. Quinn, Department of Biochemistry, Chelsea College University of London, U.K., for revising the English version of the manuscript.

References

Adamson, J.W. (1968). The erythropoietin/hemotocrit relationship in normal and polycythemic man. Implication of marrow regulation. *Blood 32,* 597–609.

Adamson, J.W. and Finch, C.A. (1960). Erythropoietin and the polycythemias. *Ann. N.Y. Acad. Sci. 149,* 560–563.

Alexanian, N. (1977). Increases erythropoietin production in man. In *Kidney Hormones II. Erythropoietin,* Fisher, J.W., ed. Acad. Press, London, New York, San Francisco, pp. 531–569.

Baciu, I. (1957). Despre stimularea eritropoezei prin hipoxie. *Stud. Cercet. Med. (Cluj) 8,* 255–260.

Baciu, I. (1974). Kontrollmechanismen während der anpassung an hypoxie. *Ergeb. Exp. Med. 15,* 195–205.

Baciu, I. (1977). La régulation de l'érythropoïèse. Arch. de l'Union. *Méd. Balkanique 15,* 12–17.

Baciu, I. (1980). *Homeostaxia Oxigenului,* Cluj-Napoca, D., ed.

Baciu, I. and Ivanof, L. (1983). Erythropoietin interaction with the mature red cell membrane. *Ann. N.Y. Acad. Sci.* in press.

Baciu, I., Ivanof, L., and Muresan, A. (1977). Influenţa eritropoetinei asupra absorbţiei de riboză de către celulele măduvei osoase. *Clujul Medical 50,* 266–271.

Baciu, I., Ivanof, L., and Pavel, T. (1980). Efectul eritropoetinei asupra consumului de riboză eritrocitar. *Clujul Medical 53,* 47–54.

Baciu, I., Stoica, N., and Raducanu, N. (1968). Rolul hipotalamusului posterior în reglarea eritropoezei. *Stud. Cerc. Fiziol. 13,* 15–23.

Baciu, I., Dorofteiu, M., Olteanu, A., and Pavel, T. (1971). Die erythropoietinsekretion während der postnatalenentwicklung. *Rev. Roum. Physiol. 8,* 37–45.

Baciu, I., Dorofteiu, M., Rusu, M., and Tomus, L. (1960). Stimularea eritropoezei prin hipoxie. IV. Despre acţiunes directă şi reflexă a hipoxeiei în reglarea nervoasă şi umorală a eritropoezia. *Stud. Cerc. Med. (Cluj) 11,* 239-251.

Baciu, I., Soltuz, V., Stoica, N., and Raducanu, N. (1965). Sur leţ centres régulateurs de l'érythropoïèse. *Rev. Roum. Physiol. 2,* 123-133.

Baciu, I., Derevenco, P., Sovrea, I., Pereni, N., and Popa, V. (1969). Semnificaţia răspunsului simpato-adrenal la hipoxie în raport cu reacţia eritrocitară. *Fiziol. Norm. Patol. 15,* 491-498.

Baciu, I., Dorofteiu, M., Olteanu, A., Pavel, T., and Daniello, R. (1973). Biogeneza eritropoetinei. *Fiziol. Norm. Patol. 19,* 127-136.

Baciu, I., Ivanof, L., Muresan, A., Pavel, T., and Benga, G. (1978). Efectul hipoxiei hipobare asupra lizozomilor renali de şobolan. *Clujul Medical 51,* 247-254.

Baciu, I., Dorofteiu, M., Rosenfeld, E., Don, F., Raducanu, N., and Timar, N. (1970). Sur le lieu de formation de l'érythropoïètine. *J. Physiol. (Paris) 62,* 231-243.

Baciu, I., Oprisiu, C., Dorofteiu, M., Rosenfeld, E., Rusu, M., and Cherebetiu, C. (1963). Le rein et la production de l'érythropoïètine. *J. Physiol. (Paris) 55,* 551-567.

Carnot, P. and DeFlandre, C.L. (1906). Sur l'activité hématopoïètique du sérum au cours de la régénération du sang. *C.R. Acad. Sci. (Paris) 143,* 384-386.

Devenuto, F. (1967). Interaction of progesterone and aldosterone with red blood cells of the rat. *Proc. Soc. Exp. Biol. Med. 124,* 478-483.

Devenuto, F., Ligon, D.F., Fridricksen, D.H., and Wilson, H.L. (1969). Human erythrocyte membrane uptake of progesterone and chemical alterations. *Biochim. Biophys. Acta 193,* 36-47.

Erslev, A.J., Solit, R.W., Damishon, R.C., and Ballinger, W.F. (1965). Erythropoietin in vitro. III. Perfusion of a lung-kidney preparation. *Am. J. Physiol. 208,* 1153-1157.

Fink, G.D. and Fisher, J.W. (1974). The role of the adrenergic nervous system in erythropoietin production. *Fed. Proceed. 23,* Abstracts 1879.

Gordon, A.S., Zanjani, E.D., Wong, K.K., and Cooper, G.W. (1968). Recent studies on the biogenesis of erythropoietin. Atti del simposio internationale sui fattori della eritropoiesis. Pavia, pp. 19-26.

Grant, W.C. and Root, W.S. (1952). Fundamental stimulus for erythropoiesis. *Phys. Rev. 32,* 449-498.

Halvorsen, S. (1961). Plasma erythropoietin levels following hypothalamic stimulation in the rabbit. *Scand. J. Clin. Lab. Invest. 13,* 564-575.

Hammond, G.D., Ishikawa, A., and Keighely, G. (1962). *The Relationship Between Erythropoietin and Severity of Anemia in Hypoplastic and Hemolytic States. Erythropoiesis,* Jacobson, L.O. and Doyle, M., eds. Grune & Stratton, New York, London, pp. 351-358.

Hammond, D.G., Shore, N., and Movassaghi, N. (1968). Production, utilization and extraction of erythropoietin. I. Chronic anemias. II. Aplastic crisis. III. Erythropoeitic effects of normal plasma. *Ann. N.Y. Acad. Sci. 149,* 516-527.

Hoff, F. (1933). Beiträge zur frage der blutregulation. Verhl deutsch. gesellsch. *Inn. Med. 45,* 124-133.

Jacobson, L.O., Goldwasser, E., Fried, W., and Plzak, L. (1957). Role of the kidney in erythropoiesis. *Nature 179,* 633-634.

Kuratowska, Z., Lewartowski, B., and Michalak, E. (1961). Studies on the production of erythropoietin by isolated perfused organs. *Blood 18,* 527-534.

Libbin, R.N. (1974). Renal lysosomes: Role in biogenesis of erythropoietin. *Science 185,* 1174-1176.

Mirand, E.A., Grace, J.T., Johnston, G.S., and Murphy, G.F. (1964). Effects of hypothalamic stimulation on the erythropoietic response in the rhesus monkey. *Nature 204,* 1163-1165.

Natscheff, N. (1961). Veränderungen in der hámopoese bei hunden nach reizung im bereich des hypothalamus. *Dokl. Bolg. Acad. Nauck. 14,* 321-324.

Nečas, E. and Thorling, E.B. (1972). Unresponsiveness of erythropoietin producing cells to cyanide. *Am. J. Physiol. 222,* 1187-1190.

Paulo, L.G., Fink, G.D., Roh, B.L., and Fisher, J.W. (1973). Influence of carotid body ablation on erythropoietin production in rabbits. *Am. J. Physiol. 224,* 442-444.

Prentice, T.C. and Mirand, E.A. (1962). *Effect of Various Hematocrit Levels on Plasma Erythropoietin During In Vitro Incubation with RBCs. Erythropoiesis,* Jacobson, L.O. and Doyle, M., eds. Grune-Stratton, New York, London, pp. 93-97.

Robinson, T.J., Archer, Z.A., Gambhir, K.K., Hollis. V.W., Carter, L., and Bradley, C. (1979). Erythrocytes: A new cell type for the evaluation of insulin receptor defects in diabetic humans. *Science 205,* 200-202.

Schulhof, K. and Mathies, M.M. (1927). Polyglobulia induced by cerebral lesions. *J.A.M.A. 89,* 2093-2094.

Seip, M. (1953). Reticulocyte studies. *Acta Med. Scand. 282* (Suppl.).

Shibko, S. and Tappel, A.L. (1965). Rat kidney lysosomes: Isolation and properties. *Biochem. J. 95,* 731-741.

Smith, R.J. and Fisher, J.W. (1973). Effects of cobalt on the renal erythropoietic factor and kidney hydrolase activity in the rat. *Blood 42,* 893-905.

Stohlman, J.F., Rath, C.E., Rose, J.C. (1954). Evidence for a humoral regulation of erythropoiesis: Studies on a patient with polycythemia secondary to regional hypoxia. *Blood 9,* 721-733.

Index

MIX
Papier aus verantwortungsvollen Quellen
Paper from responsible sources
FSC® C105338

If you have any concerns about our products,
you can contact us on
ProductSafety@springernature.com

In case Publisher is established outside the EU,
the EU authorized representative is:
Springer Nature Customer Service Center GmbH
Europaplatz 3, 69115 Heidelberg, Germany

Printed by Libri Plureos GmbH
in Hamburg, Germany